W9-BYU-731

THE PRE-EXCITATION SYNDROME: FACTS AND THEORIES

THE PRE-EXCITATION SYNDROME: FACTS AND THEORIES

Printed in the United States of America

First Edition

Library of Congress Catalog Card Number: 78-56930

International Standard Book Number: 0-914316-13-3

THE PRE-EXCITATION SYNDROME: FACTS AND THEORIES

Libi Sherf, MD
Associate Professor of Cardiology
Heart Institute, Chaim Sheba Medical Center, Tel Hashomer
University of Tel Aviv, Sackler School of Medicine, and
Visiting Associate Professor, Department of Medicine
University of Alabama School of Medicine, Birmingham, Alabama

Henry N. Neufeld, MD
Professor of Medicine and Chaim Sheba Professor of Cardiology
Chief, Heart Institute, Chaim Sheba Medical Center, Tel Hashomer
University of Tel Aviv, Sackler School of Medicine

YORKE MEDICAL BOOKS
666 FIFTH AVENUE, NEW YORK, NEW YORK

Acknowledgements

It is a pleasure to thank our colleagues, Drs. I. Drori and H. Herzeanu, who helped in the preparation of the Chapter dealing with exercise tests in pre-excitation and ambulatory monitoring.

We are also grateful to Mrs. Ruth Olinger, Mrs. Sara Mishor-Millerizky, Mrs. Ruth Redmont and Mrs. Francis Zetland for their secretarial, library and editing help, and to Miss Simcha Chaim and Miss Evi Molcho for helping in the arrangement of the illustrations.

This review has been supported by a grant from the United States-Israel Binational Science Foundation (BSF), Jerusalem, Israel, in continuation of the collaborative health research communications program between the Israel Journal of Medical Sciences and the National Library of Medicine, Bethesda, Maryland, USA.

To my wife

CHAIA

for her help, encouragement and patience

L.S.

FOREWORD

Having personally known and worked with Libi Sherf for well over a decade, I find it a special pleasure to write these few words about this remarkable book by Sherf and Neufeld. The Wolff-Parkinson-White syndrome and its fullest understanding have been Sherf's intellectual obsession for much of his professional life. His pursuit of all available information on the subject has been with a persistence and avidity comparable only to those legendary knights who sought the Holy Grail.

What is offered richly in this book is a methodical and comprehensive analysis of virtually all currently available knowledge on the subject. This is done with an admirable even-handedness, dealing objectively with all viewpoints but at the same time evaluating the relative strengths and weaknesses of each observation and hypothesis. Wisdom is brought to this book which only time and experience can provide.

The encyclopedic references alone would make the book invaluable to anyone seriously interested in the Wolff-Parkinson-White syndrome. But when knowledgeable discussion and personal observations are added as they have been, it is genuinely a monumental contribution to clinical cardiology and cardiac electrophysiology.

Thomas N. James, MD
The Mary Gertrude Waters Professor of Cardiology and Chairman of the Department of Medicine, University of Alabama in Birmingham, and Physician-in-Chief, University of Alabama Hospitals

CONTENTS

PART I: FACTS

I General Considerations 1

Terminology 3
Prevalence 5
Distribution by Sex 7
Distribution by Age 8
Familial Occurrence 9
The Pre-excitation Syndrome Associated with Congenital Heart Disease 12
The Pre-excitation Syndrome Associated with Cardiac Diseases Other than Congenital 16
The Pre-excitation Syndrome Associated with Noncardiac Diseases 22

II The Electrocardiogram in the Pre-excitation Syndrome: Complexes and Configurations 36

P-QRS 36
The P-R Interval 36
P Waves 41

The QRS Complex 42
Characteristic Patterns 45
Duration of the QRS 47
Classification of QRS Patterns 48
Variations in QRS Form in the Same Patient 55

III The Electrocardiogram in the Pre-excitation Syndrome: Disturbances of Rate and Rhythm 61

Premature Beats 61
Tachycardias 63
Supraventricular Tachycardias 65
Pseudoventricular Tachycardias 69
True Ventricular Tachycardias 72
Other Arrhythmias 72
Sinoatrial Conduction Anomalies 72
Atrioventricular Conduction Anomalies 78

IV Pathology 89

Nomenclature 90
Postmortem Findings in 49 Pre-excitation Cases 90
Analysis of the 26 Most Thoroughly Studied Cases 111
Animal Studies 111
Pathological Studies in Cases in which Pre-excitation was Not Found during Lifetime 112

V Noninvasive Methods of Investigation 117

Vectorcardiography 117
The Tel Hashomer Series 117
Order of Depolarization in the Ventricles during Pre-excitation Conduction 123
Classification of Pre-excitation Cases 124
Use of VCG in Determining the Presence of Bundle Branch Blocks or Other Additional Pathological Conditions 124
Use of VCG in Determining the Location of an Accessory Pathway 125
Long-term Ambulatory Monitoring 127
Incidence of Supraventricular Paroxysmal Tachycardia 128
Correlation of Symptoms Suggestive of Paroxysmal Supraventricular Tachycardia with Actual ECG Events 128

Occurrence of Asymptomatic Episodes in Patients Without a History of Previous Symptoms 129
Nature of Pre-excitation, Persistent or Intermittent, during Daily Normal Activity 129
Detection of Other Electrocardiographic Abnormalities 130
Methods for Detection of Pre-contraction of the Ventricular Muscle 130
Exercise Tests 133
Work Capacity and Physical Fitness 134
ECG Signs of Coronary Insufficiency 134
Relationship between Effort and Attacks of Paroxysmal Tachycardia 136
Physical Effort and Normalization of the Conduction 137
Additional ECG Findings 137
Body Surface Isopotential Mapping 138

VI Invasive Methods of Investigation 143

Intracardiac Recordings and Stimulations 143
Determination of the Type of Anomalous AV Connection 143
Location of the Accessory Pathway 149
Participation of the Assumed Anomalous Pathway in the Tachycardias 151
Determination of Effective Refractory Period of Anomalous AV Connection Compared to Normal Conduction System 152
Assessment of the Effect of Drugs 153
Identification of High Risk Patients 153
Postoperative Evaluation 154
Epicardial Mapping 154

VII Prognosis 162

Morbidity 163
Without Tachycardias 163
Isolated Pre-excitation 163
With Associated Heart Disease 163
With Tachycardias 164
Age at Onset of Paroxysmal Tachycardias 165
Duration of Suffering from Tachycardias 167
Frequency and Duration of Attacks of Tachycardia 168
Types of Tachycardia 170

Symptoms and Signs during Tachycardia 170
"Burned Out" Cases 172
Mortality 172
Deaths Associated with Tachycardia 173
Sudden Death 174

VIII Treatment **179**

Medical Management 179
During Paroxysmal Attacks of Tachycardia 179
Preventive Treatment to Avoid Tachycardia 182
Electrical Treatment 184
Electrical Cardioversion 184
Artificial Pacemaker 185
Surgery 185
Surgical Techniques 185
Section of an Accessory Bundle 186
Division of His Bundle 187
Results of Surgery 187
Indications and Contraindications for Surgery 190

IX Problems of Diagnosis: Some Illustrative Cases **198**

Pre-excitation Patterns Mimicking or Obscuring Other ECG Abnormalities 198
Pitfalls in the Diagnosis of Myocardial Infarction 198
Pitfalls in the Diagnosis of Ventricular Hypertrophy 202
Right and Left Bundle Branch Block in the Presence of Pre-excitation 204
Right Bundle Branch Block, Left Axis Deviation and Coronary Insufficiency 205
Pre-excitation Mimicking Arrhythmias: Ventricular Premature Beats and Ventricular Tachycardia 207
Diagnostic Clues for Discovering Myocardial Infarction in the Presence of Pre-excitation 208
Atypical Cases 211
The Short P-R Normal QRS Syndrome 212

PART II: THEORIES

X A Summary of the Findings and Theories of the Pre-excitation Syndrome **227**

Findings ... 227
Electrophysiological Phenomena ... 227
Anatomical and Structural Findings ... 228
Histopathological Findings ... 228
Clinical Findings ... 229
Recent Popular Theories ... 229
Structural Theories ... 230
The "Kent Bundle" Theory ... 230
Other Accessory AV Bypasses ... 230
Postnatal Remnants of AV Nodal and His Bundle Tissues ... 231
Pathophysiological Theories ... 232
"Excitable Center" Theory ... 232
"Accelerated Conduction" Theory ... 233
"Synchronized Sinoventricular Conduction" Theory ... 233

XI A Concept of Interacting Structural and Functional Factors **236**

Dual or Multiple Atrioventricular Conduction Pathways ... 236
Electrical Instability in the Heart in Pre-excitation ... 238

XII Explanation of Some Phenomena of the Pre-excitation Syndrome by the Proposed Concepts **244**

Congenital and Postnatal Pre-excitation ... 244
Acquired Pre-excitation ... 244
Genesis of Tachycardias in Pre-excitation ... 246
Concomitant P and QRS Changes ... 247
Correlation of the P-R Interval during Normal and Anomalous Conduction in a Single Patient ... 250
First Degree, Second Degree and Complete AV Block ... 251
Bradycardia-Tachycardia Syndrome ... 253

XIII Various QRS Patterns in the Different Forms of Pre-excitation **255**

TABLES

		PAGE
I	Pre-excitation Syndrome: Prevalence	4
II	Pre-excitation Syndrome: Distribution by Sex	6
III	The Tel Hashomer Sample: Distribution by Sex and Age	7
IV	Familial Occurrence	8
V	Pre-excitation with Associated Congenital Heart Disease	13
VI	Pre-excitation and Cardiovascular Diseases	17
VII	The Pre-excitation Syndrome Associated with Noncardiac Diseases in the Tel Hashomer Series (215 Patients)	24
VIII	The P-R Interval	37
IX	The P-R Interval by Age Groups in the Tel Hashomer Series (215 Patients)	38
X	The P-R Interval Exhibiting Both Normal and Anomalous Ventricular Complexes in the Same Patient	39
XI	Cases with Changes in Both P Wave Form and Shape of the QRS Complex	46
XII	Duration of the QRS Complex in Pre-excitation	48
XIII	Proposed Descriptive Classification of QRS Patterns	54
XIV	Incidence of Tachycardia	64
XV	Accessory AV Connections and Other Anatomical and Anatomopathological Findings in Cases with Pre-excitation Syndrome	92
XVI	Proposed Nomenclature for Normal and Abnormal Muscular Connections in the Pre-excitation Syndrome and Additional Pathological Findings	112
XVII	Correlation of VCG and ECG Findings in 25 Patients	120
XVIII	Studies with Long-term Follow-up	164
XIX	Age at First Attack of Tachycardia in the Tel Hashomer Series	165
XX	Number of Years of Follow-up for Paroxysmal Tachycardias in the Tel Hashomer Series	167
XXI	Frequency of Paroxysms: Patients Suffering 1-5 Paroxysms of Tachycardia in the Tel Hashomer Series	168
XXII	Duration of Attacks of Paroxysmal Tachycardia in the Tel Hashomer Series	168
XXIII	Patients' Assessment of Subjective Feeling in the Tel Hashomer Series	169
XXIV	Type of Tachycardia in the Tel Hashomer Series	170

PAGE

XXV Preventive Treatment of Tachycardias in the Tel Hashomer Series ... 183
XXVI Results of Surgery ... 188
XXVII Electrocardiographic Values in 24 Cases during Pre-excitation and Normal Conduction in the Tel Hashomer Series ... 258
XXVIII Pre-excitation with Intermittency in 4 Studies from the Literature and the Tel Hashomer Series ... 259

PART ONE

FACTS

I

General Considerations

The pre-excitation syndrome was first recognized as a distinct clinical and electrocardiographic entity in 1930 with the publication of Wolff, Parkinson and White's first paper on the subject.[1] The story began, however, in 1928 when Paul Dudley White encountered a puzzling case of paroxysmal atrial fibrillation in a healthy 35-year-old man whose resting ECG showed abnormal ventricular complexes, a P-R interval of 0.10 second and normal P waves. The response to exercise was even more perplexing: the ventricular complexes became normal and the P-R interval increased to 0.16 second, although the heart rate rose to 140 beats per minute. After finding three similar cases in the hospital files, White and Wolff concluded that they were dealing with a definite but unrecognized clinical entity. A painstaking search of the literature failed to yield other descriptions of the syndrome, although some ECG tracings were found which displayed similar features.[2-5] All of them, however, were erroneously attributed to an AV nodal rhythm, and the paroxysmal tachycardias were disregarded. White added 7 cases found by Sir John Parkinson to his 4, and eventually the 11 cases were published in 1930 under the joint authorship of Wolff, Parkinson and White.[1]

Two years later Holzman and Scherf[6] stated that the ventricular depolarization in this syndrome was dissimilar from bundle branch block and was associated with the short P-R interval, which they considered to be sinus rhythm and not an AV nodal rhythm. They proposed that an anomalous AV connection might be responsible for the syndrome, based partly on the anatomic studies of Kent.[7-13] Their suggestion of an alternative physiopathological rather than anatomicopathological mechanism—the existence of a hypersensitive ectopic ventricular focus triggered by an obscure mechanical or electrical phenomenon which originates in the atria—was not adopted. However, their postulation of an accessory muscle bundle directly connecting the atria and the ventricles has been accepted as the most probable pathological mechanism of the syndrome.[14,15]

Supportive data for this theory was soon forthcoming. Butterworth and Poindexter[16-18] artificially produced an electrical connection between atrium and ventricle in the cat heart, and reproduced all forms of ventricular pre-excitation. Wood, Wolferth and Gekeler[19] found a muscular bridge connecting the right atrium with the right ventricle in a patient with pre-excitation. At this point, the problem was generally regarded as "neatly solved," and the Wolff-Parkinson-White (WPW) syndrome became synonymous in some circles with "bundle of Kent syndrome."[8]

During the next 40 years more than one thousand papers appeared dealing with this syndrome: case reports,[20-42] series of cases,[43-59] critical reviews of the literature[60-119] and new methods of investigation.[120-126] Various opinions have been expressed, as well as some original and provocative ideas. In 1944[53] Ohnell introduced the term "pre-excitation," and Prinzmetal[127] postulated an "accelerated AV conduction" which attributed the entire disorder to a lesion inside the AV node.[128] Sherf and James[110,129] proposed the existence of an ectopic electrical focus located in the right atrium, which uses the posterior internodal pathway and the "AV nodal bypass fibers" of James[130] for the conduction of the electrical stimulus; bypassing the bulk of the AV node could produce a short P-R interval, and input of the stimulus in the lower pole of the AV node could cause an asynchronization of the normal depolarization front in the His bundle and the ventricles, and lead to pre-excitation.

During this period most investigations of the disorder were limited to ECG tracings, VCG studies[120] and occasional esophageal ECG[131,132] or intracardiac tracings.[133,134] Beginning in 1967, a new era opened up with the introduction of various electronic methods. Durrer and Roos,[135] using epicardial mapping of the human heart during cardiac surgery, were able to show an early point of ventricular activation at the right lateral portion of the AV sulcus in a patient with pre-excitation. In 1970 Castellanos et al[121] introduced His bundle electrograms and electronic pacing. Surgeons began to employ these new sophisticated techniques of investigation in an attempt to locate the anomalous pathway, cut the suspected muscular bridge[136] and hopefully abolish the abnormal ECG pattern and the tachycardias. Although the first attempt was unsuccessful,[136] the door to cardiac surgery in pre-excitation was opened, and by 1976 more than 80 patients underwent surgery[122,137] (see Chapter VIII, Table XXVI).

Physicians, pathologists and physiologists displayed great interest in this relatively rare and mostly benign syndrome.[138] This may be explained by the deep conviction of many researchers that the pre-excitation syndrome, with its multiple and strange facets, presented a kind of "Rosetta stone" of electrocardiography, whose deciphering may explain many of the still obscure electrophysiological phenomena occurring in the human and the animal heart.

Terminology

Often the more obscure the etiology or pathological mechanism of a clinical disorder, the more names it is given and the less clear-cut and firm is its definition. This seems particularly relevant in the case of pre-excitation.

The classic Wolff, Parkinson and White paper[1] spoke of a "bundle branch block with short P-R interval." It has since been referred to as "WPW syndrome" (Ohnell[53]), "bundle of Kent syndrome" (Scherf and Schoenbruner[139]), "pre-excitation syndrome" (Ohnell[53]), "anomalous atrio-ventricular excitation" (Rosenbaum et al[140]), "aberrant atrioventricular conduction" (Fox[141]), "bypass AV conduction" (Wiggers[142]), "antesystolie" (Holzman[6,143]), "accelerated atrioventricular conduction" (Prinzmetal[127]), "non-delayed conduction" (Zao et al[144]) and "short P-R bundle block syndrome" (Hunter et al[145,146]). The most commonly used terms today are the "WPW syndrome" and the "pre-excitation syndrome."

While the number of names describing the disorder has been considerably reduced, almost every aspect of the syndrome has been a point of controversy. Wolff, Parkinson and White[1] outlined the clinical entity as follows: 1) a functional bundle branch block; 2) abnormally short P-R interval occurring mostly in otherwise healthy young people; 3) paroxysms of tachycardia or of auricular fibrillation; and 4) reversion of the ventricular complexes to normal physiological form and lengthening of the P-R interval to become normal spontaneously or following release of vagal tone by exercise or atropinization. Over the years it became clear that these clinical and electrocardiographic criteria were far too narrow to cover cases which exhibited additional features or some if not all the characteristics of the syndrome. Cases were observed with normal[147] and even prolonged P-R intervals,[148] in the elderly[149] as well as in the young,[47] and in patients suffering from congenital[28,150-152] or acquired cardiovascular diseases.[49,153] The deformed QRS complexes were shown to be neither "bundle branch block"[6] nor necessarily prolonged.[154] Palpitations were described in only about half of the cases;[124,155] atropine[156] or exercise did not always result in the normalization (intermittency) of the pathological ECG features.

Ohnell[53] in 1944 described the abnormal QRS as the most characteristic feature, particularly the slope or slurring which is always seen at the beginning of the QRS complex (later termed "delta wave" by Segers et al[157,158]). This initial slope, according to Ohnell, was the ECG counterpart of an early and slow depolarization occurring in a part of the ventricles before the electrical stimulus descends on the ordinary AV node–His bundle axis. He called this phenomenon "pre-excitation," and built around it a complex classification of ten different types of cases based on short or normal P-R interval and normal or prolonged QRS duration. He had a special group, "type WPW," for cases with positive P waves in leads I and II, a maximum P-R interval of 0.10 second, and QRS duration exceeding 0.10 second (taken verbatim from Wolff, Parkinson and White's original observations). This detailed classification is no longer accepted.

TABLE I
Pre-excitation Syndrome: Prevalence

Study	Source of Material	Subjects Studied	Cases of Pre-excitation #	%
Age 0-16 years				
Tamm[162]	random	5,500	17	0.31
Landtman[163]	random	5,600	2	0.04
Kupatz[84]	random	3,000	4	0.13
Joseph[164]	random	2,000	1	0.05
Total		16,100	24	Mean 0.15
Swiderski[165]	heart clinics	10,000	48	0.5
Schiebler[166]	heart clinics	5,400	27	0.5
Total		15,400	75	Mean 0.5
Age 16-50 years				
Manning[167]	random	17,000	33	0.18
Hiss[168]	random	122,043	187	0.15
Total		139,043	223	Mean 0.16
All ages				
Katz[169]	hospital	50,000	81	0.16
Hejtmancik[49]	hospital	55,000	80	0.15
Tranchesi[170]	hospital	32,000	27	0.08
Nasser[171]	hospital	20,000	29	0.15
Mortensen[155]	hospital	17,174	25	0.14
Frau, Maggi[75]	hospital	8.592	19	0.22
Hunter[146]	hospital	33,000	11	0.03
Tel Hashomer	hospital	43,000	72	0.15
Total		258,766	344	Mean 0.13
Grand Total		429,309	666	Mean 0.155

Durrer et al[14] and Wallace et al[159] considered "classic WPW" those cases which showed a P-R interval of 0.12 second or less, a well developed delta wave, and a QRS duration equal to or longer than 0.12 second; all other cases were labeled "variants of the WPW syndrome." These authors referred to all the cases by the term "pre-excitation," whether tachycardia was present or not, including those with the Lown-Ganong-Levine syndrome. James,[160] on the other hand, did not stress P-R interval or QRS duration in pre-excitation, but insisted that the presence of paroxysmal tachycardias is necessary in "WPW." Schiebler et al[153] ignored both the P-R interval and the tachycardias and concentrated on Wolff, Parkinson and White's criterion of "young, healthy people"; on this basis they excluded part of the WPW patients with associated heart disease from the "WPW syndrome group," and listed them as "nontraditional" cases.

Thus, there are several definitions for the syndrome, and many names. However the one feature about which there seems to be universal agreement is that a part of a ventricle is prematurely excited by an electrical stimulus arriving from the atrium. This premature stimulus starts from an unusual ventricular location and triggers the depolarization of the ventricles (pre-excitation). It is represented by a very characteristic slope at the beginning of the deformed QRS complex, and is explained by slow ventricular depolarization in the ordinary working myocardium (delta wave). Based on this universally accepted explanation, this condition will henceforth be called the "pre-excitation syndrome," and the presence of a clearly defined delta wave will be the *sine qua non* for establishing the diagnosis. In this way, short, normal or even prolonged P-R intervals, more or less deformed QRS complexes, and the presence or absence of intermittency and/or tachycardias, are understood to represent only different aspects of the syndrome.

Prevalence

The prevalence of pre-excitation is difficult to determine because the typical ECG patterns may be absent in cases of intermittent pre-excitation at the time of recording.[161] The diagnosis is usually made: 1) by the incidental finding of the typical pattern on a routine ECG tracing, in the absence of arrhythmias; 2) in 5-10% of cases suffering from paroxysmal tachycardias; or 3) in cases admitted to the hospital because of misinterpretation of a routine ECG revealing this anomaly.

Table I summarizes the prevalence of the disorder in 16 reports in the literature and the Tel Hashomer series. (The 72 cases in the latter series were gleaned from 10 years of case records [1953-1963] from the medical departments and the Heart Institute Out-Patient Clinic; an-

TABLE II
Pre-excitation Syndrome: Distribution by Sex

	Total	Male	Female
Ohnell, 1944[53]	70	45	25
Longhini, 1973[173]	24	17	7
Wellens, 1974[172]	26	20	6
Ueda, 1966[174]	30	21	9
Hunter, 1940[146]	22	13	9
Mortensen, 1944[155]	45	22	23
Grishman, 1950[175]	7	5	2
Giraud, 1956[176]	8	6	2
Wolff, 1930[1]	11	8	3
Willius, 1946[59]	65	35	30
Hejtmancik, 1957[49]	80	48	32
Bleifer, 1959[120]	38	22	16
Flensted-Jensen, 1969[47]	47	24	23
Chung, 1965[45]	40	29	11
Burch, 1959[177]	15	11	4
Rosenkranz, 1965[178]	20	15	5
Swiderski, 1962[165]	48	30	18
Schiebler, 1958[166]	28	19	9
Wolff and White, 1948[179]	52	36	16
Tranchesi, 1959[170]	27	15	12
Friedman, 1969[180]	31	19	12
Giardina, 1972[181]	62	40	22
Tel Hashomer, 1974	215	157	58
Total	1,011	657 (65%)	354 (35%)

other 142 cases in the series were specially referred by other physicians and therefore could not be included in calculation of prevalence.) A total of 666 cases of pre-excitation were discovered among 429,309 subjects (0.155%). The average prevalence of pre-excitation among hospital in-patients was similar to that among healthy individuals undergoing routine examinations. The under 16 age group did not differ from the over 16 age group. The only outstanding finding was that children below the age of 16 under care for different cardiac conditions had a prevalence of pre-excitation approximately three times higher than any of the other groups.[166,172]

This higher incidence in the child heart clinics may be due to the association of congenital heart disease with the pre-excitation syndrome, and is not carried over into the general population since many of these children die during childhood.

Distribution by Sex

In Wolff, Parkinson and White's classic paper,[1] 8 of their 11 patients were men and 3 were women. Every series of cases published since then has shown a similar male preponderance (60-70%). Table II summarizes the distribution by sex in 22 reports from the literature. Of a total of 1,011 cases of pre-excitation where the sex was mentioned, 657 were men (65%) and 354 were women (35%).

Table III presents the sex and age distribution among the 215 cases in the Tel Hashomer series (the age refers to the time when a case was first seen). Men predominate in all age groups. Analysis of our patients by sex and associated heart disease showed that the male-female ratio was the same as for the entire series in the groups of atherosclerotic heart disease, coronary heart disease and hypertension. This ratio was reversed in rheumatic valvular diseases (3 men: 6 women) and equal in congenital heart disease (3 men: 3 women). An attempt to correlate these findings with similar cases in the literature failed because there were too few reports of patients with rheumatic valvular disease and pre-excitation or no indication of the sex. There was, however, a sufficient number of reports on the sex distribution in congenital heart disease and cardiomyopathies,[147,165,181,182] and the findings were very similar to our own. Swidersky et al,[165] in a study of 48 children with pre-excitation, found 28 without associated heart diseases with the usual sex distribution (20 boys: 8 girls), while the ratio in the group with associated con-

TABLE III

The Tel Hashomer Sample: Distribution by Sex and Age

Age*	Total	Male	Female
0-11 months	2	2	0
1-10 years	6	4	2
11-16 years	16	8	8
17-20 years	50	43	7
21-30 years	38	27	11
31-40 years	30	17	13
41-50 years	26	21	5
51-60 years	29	24	5
61 + years	18	11	7
Total	215 (mean 26.8)	157	58

*Age at which patient was first seen

genital and primary myocardial diseases (19 cases) was almost equal (9 boys: 10 girls). Schiebler et al[147] described 6 cases of pre-excitation among his 23 cases of Ebstein's anomaly, 3 boys and 3 girls. Hostreiter et al[151] reported 5 cases suffering from discrete and/or hypertrophic subaortic stenosis, 2 boys and 3 girls. The meaning of these unusual findings is not clear at present.

Distribution by Age

Although Wolff, Parkinson and White[1] assumed that the pre-excitation syndrome is a disorder of young adults only, it has by now been reported in all ages, from early childhood[165,166,183] including neonates, to old age.[149] His and Lamb[168] did not find any significant variation in the incidence among the different age groups.

TABLE IV
Familial Occurrence

Author	Cases in One Family	Male	Female	Associated Familial Cardio-myopathy
Westlake[138]	6	3	3	+
Harnischfeger[188]	5	3	2	–
Wolff (quoted by Montemurro[189])	5	5	—	
Ohnell[53]	4	1	?	+?
Willius[59]	4	2	2	
Montemurro[189]	3	—	3	—
Urban[190]	3	1	2	+
Massumi[191]	3			+
Schiebler[182]	2	1	1	+
Soulié[192]	2	–	2	+
Ohnell[53]	2	1	1	–
Schiebler[166]	2	?	?	–
Braunwald[193]	2	2	—	+
McIntyre[194]	2	1	1	–
Soekmen[195]	2	2	—	–
Doumer[196]	2	1	1	–
Averill[197]	2	2	—	–
Rosenbaum[132]	2	1	1	–
Schneider[198]	2	1	1	–
Homola[199]	2	?	?	?
Swiderski[165]	2	?	?	+
Tel Hashomer	2	1	1	–

The age distribution in the 215 Tel Hashomer cases is shown in Table III. Our youngest patient was 3 days old and the oldest 80 years. There were relatively few cases at the two extremes of the age spectrum: 24 children under the age of 16, and 18 adults over 60. Patients from 11 to 20 years of age were twice as numerous (66) as the mean (26.8). This may be due to the fact that this group included many children and young adults sent for cardiac investigation because of a functional murmur or nonspecific chest pain discovered by the school or army physician.

Because of the intermittent nature of the disease, normal ECG tracings at a young age and later ones showing pre-excitation do not exclude the possibility that the patient had the disorder even from birth. Therefore, the age figures in all series present, *de facto,* only the age at which the disorder was discovered. This is true also for patients where the presenting symptom is an attack of paroxysmal tachycardia (see Chapter VII, Prognosis, Tachycardias).

Familial Occurrence

Pre-excitation syndrome in two or more members of the same family, although rare, does occur.[184-187] Table IV summarizes the findings in 21 families from the literature and one from our series. The number of affected members in any family may vary from 2 (the majority) to 6.[138] Harnischfeger[188] reported a particularly interesting family with 5 cases, 2 of which were monozygotic twins. Westlake[138] and Ohnell[53] reported families where some members had classic pre-excitation ECG's while others were borderline and suggestive but not conclusive. Pre-excitation in our family occurred in a mother and son (Figure 1 and 2).

While the combination of familial cardiomyopathy (FCM) and pre-excitation has been reported[200] (our material includes one such case, Figure 3), the findings of more than one case among the offspring of a family with FCM is rare. As shown in Table IV, this combination was found in 7 out of 21 families, and in an additional one (Ohnell[53]), where it was suggested from the described symptomatology but not specifically stated by the author. Ten years ago, Scherf and Cohen[201] noted that nothing definite was known about hereditary factors governing the appearance of the pre-excitation syndrome; this statement still applies. We consider a genetic background for the syndrome as probable, based on the following: 1) the finding of pre-excitation without any other associated heart disease in 2 or even 3 generations of the same family; 2) the occurrence of more than one case in families where the offspring have, in addition to pre-excitation, a proven genetic disorder such as familial

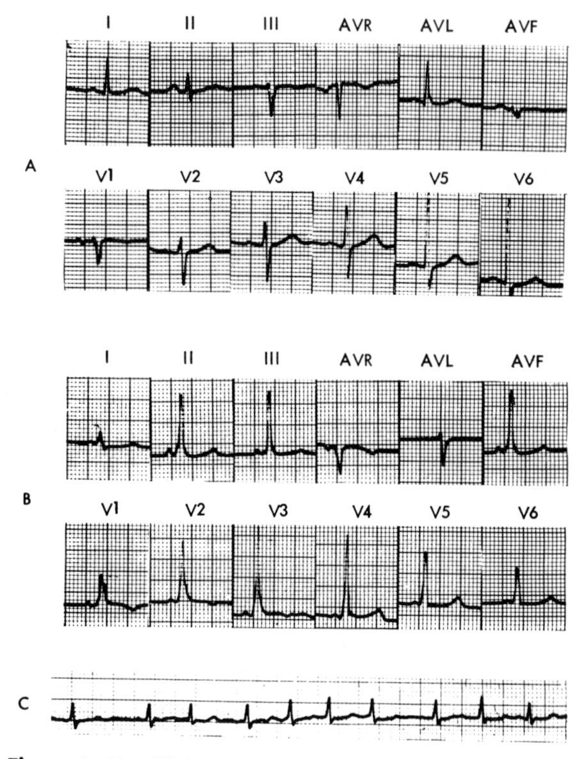

Figure 1. *Familial pre-excitation in a 50-year-old woman.* **a.** *Normal conduction.* **b.** *Pre-excitation conduction.* **c.** *Paroxysm of atrial fibrillation with narrow QRS complexes.*

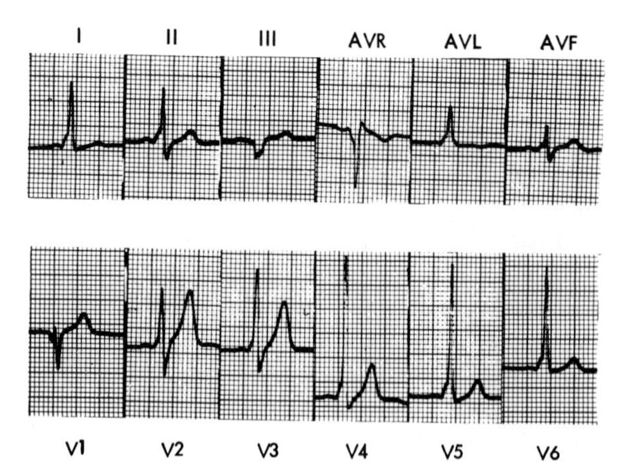

Figure 2. *Familial pre-excitation in the 28-year-old son of the woman in Figure 1. Pre-excitation type B.*

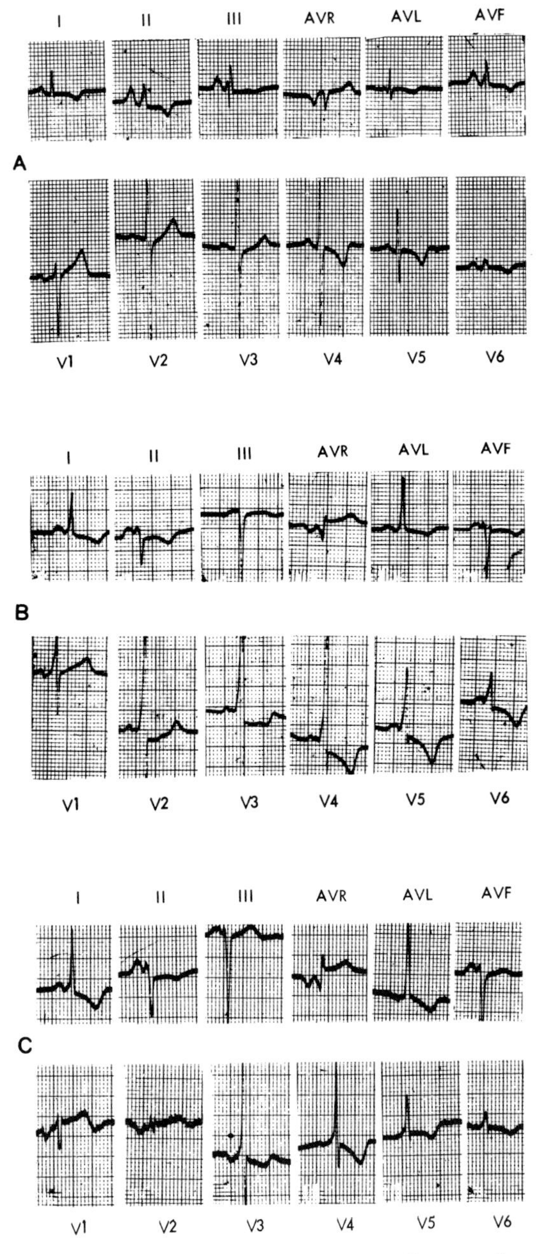

Figure 3. *Pre-excitation in familial cardiomyopathy.* **a.** *During normal conduction: P pulmonale, negative T waves in leads II, aVF, $V_{4\text{-}6}$.* **b, c.** *During pre-excitation. Note marked P wave changes in leads II, III, aVF.*

cardiomyopathy; 3) the finding of pre-excitation in 2 female monozygotic twins. However, other authors reporting only one member of a family with a classic pre-excitation syndrome did not find additional incidences of the disorder. This led Warner and McKusick[202] to state "tentatively" that heredity plays little part in the etiology of this syndrome.

Heredity may be considered one of the many working hypotheses of the etiology of pre-excitation, and will be discussed further in Theories (see Chapter X).

The Pre-excitation Syndrome Associated with Congenital Heart Disease

The association of pre-excitation and congenital heart disease was first noted in the early 1940's and subsequently reported by Nadray,[203] Pohlmann,[204] Ohnell,[53] Bodlander,[205] Moore[206] and others.[207-220] In these and other publications the association was generally considered coincidental. In 1955 Sodi-Pallares[221] first focused attention on the concomitant appearance of the ECG disorder and Ebstein's anomaly of the tricuspid valve. In 1959 Schiebler et al[166] reviewed the literature on congenital heart disease and pre-excitation and presented 68 cases with both; however, he noted that paucity of autopsies and other specific information limited their value and made generalizations tenuous.

Table V summarizes the findings of 3 studies on congenital heart disease and pre-excitation, chosen for the random manner in which congenital heart disease was found among the pre-excitation cases.[165,166,181] The data from these studies were very similar to Schiebler's findings[166] in his review of the literature. Analyzing the two groups together, Schiebler's 68 cases and the 54 summarized in Table V, the following emerges: 1) the pre-excitation syndrome occurs in many types of congenital heart disease; 2) the number of cases with ventricular septal defect (as an isolated malformation, 20 of 122 cases) and of transposition of the great vessels (not always specified if corrected or complete, 7 of 122 cases) exceeds the expected numbers; 3) the association of pre-excitation and Ebstein's anomaly of the tricuspid valve is far greater than what might be expected on the basis of the reported incidence of the latter malformation (31 of 122). If we add cases with tricuspid atresia (4 of 122) to this last group, it may be suggested that the pre-excitation syndrome has some relationship to lesions of the tricuspid valve.

There are also case reports[222,223] and smaller series[224,225] which concentrate on only one specific kind of congenital heart disease: Hostreiter et al's[151] 5 cases of pre-excitation syndrome with subvalvular aortic stenosis (discrete and/or muscular), Sealy et al's[226] 5 cases of "balloon mitral

TABLE V
Pre-excitation with Associated Congenital Heart Disease

Congenital Heart Disease	Schiebler[166]	Swiderski[165]	Giardina[181]	Total
Ebstein's Anomaly	9	4	2	15
Ventricular Septal Defect	(1)*	2 (1)	5 (2)	7 (4)
Atrial Septal Defect	1		1	2
Transposition	1	3	1	5
Patent Ductus Arteriosus	1 (1)	(1)	3	4 (2)
Tetralogy of Fallot	1		3	4
Dextroversion	1			1
Undiagnosed	1	2	1	4
Primary Cardiomyopathy	—	2	—	2
Cardiomyopathy	—	3	—	3
Coarctation of the Aorta	—	1	—	1
Pulmonary Stenosis	—	1	2 (2)	3 (2)
Tricuspid Atresia	—	1	—	1
Mitral Atresia	—	—	1	1
Double Outlet	—	—	1	1
Total	15	19	20	54 (8)

*associated with other congenital cardiac malformations

valve" among patients undergoing surgery, and Westlake et al's[138] 6 cases in a family with primary heart disease (cardiomyopathy). These papers, interesting in themselves, are highly specific and do not contribute information about the frequency of heart disease associated with pre-excitation in the general population.

Among our own 6 cases of pre-excitation with congenital heart disease, there was one of each of the following: Ebstein's anomaly of the tricuspid valve (Figure 4); patent ductus arteriosus, associated with pulmonary hypertension (Figure 5a and b); discrete and muscular subaortic stenosis; familial cardiomyopathy (Figure 3); endocardial fibroelastosis (Figure 6); and an anomalous origin of both great vessels from the right ventricle (Figure 7a and b).

All in all, there is little information regarding the incidence of the pre-excitation syndrome in congenital heart disease. Donzelot et al[227] reported 3 cases among 1,100 patients (0.27%), while Hecht[228] reported 3 among 350 consecutive patients (0.86%). However, when pre-excitation is taken as the point of reference, the following is seen: 15 cases of congenital heart disease among 28 patients (53%) with pre-excitation

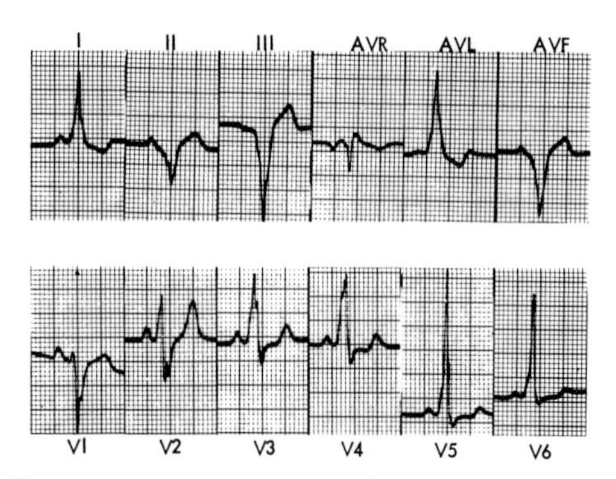

Figure 4. *Pre-excitation type BS in Ebstein's anomaly in a 17-year-old man (see Chapter II and Table XIII for abbreviations).*

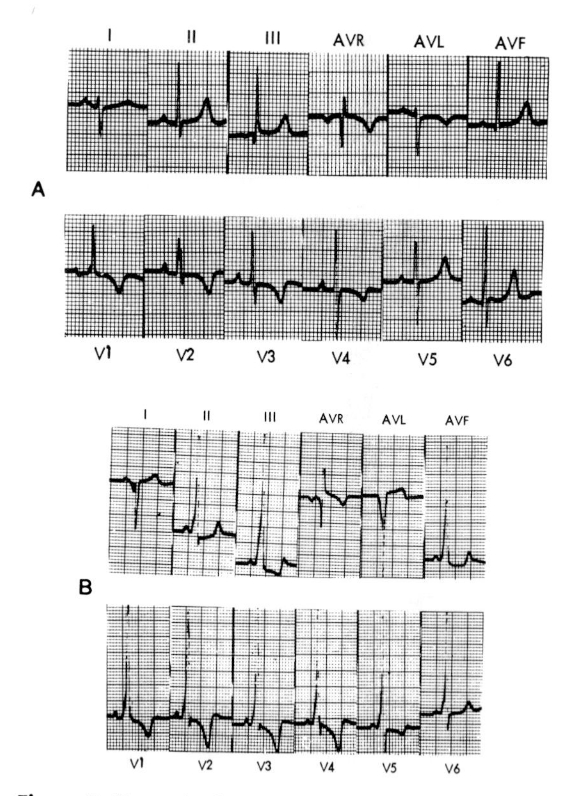

Figure 5. *Pre-excitation in patent ductus arteriosus (PDA) with pulmonary hypertension in a 13-year-old girl.* **a.** *During normal conduction: right ventricular hypertrophy, right axis deviation.* **b.** *During pre-excitation type AI (see Chapter II and Table XIII for abbreviations).*

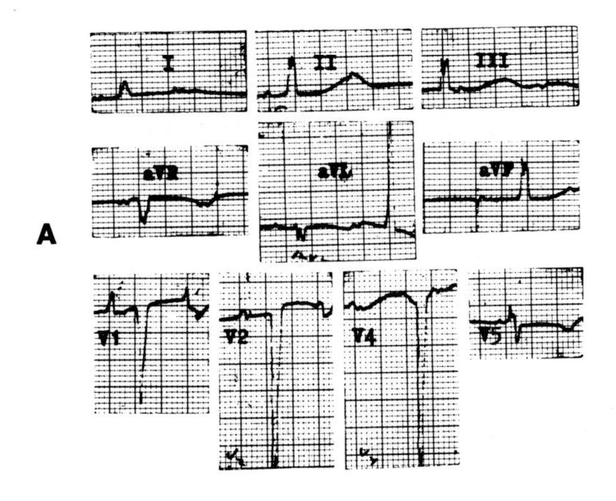

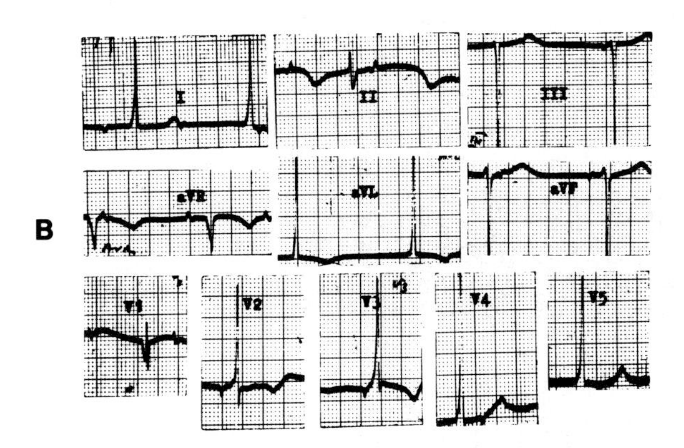

Figure 6. *Pseudo–pre-excitation in fibroelastosis.* **a.** *Complete AV block with idioventricular rhythm.* **b.** *Complete AV block with QRS complexes exhibiting all the characteristics of pre-excitation (see Chapter III).*

(Schiebler et al[166]); 19 cases of congenital heart disease among 48 children (40%) with pre-excitation (Swidersky et al[165]); and 20 cases of congenital heart disease among 62 patients (33%) with pre-excitation (Giardina[181]). If these figures are considered definitive, we can expect congenital cardiac malformation in one-third to one-half of all children with pre-excitation. The high mortality rate from congenital heart disease in these children (8 out of 20 died during Giardina's follow-up study[181]) reduces the number of adults with concurrence of the two disorders.

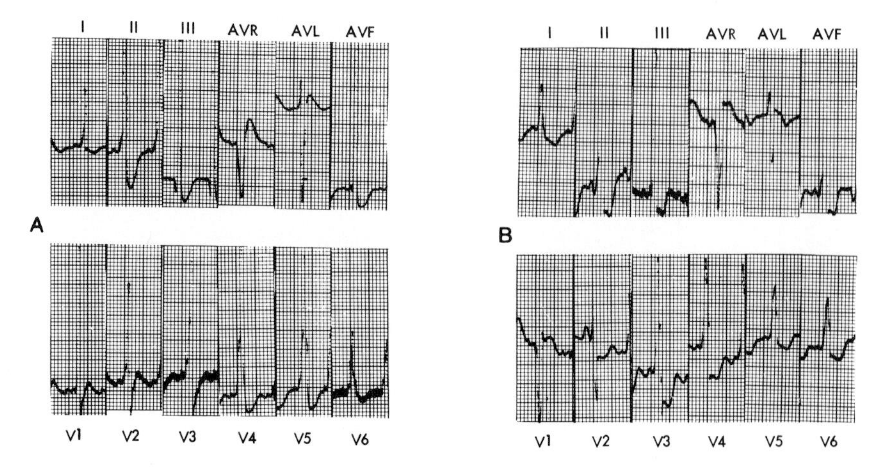

Figure 7. *Pre-excitation in anomalous origin of both great vessels from the right ventricle in a one-week-old male infant.* **a.** *Note Rs pattern of the QRS in $V_{1\text{-}2}$ (type A).* **b.** *Qr pattern in V_1 (type B).*

The Pre-excitation Syndrome Associated with Cardiac Diseases Other than Congenital

The combination of congenital heart disease and the pre-excitation syndrome was considered separately, out of a belief that it is mainly a pediatric problem. The high percentage of congenital malformations among children suffering from pre-excitation, the high mortality rate in these children and the affinity of the syndrome to some specific cardiac malformations are not reflected in the series of heart disease in the general population with pre-excitation. Table VI summarizes 11 series covering 593 cases of pre-excitation collected from the literature and from our files, in which the incidence and details of associated heart disease were available. No cardiac involvement was detected in the vast majority (68.3%),[168] and these cases are considered "isolated pre-excitation." In the remaining 188 cases (31.7%), the percentage of heart disease ranged from 18.4% (Bleifer[120]) to 62.2% (Otto[149]).

Rheumatic Disease. Thirty-two of the 593 patients had associated rheumatic valvular disease. All but one of the authors reported such an association. We found it in 9 of our 215 patients (Figure 8a, b and c, 9a, b and c, 10).

Atherosclerotic Heart Disease. The predominant associated heart disease was atherosclerosis, which included anginal syndrome, coronary insufficiency, myocardial ischemia and myocardial infarction.[229-234]

TABLE VI
Pre-excitation and Cardiovascular Diseases

Study	Cases with Pre-excitation	Associated Heart Disease	Rheumatic Valve Disease	Athero-sclerosis*	Hyper-tension	Myocarditis (+ Rheumatic Fever)	Uncertain Cardiac Involvement	Congenital Heart Disease**
Otto, 1967[149]	37	23 (62.2%)	3	3	5	12	—	—
Nasser et al, 1971[171]	29	15 (51.7%)	3	7	2	—	—	3
Hejtmancik, 1957[49]	80	32 (40%)	2	16	6	—	2	6
Mortensen, 1954[155]	45	12 (26.7%)	—	1	5	—	6	—
Burch, 1961[177]	15	8 (53.3%)	2	4	2	—	—	—
Longhini, 1973[173]	24	8 (33.3%)	1	2	3	1	—	1
Hunter, 1940[146]	19	6 (31.6%)	2	—	1	2	1	—
Willius, 1946[59]	65	19 (29.2%)	5	4	9	—	1	—
Bleifer, 1959[120]	38	7 (18.4%)	2	1	—	—	—	4
Wellens, 1974[172]	26	8 (30.8%)	3	2	—	—	3	—
Tel Hashomer, 1974	215	50 (23.3%)	9	14	12	2	7	6
Total	593	188 (31.7%)	32	54	45	17	20	20

*Atherosclerotic heart disease includes anginal syndrome, coronary insufficiency, myocardial ischemia and myocardial infarction
**Includes the number observed in all age series

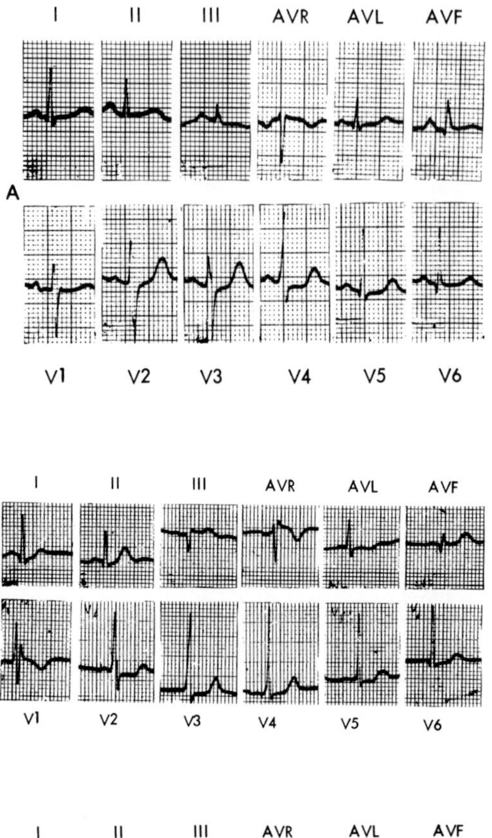

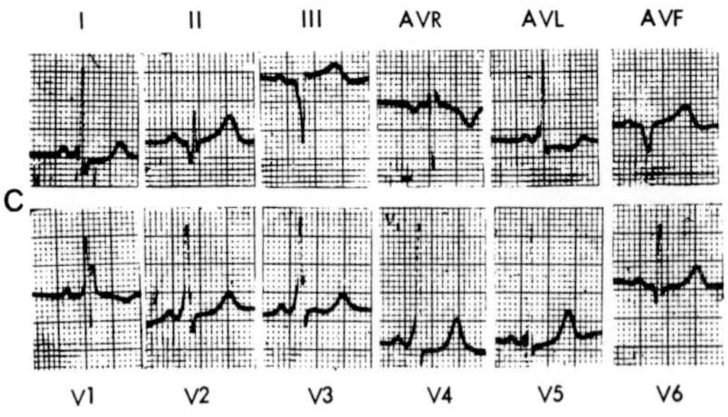

Figure 8. *Pre-excitation in mitral stenosis in a 33-year-old woman.* **a.** *Normal AV conduction.* **b.** *Pre-excitation conduction. Note P wave changes, especially in leads II, III, aVF. Relatively narrow QRS complexes. Rsr' pattern in V_1. QS in III and qr in aVF.* **c.** *A second form of pre-excitation QRS pattern. Note additional P wave changes.*

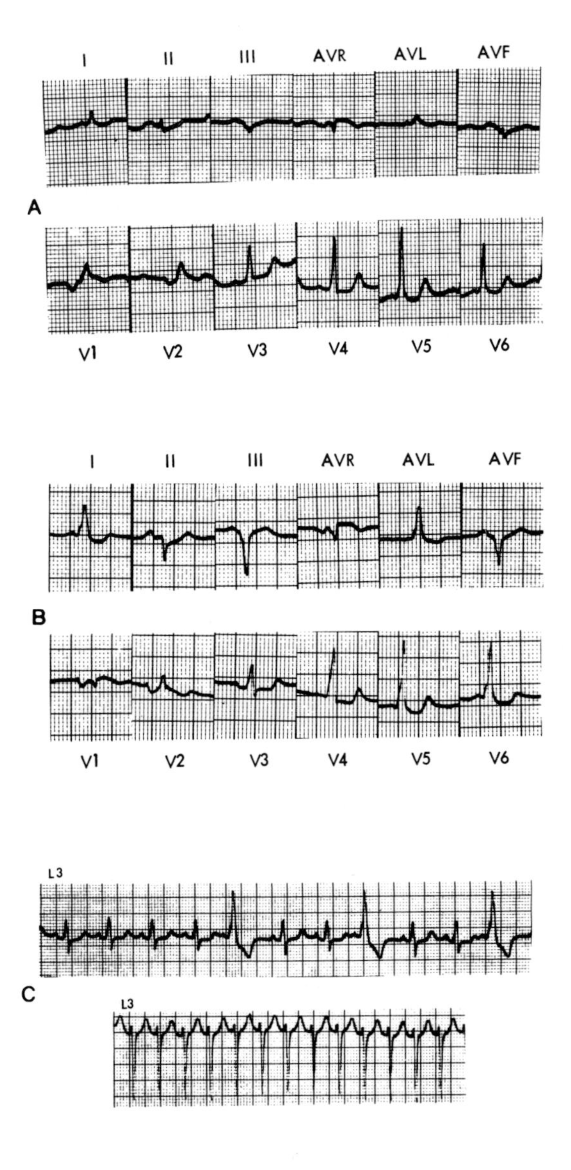

Figure 9. *Pre-excitation in mitral stenosis in a 45-year-old woman.* **a.** *Pre-excitation conduction. Note marked P mitral.* **b.** *Pre-excitation conduction in the same patient 2 months after mitral commissurotomy. Note disappearance of R in V_1 and more left axis deviation.* **c.** *Intermittent pre-excitation, lead III in the upper panel and paroxysmal atrial tachycardia (PAT) in the lower panel.*

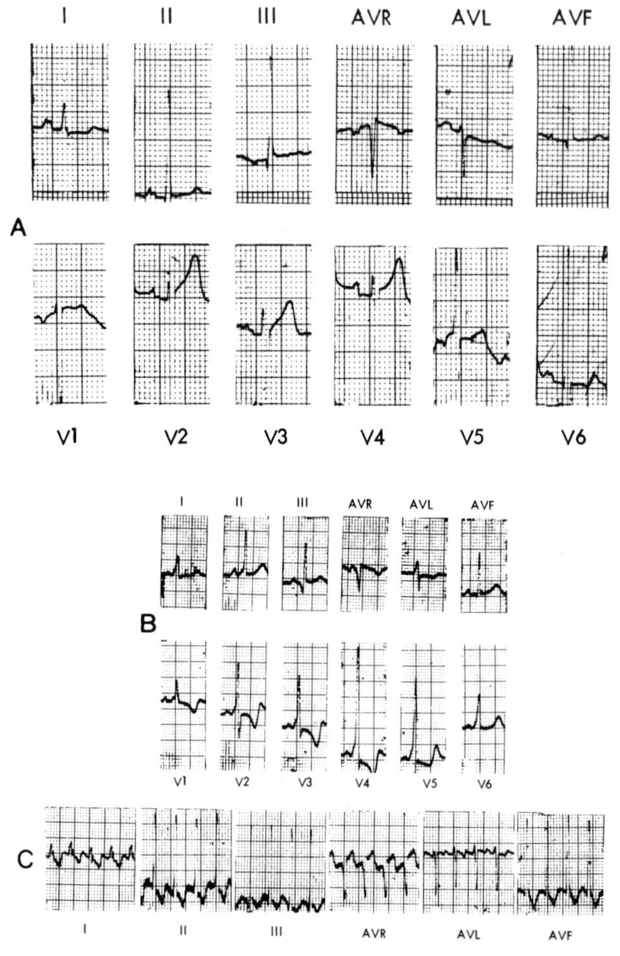

Figure 10. *Pre-excitation in aortic insufficiency in a 10-year-old boy.* **a.** *During normal AV conduction. Note deep S waves in V_1 and high R in V_6 (left ventricular hypertrophy).* **b.** *Pre-excitation conduction, type AI. Note the disappearance of deep S in V_1 and high R in V_6. Note also P wave changes when comparing a with b.* **c.** *Paroxysmal atrial tachycardia (PAT).*

Again, all but one author[146] found cases of atherosclerosis among their patients, and we found 14 (Figure 11a, b and c, 12).

Hypertensive Cardiovascular Disease. This was another relatively large group, with 45 cases; many of them also had atherosclerotic heart disease but were classified under the heading of hypertension

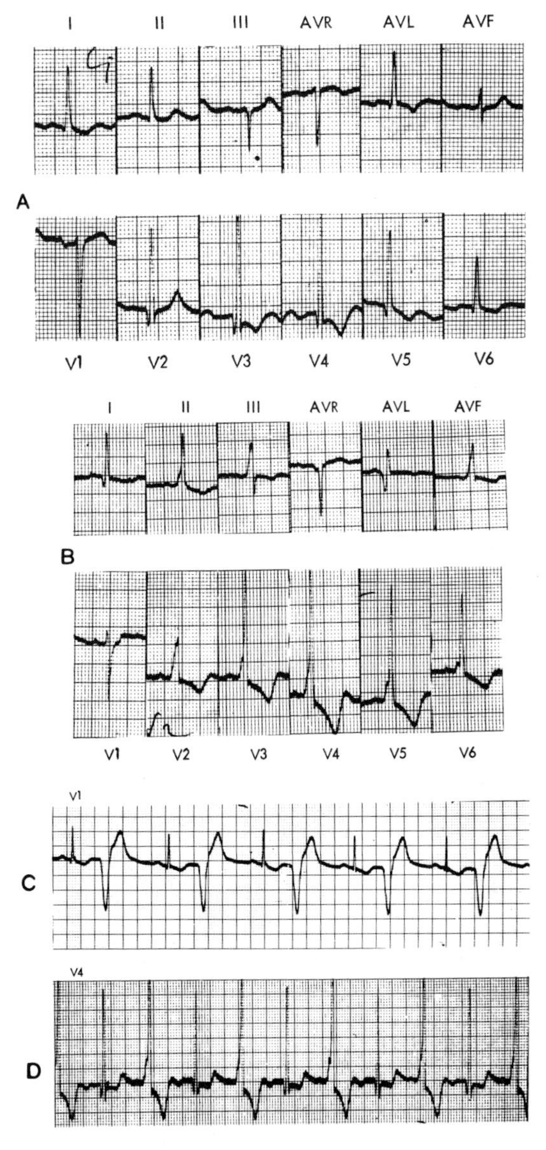

Figure 11. *Pre-excitation in atherosclerotic heart disease in a 55-year-old man with anginal syndrome.* **a.** *Normal AV conduction. Left heart hypertrophy and negative T waves in leads I, aVL, $V_{3\text{-}6}$. P-R = 0.24 second (V_4).* **b.** *Pre-excitation conduction. The P-R is reduced to 0.16 second (normal P-R with pre-excitation).* **c.** *Ventricular premature beats during normal AV conduction, bigeminy.* **d.** *Intermittent pre-excitation imitating ventricular bigeminy. Note the differences in form of the P waves (lead V_4) as well as in the P-R intervals of normal and anomalous conducted beats.*

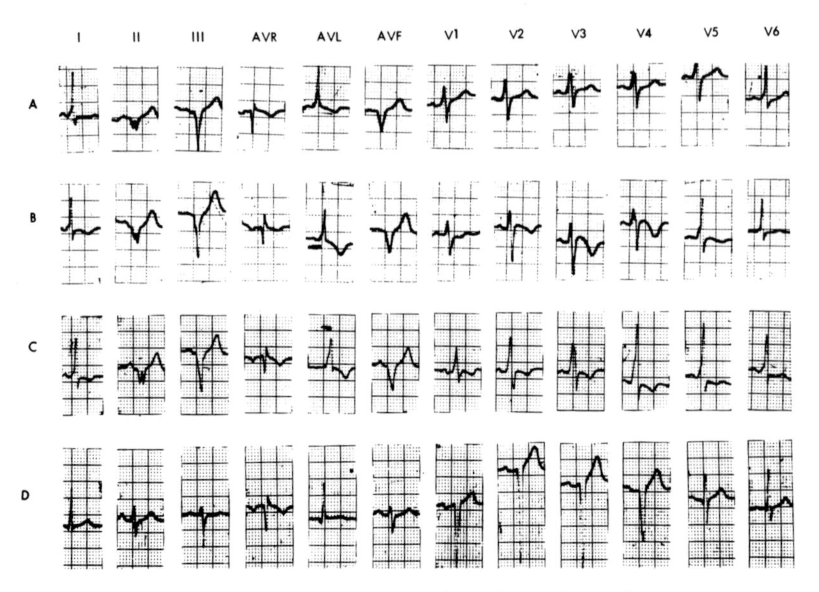

Figure 12. *Pre-excitation in atherosclerotic heart disease in a 52-year-old man who suffered an attack of acute mycoardial infarction.* **a, b, c.** *Pre-excitation conduction. Note QS pattern in leads II, III, aVF and appearance of negative T waves in leads $V_{2\text{-}5}$ in b.* **d.** *During normal AV conduction. Note the change in QRS pattern in limb and chest leads (see Chapter IX).*

only. Twelve were from our series (Figure 13a and b) including one case of juvenile hypertension.

Myocarditis and/or "Acute Rheumatic Fever." There were 17 patients in this group, two of which were ours.

Uncertain Cardiac Involvement. In 20 patients there was no definite diagnosis of heart disease, although the authors felt that there was some type of cardiac involvement. They did not fit into the framework of known cardiac disorders.

The Pre-excitation Syndrome Associated with Noncardiac Diseases

Noncardiac diseases associated with pre-excitation have been reported in our series, as well as others. In some cases, pre-excitation was discovered during investigations of noncardiac diseases, and vice versa. While there is a rather high degree of overlap between the two, this would be expected in any large series of patients investigated and was thought to be purely coincidental.[201,235] The association of the two be-

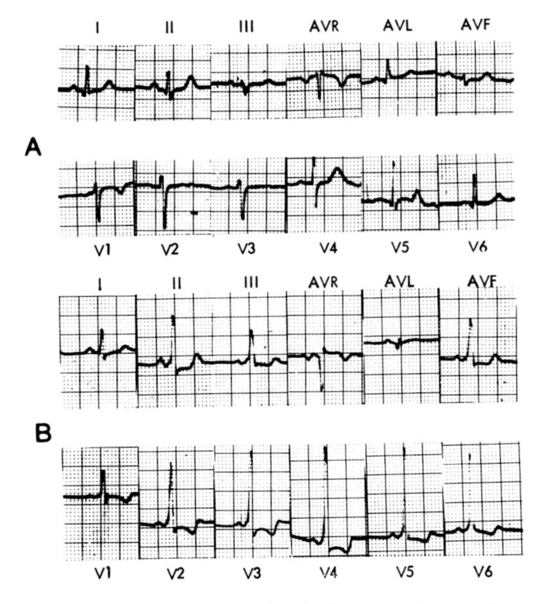

Figure 13. *Pre-excitation in hypertensive cardiovascular disease in a 47-year-old man.* **a.** *Normal AV conduction.* **b.** *Pre-excitation conduction.*

came important, however, with the increased controversy over the existence of an acquired pre-excitation syndrome. Some authors believe that a particular disease, *eg,* allergic diseases[236] and other conditions,[2,53,155,237] may produce the electrical disorder in the heart. Prinzmetal et al[127] observed the syndrome after a respiratory infection or tonsillitis (some even saw it disappear after tonsillectomy). Similar observations were reported in cases of thyrotoxicosis,[59,124,238-245] especially when the anomalous conduction was no longer found after successful treatment of the hyperthyroidism. Central nervous system disturbances have been found in some series,[19,49,132,155,162,246-250] and in an especially large number in children (Swiderski et al,[165] 9 of 48; Schiebler,[153] 5 of 27).

Acute rheumatic fever and myocarditis may also be considered here, as they are systemic as well as cardiac diseases, and because many authors stressed their appearance before or at the time pre-excitation was found.[251-264]

Doumer and Merle[265] described the disorder together with various other noncardiac diseases.[266-281]

TABLE VII
The Pre-excitation Syndrome Associated with Noncardiac Diseases in the Tel Hashomer Series (215 Patients)

Noncardiac Diseases	# Cases
Respiratory diseases	10
Orthopedic disorders and accidents	8
Urogenital diseases	7
Metabolic diseases (diabetes, thyrotoxicosis)	5
Psychiatric disturbances	4
Tumors	4
Gastrointestinal disorders	3
Other diseases	5
Total	46 (21.39%)

In our series of 215 cases of pre-excitation, 46 patients (21.39%) were found to have associated noncardiac diseases (Table VII). The particular diseases of these patients were randomly distributed and did not confirm the associations between pre-excitation and diseases reported in the literature. Similar random distribution was described in other series.[149,282]

The general problem of acquired pre-excitation will be discussed in Theories (see Chapter X).

REFERENCES

1. **Wolff L, Parkinson J, White PD:** Bundle branch block with short P-R interval in healthy young people prone to paroxysmal tachycardia. *Am Heart J* 5:685, 1930.
2. **Hamburger WW:** Bundle branch block. Four cases of intraventricular block showing some interesting and unusual clinical features. *Med Clin North Am* 13:343, 1929.
3. **Sigler LH:** Functional bundle-branch block (partial) paradoxically relieved by vagal stimulation. *Am J Med Sci* 185:211, 1933.
4. **Wedd AM:** Paroxysmal tachycardia with reference to nomotopic tachycardia and the role of the extrinsic cardiac nerves. *Arch Intern Med* 27:571, 1921.
5. **Wilson FN:** A case in which the vagus influenced the form of the ventricular complex of the electrocardiogram. *Arch Intern Med* 16:1008, 1915.
6. **Holzmann M, Scherf D:** Über Elektrokardiogramme mit verkürtzter Vorhof-Kammer-Distanz und positiven P-Zacken. *Z Klin Med* 121:404, 1932.
7. **James TN:** The Wolff-Parkinson-White syndrome. *Ann Intern Med* 71:399, 1969.
8. **James TN:** The Wolff-Parkinson-White syndrome: evolving concepts of its pathogenesis. *Prog Cardiovasc Dis* 13:159, 1970.

9. **Kent AFS:** Researches on the structure and function of the mammalian heart. *J Physiol* 14:233, 1893.
10. **Kent AFS:** The structure of the cardiac tissue at the auriculoventricular junction. *J Physiol* 47:17, 1913.
11. **Kent AFS:** A lecture on some problems in cardiac physiology. *Br Med J* 2:105, 1914.
12. **Kiss T, Tóth J:** Intermittent WPW syndrome in myocardial infarcts. (Hun) *Orv Hetil* 113:395, 1972.
13. **Lev M, Lerner R:** The theory of Kent. A histologic study of the normal atrioventricular communications of the human heart. *Circulation* 12:176, 1955.
14. **Durrer D, Schuilenburg RM, Wellens HJ:** Pre-excitation revisited. *Am J Cardiol* 25:690, 1970.
15. **Gallagher JJ, Gilbert M, Svenson RH, et al:** Wolff-Parkinson-White syndrome. The problem, evaluation, and surgical correction. *Circulation* 51:767, 1975.
16. **Butterworth JS:** The experimental production of the syndrome of apparent bundle branch block with short P-R interval. *J Clin Invest* 20:458, 1941.
17. **Butterworth JS, Poindexter CA:** Short PR interval associated with a prolonged QRS complex. *Arch Intern Med* 69:437, 1942.
18. **Butterworth JS, Poindexter CA:** Fusion beats and their relation to the syndrome of short P-R interval associated with a prolonged QRS complex. *Am Heart J* 28:149, 1944.
19. **Wood FC, Wolferth CC, Geckeler GD:** Histologic demonstration of accessory muscular connections between auricle and ventricle in a case of short P-R interval and prolonged QRS complexes. *Am Heart J* 25:454, 1943.
20. **Burchell HB:** Atrioventricular nodal (reciprocating) rhythm. Report of a case. *Am Heart J* 67:391, 1964.
21. **Chia BL:** Atrial flutter, supraventricular tachycardia, junctional rhythm and complete right bundle branch block in a patient with the Wolff-Parkinson-White syndrome (type B). *J Electrocardiol* 6:353, 1973.
22. **Ciaramella S:** 2 cases of Wolff-Parkinson-White syndrome in pregnancy. (Ital) *Arch Ostet Ginecol* 75:254, 1970.
23. **Cossio P, Berconsky I, Kreutzer R:** P-R acortado con QRS ancho y mellado tipo bloqueo de rama intra-ventricular de menor grado. *Rev Argent Cardiol* 2:411, 1936.
24. **de Mey D:** Un cas de Wolff-Parkinson-White du type rare. Action pharmacodynamique. *Arch Mal Coeur* 47:688, 1954.
25. **Fontaine G, Plagne A, Vachon JM, et al:** Complete paroxysmal arrhythmia in Wolff-Parkinson-White syndrome (study of a case with multicanal analyser). *Ann Med Interne* (Paris) 121:687, 1970.
26. **Gordon MK:** Some remarks concerning a case of Wolff-Parkinson-White syndrome. (Rus) *Kardiologiia* 4:84, 1964.
27. **Grasso A, Artese D, Ferro-Luzzi M, et al:** Pathogenetic considerations on the Wolff-Parkinson-White syndrome. A propos of a case. (Ital) *Sicilia Sanit* 20 (suppl):229, 1967.
28. **Green JR Jr, Bartley TD, Schiebler GL:** The Wolff-Parkinson-White syndrome and ventricular septal defect. A case report of successful surgery in an 18-year-old man. *Dis Chest* 47:659, 1965.
29. **Hannington-Kiff JG:** The Wolff-Parkinson-White syndrome and general anesthesia. *Br J Anesth* 40:791, 1968.
30. **Heikkilä J, Jounela A:** Intra-A-type variation of WPW syndrome. *Br Heart J* 37:767, 1975.
31. **Kuppardt H, Israel G, Kuppardt B, et al:** Case report on WPW syndrome. (Ger) *Z Gesamte Inn Med* 27:748, 1972.
32. **Lange A:** An unusual case of Wolff-Parkinson-White syndrome with features of the pre-excitation complex of Ohnell. *Bull Pol Med Sci Hist* 13:73, 1970.
33. **Levine H:** Electrocardiograms of the month. *Conn Med* 33:283, 1969.
34. **Page HL, Campbell WB:** EKG of the month: Wolff-Parkinson-White syndrome. *J Tenn Med Assoc* 66:43, 1973.

35. **Pearson JR, Wallace AW:** The syndrome of paroxysmal tachycardia with short P-R interval and prolonged QRS complexes, with report of two cases. *Ann Intern Med* 21:830, 1944; **correction** 22:294, 1945.
36. **Pentimone F, Cinotti G, Pesola A, et al:** On the correlation between ventricular activation and changes of the 1st and 2nd tone in the Wolff-Parkinson-White syndrome. (Ital) *Boll Soc Ital Cardiol* 13:116, 1968.
37. **Poumailloux M:** A propos d'une observation du syndrome de Wolff-Parkinson-White. *Arch Mal Coeur* 39:229, 1946.
38. **Rubin IL, Frieden J:** Electrocardiogram of the month: Wolff-Parkinson-White syndrome. *NY State J Med* 73:689, 1973.
39. **Shamroth L, Krikler DM:** Location of the pre-excitation areas in the Wolff-Parkinson-White syndrome. *Am J Cardiol* 19:889, 1967.
40. **Soulié P, Combet J, Brial M:** Syndrome de Wolff-Parkinson-White et dérivations précordiales. *Arch Mal Coeur* 40:455, 1957.
41. **Zakopoulos KS, Tsatas AT, Liokis TE:** Type B Wolff-Parkinson-White syndrome associated with right bundle branch block. *Dis Chest* 46:346, 1964.
42. **Zuckermann R:** Ein Fall von Wolff-Parkinson-White. *Kreislaufforsch* 45:712, 1956.
43. **Bioerck G:** Five cases with pre-excitation electrocardiograms. *Acta Med Scand* 125:465, 1946.
44. **Boyer NH:** The Wolff-Parkinson-White syndrome. *NEJM* 234:111, 1946.
45. **Chung KY, Walsh TJ, Massie E:** Wolff-Parkinson-White syndrome. *Am Heart J* 69:116, 1965.
46. **Doschitsin VL, Eremina GA:** Clinical significance of WPW and CLC syndrome. (Rus) (English Abstr) *Kardiologiia* 13:87, 1973.
47. **Flensted-Jensen E:** Wolff-Parkinson-White syndrome. A long-term follow-up of 47 cases. *Acta Med Scand* 186:65, 1969.
48. **Guasi Gené C, Pulido SP, Vela YE, et al:** Wolff-Parkinson-White syndrome. Study of 70 cases. (Sp) *Arch Inst Cardiol Mex* 36:74, 1966.
49. **Hejtmancik MR, Herrman GR:** The electrocardiographic syndrome of short P-R interval and broad QRS complex. A clinical study of 80 cases. *Am Heart J* 54:708, 1957.
50. **Ishikawa H, Morisato F, Nonaka H, et al:** Studies on the Wolff-Parkinson-White syndrome. Clinical history and symptoms of 45 cases. *J Nara Med Assoc* 23:319, 1972.
51. **Jouve A:** Raccourissement permanent de PR avec déformation du ventriculogramme. Action des épreuves clinico-physiologiques. *Arch Mal Coeur* 37:133, 1944.
52. **Kuramitsu H, Obara Y, Sasaki T, et al:** On the WPW syndrome. Report of 22 cases. (Jap) *Jpn Circ J* 27:668, 1963.
53. **Ohnell RF:** Pre-excitation, a cardiac abnormality. *Acta Med Scand* (suppl 152), 1944.
54. **Pajaron Lopez A, Povedasierra J, Medrano GA:** Pre-excitation syndrome (review of 235 cases). *Arch Inst Cardiol Mex* 43:826, 1974.
55. **Ray D, Danino EA:** Wolff-Parkinson-White syndrome. A review of 42 cases. *Indiana Heart J* 27:13, 1975.
56. **Reed JC, Langley RW, Utz DC:** A review of eight cases of Wolff-Parkinson-White syndrome including four cases with initial left ventricular activation as demonstrated by multiple precordial leads. *Am Heart J* 33:701, 1947.
57. **Reinikainen M:** Occurrence of the WPW (Wolff-Parkinson-White) syndrome in a series from certain medical hospitals in Finland. *Ann Med Intern Fenn* 50 (suppl 34), 1961.
58. **Scherf D, Bornemann C:** A case of pre-excitation. *Am Heart J* 88:627, 1974.
59. **Willius FA, Carryer HM:** Cardiac clinics; electrocardiograms displaying short P-R intervals with prolonged QRS complexes. An analysis of sixty-five cases. *Proc Staff Meet Mayo Clin* 21:438, 1946.
60. **Aixala R:** Sindrome de Wolff-Parkinson-White (PR corto and QRS ancho y mellado). *Rev Cubana Cardiol* 4:61, 1943.
61. **Antonioli G, Pozzar C, Abrahamsohn R:** Clerc-Levy-Cristesco syndrome (clinical contribution). (Ital) *Anna Ferrara* 18:1145, 1965.

62. **Averill KH, Lamb LE:** Electrocardiographic findings in 67,375 asymptomatic subjects. I. Incidence of abnormalities. *Am J Cardiol* 6:76, 1960.
63. **Bardan V:** Wolff-Parkinson-White syndrome. (Rec progress: still unsolved). *Med Intern* (Bucur) 25:1435, 1973.
64. **Cagan S, Cagánova A, Králová G:** Electrocardiographic pictures of WPW syndrome with early delta wave. (Slo) *Bratisl Lek Listy* 55:239, 1971.
65. **Cariello M:** The Wolff-Parkinson-White syndrome and its legal and medicolegal repercussions. (Ital) *Folia Med* (Napoli) 46:491, 1963.
66. **Castellanos A Jr, Lemberg L, Claxton BW:** Wolff-Parkinson-White syndrome: generalities. *Chest* 65:307, 1974.
67. **Castellino N:** Contribution to the study of the Wolff-Parkinson-White syndrome. (Ital) *Riforma Med* 78:909, 1964.
68. **Clerc A, Levy R, Cristesco C:** A propos du raccourcisseement permanent de l'espace P-R de l'électrocardiogramme sans déformation du complexe ventriculaire. *Arch Mal Coeur* 31:569, 1938.
69. **Coelho E:** Nova contribuiçao para o estudo do sindroma de Wolff-Parkinson-White (WPW). *Amatus Lusitanus* 4:603, 1945.
70. **Coelho de Oliveira N:** Wolff-Parkinson-White syndrome. (Sp) *Prensa Med Argent* 53:181, 1966.
71. **Dassen R:** A proposito del sindrome de Wolff-Parkinson-White. *Medicina* 2:192, 1942.
72. **Doktorczyk H, Galazka A, Knapikowa D:** Analysis of our cases of Wolff-Parkinson-White syndrome from the aspect of etiology and stereocardiographic picture. (Pol) *Wiad Lek* 22:1171, 1969.
73. **Doneff D, Scheid H:** Ein Beitrag zum Wolff-Parkinson-White Syndrom. *Cardiology* (Basel) 34:199, 1959.
74. **Fox TT:** Further remarks on the syndrome of aberrant A-V conduction (Wolff-Parkinson-White). *Am Heart J* 53:771, 1957.
75. **Frau G, Maggi GC:** "La patogenesi della sindrome di Wolff-Parkinson-White (contributo sperimentale)," in *Le Congrès Mondial de Cardiologie, Paris 3-9 Sept 1950,* p 380, Paris: JB Baillère et Fils, 1951.
76. **Frau G, Maggi GC:** *La Sindrome de Wolff-Parkinson-White Contributo Sperimentale e Clinico,* Milano: A Recordati, 1954.
77. **Gilgenkrantz JM:** Wolff-Parkinson-White syndrome. General, etiologic and anatomical considerations. *Ann Cardiol Angeiol* (Paris) 21:319, 1972.
78. **von Gruber Z:** Über die Elektrokardiogramme mit scheinbar kurzem A-v-Intervall und verbreitertem QRS-Komplex. *Z Kreislaufforsch* 30:100, 1938.
79. **Hayton RC, Lopez JF:** The ventricular pre-excitation (Wolff-Parkinson-White) syndrome. *Can Med Assoc J* 98:950, 1968.
80. **Heinecker R:** The significance of the Wolff-Parkinson-White syndrome. *Dtsch Med Wochenschr* 93:357, 1968.
81. **Holzmann M:** Das Syndrom von Wolff-Parkinson-White. *Z Kreislaufforsch* 51:275, 1962.
82. **Holzmann M, Volkert M:** Clinical contributions to the Wolff-Parkinson-White syndrome. (Fr) *Actual Cardiol Angeiol Int* 14:273, 1965.
83. **Kapanjie R:** Wolff-Parkinson-White syndrome. *J Am Osteopath Assoc* 68:267, 1968.
84. **Kon'kov AV, Bogatyreva AP:** Wolff-Parkinson-White syndrome. (Rus) *Vrach Delo* 1:73, 1967.
85. **Kuramoto K, Matsushita S:** Classification and interpretation of WPW syndrome. (Jap) (Eng Abstr) *Jpn J Clin Med* 30:1770, 1972.
86. **Laham J:** *Le Syndrome de Wolff-Parkinson-White.* Paris: Libraire Maloine, 1969.
87. **Leduc G:** Wolff-Parkinson-White syndrome. (Fr) *Un Med Canada* 94:805, 1965.
88. **Limborskaia TI:** Wolff-Parkinson-White syndrome according to data from the "Pushkin" sanatorium. (Rus) *Ter Arkh* 37:113, 1965.
89. **Lirman AV, Stolbun BM:** The clinical picture of the Wolff-Parkinson-White syndrome. (Rus) (Eng Abstr) *Klin Med* (Moskva) 49:79, 1971.

90. **Lobel I, Stegaru D, Stegaru B:** Contribution to the study of the Wolff-Parkinson-White syndrome. (Rum) *Med Intern* (Bucur) 15:1129, 1963.
91. **Loew DE:** Wolff-Parkinson-White syndrome. *J Electrocardiol* 4:77, 1971.
92. **Mahaim I:** Syndrome de Wolff-Parkinson-White acquis, évolutif. *Folia Cardiol* 7:625, 1948.
93. **Martin-Noel P, Brenier RH, Grundwald D, Denis B, et al:** Wolff-Parkinson-White syndrome. (Fr) *J Med Lyon* 49:1609, 1968.
94. **Massumi RA, Vera Z, Mason DT:** The Wolff-Parkinson-White syndrome, a new look at an old problem. *Mod Concepts Cardiovasc Dis* 42:41, 1973.
95. **Messina JJ:** Wolff-Parkinson-White syndrome. *Maryland Med J* 18:105, 1969.
96. **Misiuniené N:** Premature ventricular beat (WPW) syndrome. (Lith) *Sveik Apsaug* 9:55, 1964.
97. **Narula OS:** Wolff-Parkinson-White syndrome. A review. *Circulation* 47:872, 1973.
98. **Noro L:** On pre-excitation (WPW syndrome). *Acta Med Scand* 121:563, 1945.
99. **Paereli RS:** Clinical evaluation of the nodular form of the Wolff-Parkinson-White syndrome. (Rus) *Kardiologiia* 7:78, 1967.
100. **Pattani F:** Das Syndrom von Wolff, Parkinson and White. *Cardiologia* 12:247, 1947-8.
101. **Pedemonte LE:** Wolff-Parkinson-White syndrome. (Sp) *Rev San Mil Argent* 67:52, 1968.
102. **Pick A, Langendorf R, Katz LN:** Advances in the electrocardiographic diagnosis of cardiac arrhythmias. *Med Clin North Am* 41:269, 1957.
103. **Prieto Solis JA:** Wolff-Parkinson-White syndrome (diagnosis, evaluation and treatment). (Sp) (Eng Abstr) *Rev Clin Esp* 131:399, 1973.
104. **Pruitt RD:** Ventricular pre-excitation (Wolff-Parkinson-White syndrome). *Am J Cardiol* 25:734, 1970.
105. **Rezvin IM:** Premature excitation of cardiac ventricles—Wolff-Parkinson-White syndrome (clinico-electrocardiographic observations). (Rus) *Kardiologiia* 5:61, 1965.
106. **Rosenbaum FF:** Anomalous atrioventricular excitation: panel discussion. *Ann NY Acad Sci* 65:832, 1957.
107. **Sazama KJ:** The syndrome known as WPW: a review. *J Am Med Wom Assoc* 31:56, 1976.
108. **Scherf D:** Über Elektrokardiogramme mit verkürzter Vorhof-Kammer-Distanz und positiven P-Zacken. *Z Klin Med* 121:404, 1932.
109. **Schramm E:** Beitrag zur Klinik und Ätiologie des WPW-Syndroms. *Z Kreislaufforsch* 43:875, 1954.
110. **Sherf L, James TN:** A new electrocardiographic concept: synchronized sinoventricular conduction. *Dis Chest* 55:127, 1969.
111. **Sokoll U:** Wolff-Parkinson-White syndrome. (Ger) *Z Allg Med* 46:1406, 1970.
112. **Terreni F, del Corso P, Antonelli G:** Wolff-Parkinson-White syndrome. *Cardiol Prat* 22:195, 1971.
113. **Tranchesi J, Grinberg M, Moffa P:** Current concepts of Wolff-Parkinson-White syndrome. *Arq Bras Cardiol* 24:71, 1973.
114. **Videla JG:** El sindrome de Wolff-Parkinson-White. *Med Panamericana* 9:281, 1957.
115. **Villa JG, Concepción PM, De Artaza M:** Clinical and electrocardiographic study of Wolff-Parkinson-White syndrome, analysis of 30 cases and review of the literature. (Sp) (Eng Abstr) *Rev Esp Cardiol* 38:109, 1975.
116. **Wolferth CC, Wood FC:** The mechanism of production of short P-R intervals and prolonged QRS complexes in patients with presumably undamaged hearts; hypothesis of an accessory pathway of auriculoventricular conduction (bundle of Kent). *Am Heart J* 8:297, 1933.
117. **Wolff L:** Anomalous atrioventricular excitation, panel discussion. *Ann NY Acad Sci* 65:828, 1957.
118. **Wolff L:** Wolff-Parkinson-White syndrome: historical and clinical features. *Prog Cardiovasc Dis* 2:677, 1960.
119. **Zamfir C, Efanov A:** Considerations on the Wolff-Parkinson-White syndrome. (Rum) *Med Intern* (Bucur) 20:141, 1968.

120. **Bleifer S, Kahn M, Grishman A, et al:** Wolff-Parkinson-White syndrome. A vectorcardiographic, electrocardiographic and clinical study. *Am J Cardiol* 4:321, 1959.
121. **Castellanos A Jr, Chapunoff E, Castillo C, et al:** His bundle electrograms in two cases of Wolff-Parkinson-White (pre-excitation) syndrome. *Circulation* 41:399, 1970.
122. **Gallagher JJ, Svenson RH, Sealy WC, et al:** The Wolff-Parkinson-White syndrome and the pre-excitation dysrhythmias. *Med Clin North Am* 60:101, 1976.
123. **Isaeff DM, Harper Gaston J, Harrison DC:** Wolff-Parkinson-White syndrome. Long-term monitoring for arrhythmias. *JAMA* 222:449, 1972.
124. **Sandberg, L:** Studies on electrocardiographic changes during exercise tests. *Acta Med Scand* (suppl 365): 88, 1961.
125. **Wellens HJ, Schuilenberg RM, Durrer D:** Electrical stimulation of heart in study of patients with the Wolff-Parkinson-White syndrome type A. *Br Heart J* 33:147, 1971.
126. **Zuberbuhler JR, Bauersfeld SR:** Paradoxical splitting of the second heart sound in the Wolff-Parkinson-White syndrome. *Am Heart J* 70:595, 1965.
127. **Prinzmetal M, Kennamer R, Corday E, et al:** *Accelerated Conduction. The Wolff-Parkinson-White Syndrome and Related Conditions.* New York: Grune & Stratton, 1952.
128. **Borduas JL, Rakita L, Kennamer R, et al:** Studies on the mechanism of ventricular activity; clinical and experimental studies of accelerated auriculoventricular conduction. *Circulation* 11:69, 1955.
129. **Sherf L, James TN:** A new look at some old questions in clinical electrocardiography. *Henry Ford Hosp Med Bull* 14:265, 1966.
130. **James TN:** The connecting pathways between the sinus node and A-V node and between the right and the left atrium in the human heart. *Am Heart J* 66:498, 1963.
131. **Fleischmann P:** Interpolation of atrial premature beats of intraatrial origin due to concealed A-S conduction. *Am Heart J* 66:309, 1963.
132. **Rosenbaum FF, Hecht HH, Wilson FN, et al:** The potential variations of the thorax and esophagus in anomalous atrioventricular excitation (Wolff-Parkinson-White syndrome). *Am Heart J* 29:281, 1945.
133. **Kossman CE, Berger AR, Briller SA, et al:** Anomalous atrioventricular excitation produced by catheterization of the normal human heart. *Circulation* 1:902, 1950.
134. **Kossman CE, Goldberg HH:** Sequence of ventricular stimulation and contraction in a case of anomalous atrioventricular excitation. *Am Heart J* 33:308, 1947.
135. **Durrer D, Roos JP:** Epicardial excitation of the ventricles in a patient with Wolff-Parkinson-White syndrome (type B). *Circulation* 35:15, 1967.
136. **Burchell HB, Frye RL, Anderson MW, et al:** Atrioventricular and ventriculoatrial excitation in Wolff-Parkinson-White syndrome (type B). Temporary ablation at surgery. *Circulation* 36:663, 1967.
137. **Iwa T:** Surgical experiences with the Wolff-Parkinson-White syndrome. *J Cardiovasc Surg* (Torino) 17:549, 1976.
138. **Westlake RE, Cohen W, Willes WH:** Wolff-Parkinson-White syndrome and familial cardiomegaly. *Am Heart J* 64:314, 1962.
139. **Scherf D, Schoenbrunner E:** Beitrage zum Problem der verkuerzten Hofkammerleitung. *Z Klin Med* 128:750, 1935.
140. **Rosenbaum FF:** The nature of paroxysmal tachycardia in anomalous atrioventricular excitation. *Am Heart J* 37:668, 1949.
141. **Fox TT, Weaver J, March HW:** On the mechanism of the arrhythmias in aberrant atrioventricular conduction (Wolff-Parkinson-White). *Am Heart J* 43:507, 1952.
142. **Wiggers CJ:** *Physiology in Health and Disease,* ed 5. Philadelphia: Lea & Febiger, 1949.
143. **Holzmann M:** New diagnostic and therapeutic developments in Wolff-Parkinson-White syndrome. (Ger) *Schweiz Med Wochenschr* 101:494, 1971.
144. **Zao ZZ, Herrman CR, Hejtmancik MR:** A vector study of the delta wave in "non-delayed" conduction. *Am Heart J* 56:920, 1958.
145. **Garello L, Ribaldone D:** Effects of oculocompression on the electrocardiographic patterns in the Wolff-Parkinson-White syndrome in children. (Ital) *Arch E Maragliano Patol Clin* 25:55, 1969.

146. **Hunter A, Papp C, Parkinson J:** The syndrome of short P-R interval apparent bundle branch block and associated paroxysmal tachycardia. *Br Heart J* 2:107, 1940.
147. **Schiebler GL, Adams P Jr, Anderson RC, et al:** Clinical study of twenty-three cases of Ebstein's anomaly of the tricuspid valve. *Circulation* 19:165, 1959.
148. **Pick A, Katz LN:** Disturbances of impulse formation and conduction in the pre-excitation (WPW) syndrome. Their bearing on its mechanism. *Am J Med* 19:759, 1955.
149. **Otto H:** The WPW syndrome in older people. (Ger) *Dtsch Med J* 18:723, 1967.
150. **Chen S, Fagan LF:** Wolff-Parkinson-White syndrome in Marfan's syndrome. *J Pediatr* 84:302, 1974.
151. **Hatreiter AR, Rodriguez-Coronel A, Paul MH:** Pre-excitation syndrome associated with subvalvular aortic stenosis in children. *Pediatrics* 41:1115, 1968.
152. **Plavsic C:** Une forme curieuse du syndrome de Wolff-Parkinson-White dans un cas de dextro-version du coeur. *Arch Mal Coeur* 42:1221, 1949.
153. **Schiebler GL, Adams P Jr, Anderson RC:** Wolff-Parkinson-White (pre-excitation) syndrome in infancy and childhood. *Univ Minn Med Bull* 30:94, 1958.
154. **Fox TT:** Aberrant atrio-ventricular conduction in a case showing a short P-R interval and an abnormal, but not prolonged QRS complex. *Am J Med Sci* 209:199, 1945.
155. **Mortensen V, Nielsen AL, Eskildsen P:** Wolff, Parkinson and White's syndrome. *Acta Med Scand* 118:506, 1944.
156. **Engle MA:** Wolff-Parkinson-White syndrome in infants and children. *Am J Dis Child* 84:692, 1952.
157. **Segers M, Lequime J, Denolin H:** L'activation ventriculaire précoce de certains coeurs hyperexcitables. Etude de l'onde de l'electrocardiogramme. *Cardiologia* 8:113, 1944.
158. **Segers M, Leguime J, Denolin H:** Le syndrome de Wolff-Parkinson-White. *Arch Mal Coeur* 38:57, 1945.
159. **Wallace AG, Boineau JB, Davidson RM, et al:** Symposium on electrophysiologic correlates of clinical arrhythmias. 3. Wolff-Parkinson-White syndrome: a new look. *Am J Cardiol* 28:509, 1971.
160. **James TN:** "Heuristic thoughts on the Wolff-Parkinson-White syndrome," in *Advances in Electrocardiography* (Schlant RC, Hurst JW, Eds), p 259, New York: Grune & Stratton, 1972.
161. **Lyle AM:** Latent Wolff-Parkinson-White syndrome. *Am Heart J* 46:49, 1953.
162. **Tamm RH:** Das Wolff-Parkinson-White Syndrom in Kindesalter. *Helv Paediatr Acta* 11:78, 1956.
163. **Landtman B:** Heart arrhythmias in children. *Acta Paediatr Finland* (suppl 1): 9, 1947.
164. **Joseph R, Ribierre M, Najean Y:** Le syndrome de Wolff-Parkinson-White dans la première enfance. Ses rapports avec la tachycardia paroxystique du nourrisson. *Sem Hop Paris* 34:552, 1958.
165. **Swiderski J, Lees MH, Nadas AS:** The Wolff-Parkinson-White syndrome in infancy and childhood. *Br Heart J* 24:561, 1962.
166. **Schiebler GL, Adams P Jr, Anderson RTC:** The Wolff-Parkinson-White syndrome in infants and children. A review and a report of 28 cases. *Pediatr* 24:585, 1959.
167. **Manning GW:** An electrocardiographic study of 17,000 fit, young Royal Canadian Air Force aircrew applicants. *Am J Cardiol* 6:70, 1960.
168. **Hiss RG, Lamb LE:** Electrocardiographic findings in 122,043 individuals. *Circulation* 25:947, 1962.
169. **Katz LN, Pick A:** *Clinical Electrocardiography: The Arrhythmias,* pp 100-104, Philadelphia: Lea & Febiger, 1956.
170. **Tranchesi J, Guimaraes AC, Teixeira V, et al:** Vectorial interpretation of the ventricular complex in Wolff-Parkinson-White syndrome. *Am J Cardiol* 4:334, 1959.
171. **Nasser WK, Mishkin ME, Tavel ME, et al:** Occurrence of organic heart disease in association with the Wolff-Parkinson-White syndrome. Analysis of 29 cases. *J Indiana State Med Assoc* 64:111, 1971.
172. **Wellens HJ, Durrer D:** Wolff-Parkinson-White syndrome and atrial fibrillation. Rela-

tion between refractory period of accessory pathway and ventricular rate during atrial fibrillation. *Am J Cardiol* 34:777, 1974.
173. **Longhini C, Pinelli G, Martelli A, et al:** Polycardiographic study of ventricular pre-excitation. *Minerva Cardioangiol* 21:89, 1973.
174. **Ueda H, Harumi K, Shimomura K, et al:** A vectorcardiographic study of WPW syndrome. *Jpn Heart J* 7:255, 1966.
175. **Grishman A, Kroop IG, Steinberg MF:** The course of the excitation wave in patients with electrocardiograms showing short P-R intervals and QRS complexes (Wolff-Parkinson-White syndrome). An analysis based on studies with intracardiac and esophageal leads. *Am Heart J* 40:554, 1950.
176. **Giraud G, Latour H, Puech P, et al:** Les troubles du rhythme du syndrome de Wolff-Parkinson-White. Analyse électrocardiographique endocavitaire. *Arch Mal Coeur* 49:101, 1956.
177. **Burch GE, DePasquale NP:** Electrocardiographic and vectorcardiographic detection of heart disease in the presence of the pre-excitation syndrome (Wolff-Parkinson-White syndrome). *Ann Intern Med* 54:387, 1961.
178. **Rosenkranz KA:** On the differential diagnosis of the electrocardiographic picture of the WPW syndrome. (Ger) *Z Kreislaufforsch* 54:1168, 1965.
179. **Wolff L, White PD:** Syndrome of short P-R interval with abnormal QRS complexes and paroxysmal tachycardia. *Arch Intern Med* 82:446, 1948.
180. **Friedman S, Wells RE, Amiri G:** The transient nature of Wolff-Parkinson-White anomaly in childhood. *J Pediatr* 74:296, 1969.
181. **Giardina AC, Ehlers KH, Engle MA:** Wolff-Parkinson-White syndrome in infants and children. A long-term follow-up study. *Br Heart J* 34:839, 1972.
182. **Schiebler GL, Adams P Jr, Anderson RC:** Familial cardiomegaly in association with the Wolff-Parkinson-White syndrome. *Am Heart J* 58:113, 1959.
183. **Apostolov L:** Wolff-Parkinson-White syndrome and branch block. (Ital) *Minerva Cardioangiol* 15:20, 1967.
184. **Belotserkovskiĭ ZB:** Case of Wolff-Parkinson-White syndrome in sportsmen twins. (Rus) *Klin Med* (Mosk) 53:120, 1975.
185. **Hilmer W, Lagally W:** Family studies in Wolff-Parkinson-White syndrome. (Ger) (Eng Abstr) *Z Kreislaufforsch* 60:89, 1971.
186. **Ohnell RF:** Zur Kenntnis der Tachycardia Paroxysmalis: Zwei Familien, in denen Neigung zu "Herzanfaellen" und gewisse Ekg-Ver-aenderungen gehaeuft vorkamen. *Cardiologia* 5:326, 1941.
187. **Pedich W:** Cardiogenic studies. Familial occurrence of Wolff-Parkinson-White syndrome. (Pol) *Pol Tyg Lek* 24:1232, 1969.
188. **Harnischfeger WW:** Heredity occurrence of the pre-excitation (Wolff-Parkinson-White syndrome) with re-entry mechanism and concealed conduction. *Circulation* 19:28, 1959.
189. **Montemurro G, D'Auino R, Miraldi C:** Familial Wolff-Parkinson-White syndrome. (Ital) *Boll Soc Ital Cardiol* 12:384, 1967.
190. **Urban J, Fabian J:** Familial occurrence of WPW syndrome in 2 generations. (Cz) *Vnitr Lek* 12:1198, 1966.
191. **Massumi RA:** Familial Wolff-Parkinson-White syndrome with cardiomyopathy. *Am J Med* 43:951, 1967.
192. **Soulié P, Di Matteo J, Abaza A, et al:** Cardiomégalie familiale. *Arch Mal Coeur* 50:22, 1957.
193. **Braunwald E, Andrew G, Morrow WP, et al:** Idiopathic hypertrophic subaortic stenosis: clinical hemodynamic and angiographic manifestations. *Am J Med* 29:924, 1960.
194. **McIntire M, Freed AE:** The Wolff-Parkinson-White syndrome. Report of a case occurring in a mother and infant. *Am J Dis Child* 89:743, 1955.
195. **Sökmen C:** Summary of a familial case of Wolff-Parkinson-White syndrome. *Am Heart J* 53:940, 1957.
196. **Doumer E, Dumez L:** Syndrome de Wolff-Parkinson-White familial. Syndrome de

WPW avec PR apparemment normal sur certaines dérivations. *Arch Mal Coeur* 44:1134, 1951.

197. **Averill J:** Wolff-Parkinson-White syndrome occurring in brothers. *Am Heart J* 51:943, 1956.
198. **Schneider RG:** Familial occurrence of the Wolff-Parkinson-White syndrome. *Am Heart J* 78:34, 1969.
199. **Homola D, Srnová V:** Familial occurrence of the Wolff-Parkinson-White syndrome (WPW). (Cz) *Vnitr Lek* 14:850, 1968.
200. **Noya M, del Rio A, de Oya JC:** Wolff-Parkinson-White syndrome in a family. *Lancet* 2:712, 1971.
201. **Scherf D, Cohen J:** *The Atrioventricular Node and Selected Cardiac Arrhythmias,* pp 373-447, New York-London: Grune & Stratton, 1964.
202. **Warner AO, McKusick VA:** Wolff-Parkinson-White syndrome: a genetic study. *Clin Res* 6:18, 1958.
203. **Nadrai, A:** Die Elektrokardiographie im Säuglings—und Kindesalter. *Ergen Inn Mediz u Kinderheilk* 60:688, 1941.
204. **Pohlmann F:** Beitrag zur Frage der verkuerzten Vorhof-Kammerleitung. *Z Klin Med* 140:1, 1941.
205. **Bodlander JW:** The Wolff-Parkinson-White syndrome in association with congenital heart disease coarctation of the aorta. Report of a case. *Am Heart J* 31:785, 1946.
206. **Moore EN, Spear JF, Boineau JP:** Recent electrophysiologic studies on the Wolff-Parkinson-White syndrome. *NEJM* 289:956, 1973.
207. **Annamalai A, Ananthasubramaniam G:** Intermittent type-A Wolff-Parkinson-White syndrome in a case of coartation of the aorta. *Lancet* 86:490, 1966.
208. **Ask-Upmark E:** Coarctatio aortae. *Acta Med Scand* 112:7, 1942
209. **Campbell M, Turner-Warwick M:** Two more families with cardiomegaly. *Br Heart J* 18:393, 1956.
210. **Canabal EJ, D'Ghiero J, Karlen AA, et al:** Diverticulum of the pericardium and Wolff-Parkinson-White syndrome. (Sp) *Torax* 15:110, 1966.
211. **Comberiati L:** Due casi di sindrome di Wolff-Parkinson-White in cardiopatia congenita operata. *Cuore e Circul* 39:169, 1943.
212. **Correale E, Corsini G, Di Filippo A:** The association of branch block with the Wolff-Parkinson-White syndrome, in a case of interatrial defect. (Ital) *Rass Int Clin Ter* 51:775, 1971.
213. **Estape Panellas F de A:** Situs inversus, tetralogía de Fallot y síndrome de W-P-W. *Medic Clin* (Barcelona) 14:341, 1950.
214. **Hiejima K, Tsuchiya S, Sakamoto Y, et al:** Two cases of Marfan's syndrome associated respectively with subacute bacteria endocarditis and the Wolff-Parkinson-White syndrome. *Jpn Heart J* 9:208, 1968.
215. **Kleiber EE:** Wolff-Parkinson-White syndrome with congenital heart disease. *Pediatrics* 4:210, 1949.
216. **Nomura Y, Fujigaki H, Miyakawa T:** Case of tetralogy of Fallot with Wolff-Parkinson-White syndrome. (Jap) *Jpn J Clin Med* 24:2339, 1966.
217. **Noya del Rio A, De Oya JC:** Familial myocardiopathy with Wolff-Parkinson-White syndrome. (Sp) *Rev Clin Esp* 124:299, 1972.
218. **Stein MH:** Wolff-Parkinson-White syndrome in a case of congenital heart disease. *Am Heart J* 35:140, 1948.
219. **Tsuchiya S, Okuma S, Sakamoto Y, et al:** Case of Marfan's syndrome associated with WPW syndrome. (Jap) *Jpn J Clin Med* 25:761, 1967.
220. **Vogt A:** Maladie d'Ebstein avec syndrome de Wolff-Parkinson-White. *Ann Paediatr* 187:286, 1956.
221. **Sodi-Pallares D, Marsico F:** The importance of electrocardiographic patterns in congenital heart disease. *Am Heart J* 49:202, 1955.
222. **Angel J, Armendariz JJ, Castillo HG, et al:** Idiopathic hypertrophic subaortic stenosis and Wolff-Parkinson-White syndrome. Changes of obstruction in left ventricular outflow depending on the type of ventricular activation. *Chest* 68:248, 1975.

223. **Krikler D, Curry P, Kafetz K:** Pre-excitation and mitral valve prolapse. *Br Med J* 1:1257, 1976.
224. **Grolleau R, Baissus C, Puech P:** Transposition corrigée des gros vaisseauz et syndrome de préexcitation (à propos de deux observations). *Arch Mal Coeur* 70:69, 1977.
225. **Pernod J, Ferrane J, Quinot B, et al:** Obstructive cardiomyopathy and Wolff-Parkinson-White syndrome. *Presse Med* 74:2135, 1966.
226. **Sealy WC, Hattler BG, Blumenschein SD, et al:** Surgical treatment of Wolff-Parkinson-White syndrome. *Ann Thor Surg* 8:1, 1969.
227. **Donzelot E:** *Traité des Cardiopathies Congénitales,* p 1025, Paris: Masson, 1954.
228. **Hecht HH:** Anomalous atrioventricular excitation, panel discussion. *Ann NY Acad Sci* 65:826, 862, 1957.
229. **Angelino PF, Mina PL, Gallo C:** Wolff-Parkinson-White syndrome and atherosclerotic myocardial disease at a young age. (Ital) *Minerva Med* 55:90, 1964.
230. **Das JP, Mohanty BK:** Intermittent WPW syndrome with congestive heart failure (a case report). *J Assoc Physicians India* 21:251, 1973.
231. **Goch JH:** Wolff-Parkinson-White syndrome and myocardial disease. (Pol) *Wiad Lek* 22:1941, 1969.
232. **Goel BG, Han J:** Manifestations of the Wolff-Parkinson-White syndrome after myocardial infarction. *Am Heart J* 87:633, 1974.
233. **Sibilia D:** Attachi anginosi legati a tachicardia parossistica da sindrome di Wolff-Parkinson e White in coronaritico. Alternanza ecgrafica del complesso: "P-R corto e blocco di branca apparente" *Cuore e Circol* 25:112, 1941.
234. **Verani MS, Baron H:** Myocardial infarct associated with Wolff-Parkinson-White syndrome. *Am Heart J* 83:812, 1971.
235. **Movitt ER:** Some observations on the syndrome of short P-R interval with long QRS. *Am Heart J* 29:78, 1945.
236. **Clagett AH Jr:** Short P-R interval with prolonged QRS complex: allergic manifestations and unusual electrocardiographic abnormalities: report of a case. *Am Heart J* 26:55, 1943.
237. **Malmström G:** Pre-excitation during hypoxemia test. *Acta Med Scand* 133:68, 1949.
238. **Akesson S:** Pre-excitation and auricular fibrillation. *Acta Med Scand* 120:1, 1945.
239. **Dumlao J, Brooks MH, Rosen KM:** Simultaneous cure of thyrotoxicosis and Wolff-Parkinson-White syndrome. *Chest* 66:568, 1974.
240. **Lamb LE:** Bundle branch block in hyperthyroidism. *Med Times* 61:234, 1933.
241. **Master AF, Jaffe HL, Dack S:** Atypical bundle branch block with short P-R interval in Grave's disease effect of thyroidectomy. *J Mt Sinai Hosp* 4:100, 1937.
242. **Mihalkovics T:** Vom Elektrokardiogramm des Paladino Kent-Typus. *Z Kreislaufforsch* 35:388, 1943.
243. **Scattini MC, Suarez LD, Buceta, JE, et al:** Hyperthyroidism and WPW syndrome. *Arch Inst Cardiol Mex* 44:661, 1974.
244. **Strong JA:** Thyrotoxicosis with ophthalmoplegia, myopathy, Wolff-Parkinson-White syndrome and pericardial friction. *Lancet* 1:959, 1949.
245. **Vallarino G, Poggi L:** Hyperthyroidism and Wolff-Parkinson-White syndrome in a 6-year-old child. Etiopathogenetic considerations. (Ital) *Minerva Pediatr* 15:947, 1963.
246. **Mininni G, Violante N, Fabiano M:** On a case of Wolff-Parkinson-White syndrome associated with epilepsy. (Ital) *Riv Clin Pediatr* 74:113, 1964.
247. **Pasini U, Carvalho F, Giuliano H, et al:** Electroencephalographic changes in pateints with paroxysmal cardiac arrhythmias. (Port) *Arq Bras Cardiol* 17:368, 1964.
248. **Seganti A, Varcasia E:** Quattro casi pediatrics della sindrome di Wolff-Parkinson e White. *Pediatria Internaz* 4:67, 1954.
249. **Sondergaard G:** The Wolff-Parkinson-White syndrome in infants. *Acta Med Scand* 145:386, 1953.
250. **Zych D, Solecki J:** Cerebral circulation disturbances in a case of Wolff-Parkinson-White syndrome. *Wiad Lek* 29:199, 1976.
251. **Anghelescu F, Mataranga E:** On 3 cases of Wolff-Parkinson-White syndrome with rheumatismal etiology. (Rum) *Med Intern* (Bucur) 19:989, 1967.

252. **Apostolov L, Stoencev S:** A case of rheumatism and syndrome of ventricle preexcitation (Wolff-Parkinson-White syndrome). (Bul) *Vatr Bolesti* (Sofyiya) 13:103, 1974.
253. **Apostolov L, Mironov N:** Contribution to the rheumatic etiopathogensis of the Wolff-Parkinson-White syndrome. *Folia Med* (Plovdiv) 8:339, 1966.
254. **Bain CWC, Hamilton CK:** Electrocardiographic changes in rheumatic carditis. *Lancet* 1:807, 1926.
255. **Blom PW:** The syndrome of pre-excitation with initial activation of the left ventricle in a patient with rheumatic heart disease. *Acta Med Scand* 140:85, 1951.
256. **Coumel P, Attuel P, Slama R, et al:** 'Incessant' tachycardias in Wolff-Parkinson-White syndrome. II. Role of atypical cycle length dependency and nodal-His escape beats in initiating reciprocating tachycardias. *Br Heart J* 38:895, 1976.
257. **Déchelotte J:** Syndrome de Wolff-Parkinson-White en bigéminisme puis permanent, au décours d'une maladie de Bouillaud. *Coeur Med Interne* 6:259, 1967.
258. **Mahaim I, Bogdanovic P:** Un cas mortel de syndrome de Wolff-Parkinson-White. Examen histologique du faisceau de His-Tawara. *Acta Med Yugoslav* 2:137, 1948.
259. **Meneely JK Jr:** Occurrence of the Wolff-Parkinson-White syndrome concomitant with acute rheumatic fever. *Yale J Biol Med* 21:87, 1948.
260. **Ougier J, Page G, Marc P:** Apparition d'un syndrome de Wolff-Parkinson-White au cours d'un rhumatisme articulaire aigu. *Sem Hop Paris* 40:1178, 1964.
261. **Partilla H:** Zur Klinik des WPW-Syndroms bei akuter und chronischer Myokarditis mit besonderer Beruecksichtigung der Therapie. *Cardiologia* 33:108, 1958.
262. **Robinson RW, Talmage WG:** Wolff-Parkinson-White syndrome. Report of five cases. *Am Heart J* 29:569, 1945.
263. **Slavov S:** Rheumatism as the most frequent cause of Wolff-Parkinson-White syndrome. (Rus) *Vopr Revm* 5:52, 1965.
264. **Stein MH:** Wolff-Parkinson-White syndrome with unusual features. *Am Heart J* 29:479, 1945.
265. **Doumer E, Merlen JR:** Syndrome de Wolff-Parkinson-White au cours d'une intoxication aigue par l'oxyde de carbone. *Acta Cardiol* 1:302, 1946.
266. **Agarwal RK, Hisra DN, Verma RK:** Wolff-Parkinson-White syndrome with paroxysmal atrial fibrillation in pseudo-hypertrophic muscular dystrophy (Duchenne type). *Indian Heart J* 25:346, 1973.
267. **Bensaid H, Marsaud G, Monassier JP, et al:** High degree atrioventricular block associated with Wolff-Parkinson-White syndrome, during a Duchenne de Boulogne type myopathy. *Ann Cardiol Angeiol* (Paris) 24:67, 1975.
268. **Bowers D:** Charcot-Marie-Tooth disease, Wolff-Parkinson-White syndrome and abnormal intracardiac conduction. *Am Heart J* 86:535, 1973.
269. **Glebowska H, Smolik R, Tawlas N, et al:** Case of WPW syndrome and hemolytic syndrome. (Pol) *Wiad Lek* 25:1183, 1972.
270. **Lirman AV, Apanasenko BG, Kunitsyn AI, et al:** Wolff-Parkinson-White syndrome and other conduction and rhythm disorders of the heart in experimental fat embolism according to electrocardiography data. (Rus) (Eng Abstr) *Kardiologiia* 11:130, 1971.
271. **Lustman F, Geerts L:** Electrocardiographic manifestations in 114 cases of carbon monoxide poisoning. *Acta Clin Belg* 26:131, 1971.
272. **Mashito T, Sawaguchi T, Himuro K, et al:** A case of hyperthyroidism associated with Wolff-Parkinson-White syndrome and periodic paralysis. (Jap) *Fukuoka Acta Med* 67:60, 1976.
273. **Maszkiewicz W, Kubisz R, Wojnar A:** Wolff-Parkinson-White syndrome in a newborn infant with hemolytic disease complicated by generalized infection. (Pol) *Wiad Lek* 26:1731, 1973.
274. **Niarchos AP, Finn R, Cohen HN, et al:** Association of Wolff-Parkinson-White syndrome with congenital abnormalities of hands and feet. *Br Heart J* 36:409, 1974.
275. **Seiling A:** Wolff-Parkinson-White syndrome following acute intoxication with carbon monoxide. (Ger) *Med Klin* 61:499, 1966.
276. **Seiling A:** Funnel chest and Wolff-Parkinson-White syndrome. (Ger) *Med Welt*

21:1255, 1969.
277. **Semler HJ:** Acute pericarditis and the Wolff-Parkinson-White syndrome. *Northwest Med* 63:235, 1964.
278. **Soltés L, Mayer M:** Wolff-Parkinson-White syndrome in a patient on prolonged corticoid therapy. (Cz) *Cesk Pediatr* 21:145, 1966.
279. **Spencer MJ, Cherry JD, Adams FH, et al:** Supraventricular tachycardia in an infant associated with a rhinoviral infection. *J Pediatr* 86:811, 1975.
280. **Yigibasi O, Mamaoglou K:** Un cas de lymphogranulomatose maligne a localisation rare (pancreas) s'accompagnant du syndrome de Wolff-Parkinson-White. *Presse Med* 68:1046, 1960.
281. **Yusa T, Bugiu T, Sato T:** Experience of neuroleptanesthesia in myasthenia accompanied by the WPW syndrome. (Jap) (Eng Abstr) *Jpn J Anesth* 21:498, 1972.
282. **Berkman NL, Lamb LE:** The Wolff-Parkinson-White electrocardiogram. A follow-up study of five to twenty-eight years. *NEJM* 278:492, 1968.

II

The Electrocardiogram in the Pre-excitation Syndrome: Complexes and Configurations

P-QRS

The P-R Interval

In Wolff, Parkinson and White's original paper,[1] one of the criteria for diagnosing the new entity was a short P-R interval. Exactly what is to be considered a "short P-R interval" has been a matter of controversy since the term was introduced. In the early thirties it was 0.10 second or less,[2] while later it was 0.11 second[3] and even 0.12 second or less.[4] Different methods of measurement contributed to the differences of opinion; some measured the interval in lead II only, some measured the interval in any lead where it could be clearly seen, and still others used concomitant multilead registration. Thus, the discussion of the P-R interval can proceed only after defining exactly what is meant by "short P-R interval" and deciding on a method of measurement.

It is generally agreed that the normal P-R value, which expresses the interval from the beginning of atrial depolarization to the beginning of ventricular depolarization, is 0.12-0.20 second. According to many investigators the interval increases with age, from shorter than 0.12 second in newborns and infants to as long as 0.20 second in adults.[5-7] We considered 0.12 second or more normal, and less than 0.12 second a short P-R interval. Our method of measurement was that of Ritter and Fatturoso,[8] whereby the P-R interval is taken from the lead showing the highest value. This avoids the pitfall of part of the interval being isoelec-

TABLE VIII
The P-R Interval

Author	Total cases	0.06-0.07 sec	0.08-0.09 sec	0.10-0.11 sec	0.12-0.14 sec	0.15 sec
Swiderski[9]	18 †	9	6	2	1	—
Swiderski[9]	20 ‡	2	9	6	3*	—
Schiebler[10]	13 †	3	5	5	—	—
Schiebler[10]	15 ‡	1	4	3	7**	—
Averill et al[11]	66	2	21	33	10	—
Nasser et al[12]	29	—	14	10	5	—
Lombardi[13]	41	8	17	10	6	—
Hejtmancik[14]	81	2	19	43	17	—
Rosenkranz[15]	20	—	6	11	3	—
Longhini[16]	24	—	11	8	3	2
Flensted-Jensen[3]	47	7	24	12	4	—
Tel Hashomer	215	4	43	94	71	3
Total	589	38	179	237	130	5

*1 of the 3 cases had Ebstein's anomaly
**5 of the 7 cases had Ebstein's anomaly
† without associated heart disease
‡ with associated heart disease

tric and not measurable, especially where a Q wave is the first deflection of the QRS complex. (In our opinion, the most precise method is concomitant multilead registration, but it was not used because of special equipment requirements.)

Table VIII summarizes the P-R values of 589 cases from the literature and our series (Tel Hashomer). Due to a variety of measuring techniques the quoted values give a general picture rather than precise information. Two facts, however, emerge from this Table: 1) the great majority of cases had a P-R interval of 0.08-0.11 second (416 out of 589); and 2) in all series, without exception, a number of cases had a normal P-R interval.[17,18] While the first finding was expected, the latter was surprising. However, as mentioned in Chapter I, all investigators found a normal P-R interval in otherwise typical pre-excitation cases (0.12 second or more). Ohnell,[19] who reported 5 cases of pre-excitation in which the P-R was 0.13 second or more, had a special subgroup in his classification for pre-excitation cases with normal P-R intervals. Other investigators published similar findings: Wolff and White,[20] 0.145 second;

TABLE IX
The P-R Interval by Age Groups in the Tel Hashomer Series (215 Patients)

P-R Interval	Total	0-25 Years	26-50 Years	51-80 Years
0.04-0.09	47	28 (60%)	14 (30%)	5 (10%)
0.10-0.11	94	44 (47%)	31 (33%)	19 (20%)
0.12 +	74	19 (25.6%)	32 (43%)	23 (31.5%)
Total	215	91 (42%)	77 (36%)	47 (22%)

Lamb,[21] 0.14 second; and Pick and Katz,[22] 0.24 second (first degree AV block). In our series of 215 cases of pre-excitation, 74 patients (34.4%) had normal intervals. This high percentage may be due in part to the method of measurement.

The phenomenon of lengthening of the P-R interval with age may perhaps be explained by the increase in the amount of connective tissue in the AV node, the place at which the normal delay occurs in the advance of the electrical stimulus descending from the atria to the ventricles. In the pre-excitation syndrome, where it is generally believed that the descending electrical impulse bypasses the AV node, there should be no such lengthening of the P-R interval with age. Surprisingly, however, a direct correlation between increasing length in the P-R interval and advancing age was also found in the pre-excitation syndrome.

In studies of healthy children (Swiderski,[9] Schiebler[10]) the great majority of cases was in the 0.06-0.09 second P-R interval group, while most of the patients in an all-age series[14] were in the 0.09-0.11 second group. A similar pattern was found in our series (Table IX): 60% of the cases in the group with the shortest P-R (0.04-0.09 second) were below the age of 25 and only 10% were over 50. This proportion was inverted in the group with the normal P-R interval (0.12 second and more): only 25.6% were under the age of 25 and 31.5% were 50 years or more. Thus, with respect to AV conduction and increasing age, pre-excitation cases are similar to normals.

Analysis of some cases from 16 published series (Table X) revealed a high correlation between the P-R interval during pre-excitation and during normal conduction in the same patient. According to the assumption that in pre-excitation the electrical stimulus bypasses the point of delay (the AV node), there should be no relationship between AV conduction time during normal and anomalous conduction. Nevertheless, in many patients suffering from this disorder, just such a correla-

TABLE X

The P-R Interval Exhibiting Both Normal and Anomalous Ventricular Complexes in the Same Patient

Prolonged P-R*			Normal P-R**		
Author	Pre-excitation P-R	Normal P-R	Author	Pre-excitation P-R	Normal P-R
Pick[22]	0.24	0.36	Wolff[20]	0.10	0.19
Pick[23]	0.16	0.24	Wolff[20]	0.10	0.18
Grant[24]	0.16	0.40	Wolff[20]	0.10	0.18
Tel Hashomer	0.16	0.24	Tel Hashomer	0.10	0.18
Pick[22]	0.16	0.20	Littman[32]	0.10	0.16
James[25]	0.16	0.24	Tel Hashomer	0.10	0.16
Tel Hashomer	0.16	0.20	Tel Hashomer	0.10	0.16
Averill[11]	0.15	0.20	Tel Hashomer	0.10	0.16
Wolff[20]	0.15	0.21	Tel Hashomer	0.10	0.16
Lepeschkin[26]	0.15	0.22	Tel Hashomer	0.10	0.16
Tel Hashomer	0.14	0.18	Wolff[20]	0.10	0.15
Schiebler[10]	0.14	0.22	Littman[32]	0.10	0.14
Lamb[21]	0.14	0.20	Tel Hashomer	0.10	0.14
McHenry[27]	0.14	0.32	Tel Hashomer	0.10	0.14
Schiebler[28]	0.14	0.24	Tel Hashomer	0.10	0.14
Lombardi[13]	0.13	0.25	Tel Hashomer	0.10	0.14
Friedberg[29]	0.13	0.37	Tel Hashomer	0.10	0.14
Wolff[30]	0.13	0.19	Tel Hashomer	0.10	0.14
Wolferth[31]	0.13	0.17	Wolff[20]	0.10	0.14

*More than 0.12 second during pre-excitation. These are all the cases that have been reported to our knowledge.
**There are often reported examples of intermittent pre-excitation with "normal" P-R intervals, *ie*, 0.08-0.12 second. These are the ones with P-R intervals of 0.10 second taken at random.

tion was found. As Table X shows, the short P-R interval during pre-excitation is often not a fixed value, but rather a function of the patient's P-R interval during normal conduction. For example, when the normal P-R was 0.12-0.19 second, the shortened pre-excitation P-R was usually 0.08-0.10 second; but, when the normal P-R was prolonged (0.18-0.40 second), the interval during pre-excitation was usually "normal" or "slightly prolonged" (0.13-0.24 second) (Figure 11a, b, and c, 14d). These findings are strengthened by observations in patients suffering from the pre-excitation syndrome and Ebstein's anomaly in which there is a general tendency toward a prolonged P-R interval. In 18 cases without pre-excitation reported by Schiebler et al,[10] 7 had P-R intervals of 0.20 second or more, while in 13 with both Ebstein's anomaly and the pre-excitation syndrome,[33,34] 8 had normal P-R intervals of 0.12 second or more during pre-excitation. In our case of Ebstein's anomaly with

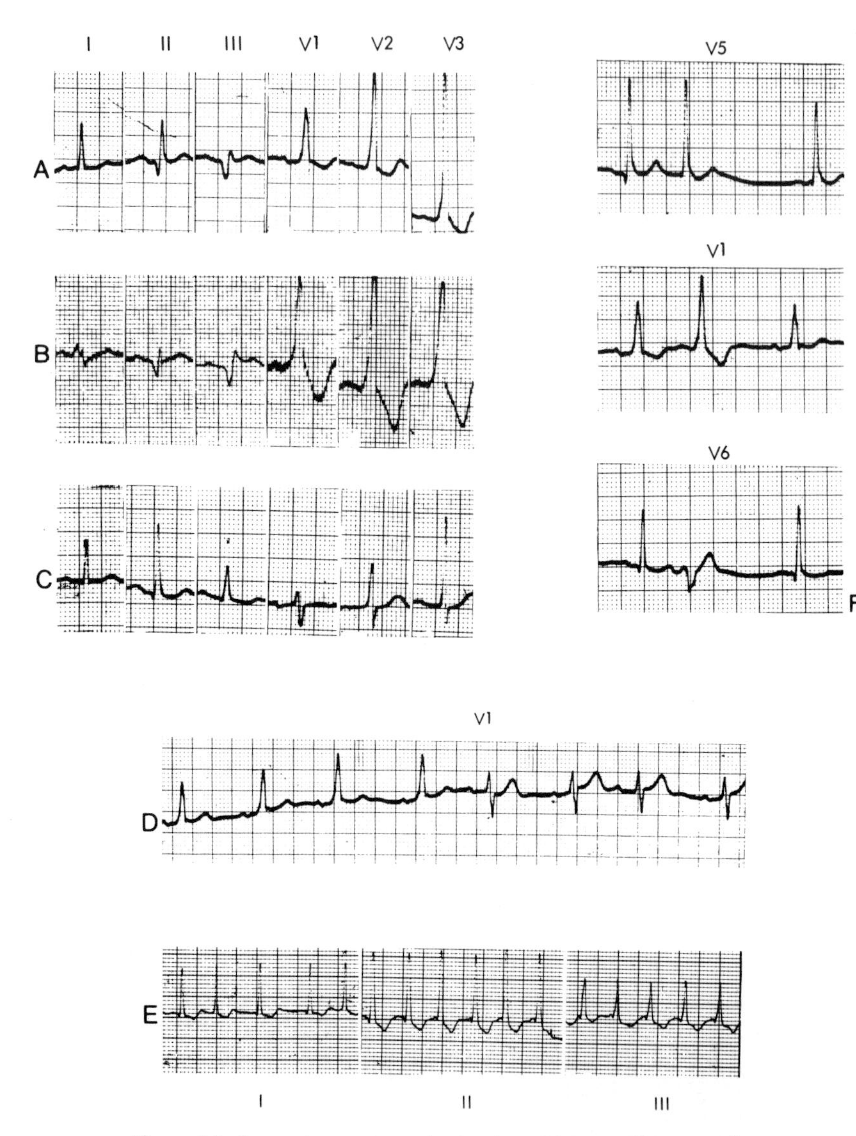

Figure 14. *Intermittent pre-excitation in a 60-year-old man.* **a, b.** *Two different patterns of pre-excitation conduction in the same patient.* **c.** *Normal AV conduction.* **d.** *Direct evidence of intermittency observed in lead V_1. Note the prolonged P-R interval during normal AV conduction and the normal P-R (0.16—0.18 second) during pre-excitation.* **e.** *Paroxysmal atrial fibrillation (with narrow QRS complexes).* **f.** *Premature beats: lead V_5 – supraventricular premature beat; lead V_1 – supraventricular beat with pre-excitation conduction; lead V_6 – ventricular premature beat.*

pre-excitation, the P-R was 0.12 second (Figure 4). Direct evidence of a prolonged P-R interval in both disorders is found in Schiebler's[28] case # 6 where the P-R interval during anomalous conduction was 0.14 second and during normal AV conduction, 0.24 second.

The relationship of the P-R interval to increasing age, and the correlation of the P-R interval during anomalous and normal conduction in the same individual will be discussed in Theories (see Chapter X).

P WAVES

When the heart rate is normal, the P wave in the pre-excitation syndrome is upright in leads I, II and III. The isoelectric segment from the end of the P to the beginning of the R is absent in most cases. Therefore the ventricular complex starts immediately after or overlaps the end of the P wave, reducing the distance from the onset of P to the onset of R (or Q) to less than 0.12 second.[35] In the early cases of pre-excitation[36-38] published prior to 1930, the pattern was considered an AV nodal rhythm because of the short P-R interval. However, Wolff, Parkinson and White were convinced from the beginning that the P wave was of sinus origin, because "frequent ventricular premature beats were followed consistently by compensatory pauses and the form of the P wave occurring with both long and short P-R intervals was identical. In several patients the P waves were distinctly notched, making identification easy and certain. Finally, the P waves were always upright in all the leads."[1] Holzman and Scherf[39] agreed that a sinus rhythm was to be considered in cases with pre-excitation. They stated that the combination of an unusual form of QRS complexes and the short P-R interval negated the possibility of a simple AV nodal rhythm, where the QRS is generally normal in configuration.

The question as to whether or not all the rhythms encountered in pre-excitation are of sinus origin arose again when changes in the form of the P waves were reported by, among others, Hunter, Papp and Parkinson[35] in 4 of 19 cases. They stated that any hypothesis explaining the mechanism of the syndrome must consider the site of the pacemaker and conduction of the impulse, and the fact that during a reversion of pre-excitation to normal conduction the P waves often change in shape. They suggested that the P waves may originate in an ectopic location near the sinus node, but not in it.

Since then, many pre-excitation cases with P wave changes have been reported, consisting of changes in configuration,[19,32,40-42] in polarity,[38,43,44] or disappearance of the P waves altogether.[45] According to Brody,[46] who favors the existence of a specific atrial conduction system, there are three possible mechanisms which singly or in combination may account for P wave changes: 1) actual translocation of the pacemaker locus; 2) temporary blocks in the atrial preferential pathways; and 3) variable exit sites from the sinus node. From a clinical point of view, spontaneous changes in P wave form which occur concomitantly with a change in heart rate are probably evidence of an active or escape shift of the primary pacemaker to a secondary site. When they occur without a concomitant change in heart rate, the diagnosis of block in one or more of the internodal or interatrial pathways is more plausible.[47,48] All these possibilities may play a role in the changes in P wave shape seen in pre-excitation. Sometimes such changes may be observed in a single lead (direct evidence).[49] They were variably thought to originate in the atria,[35] the AV node[45] or even in an accessory muscle bundle connecting the atria directly to the ventricles.[22] Recent investigations, however, have proved that the actual location of a pacemaker cannot be learned from the morphology or polarity of P waves in regular ECG tracings.[50]

In our series of 215 patients, 30 (13.95%) showed P wave changes, subtle but unmistakable in some and more pronounced in others (Figure 8a, b and c, 11a, b and c, 15a and b, 16). In most there was a change in rate when seen in "direct evidence" tracings, favoring the possibility of an ectopic (Figure 17b and c, 18) pacemaker.

In most published cases as well as in our series, the change in P wave form was accompanied by a change in the shape of the QRS complex (Figure 3, 8, 10, 11, 15, 16, 17). Thirty-eight such instances are summarized in Table XI. In some cases the QRS change represented a reversion to normal AV conduction (Figure 3a and b, 8a and b, 10a and b, 11c), and in others a change to another kind of pre-excitation pattern. Retrograde atrial conduction was observed during normal AV conduction (Figure 17a and b, 19d) in one without any QRS changes during the ectopic takeover; in the other an aberrant QRS pattern appeared dissimilar to both the normal and the pre-excitation QRS patterns observed in other tracings of the same patients.

P wave changes with and without concomitant QRS changes are assigned varying degrees of importance in the literature. All that can be said with certainty at present is that an active ectopic activity probably occurs often in patients with pre-excitation, the meaning and importance of which is unclear (see Chapter X).

THE QRS COMPLEX

By far the most striking part of the ECG tracing in pre-excitation is the QRS complex, which consists of a slurred, slow-conducting compo-

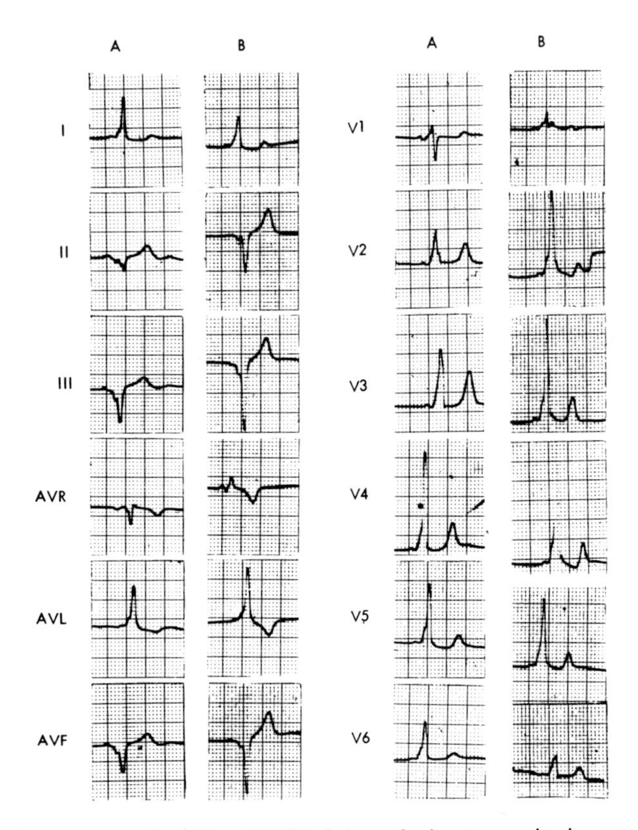

Figure 15. *Concomitant P and QRS changes during pre-excitation conduction in a 60-year-old man. Note changes in P wave configuration and polarity (leads II and III in a and b) and transformation of type B (in a) to type A pre-excitation according to Rosenbaum's classification (in b).*

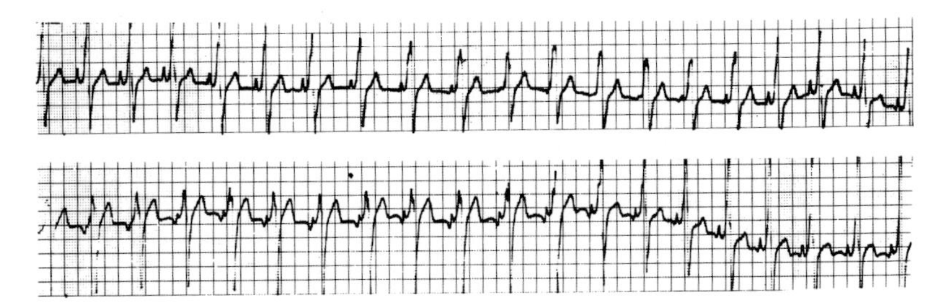

Figure 16. *Concomitant P and QRS changes during pre-excitation in a 17-year-old man. Ambulatory ECG monitoring. The P wave changes are especially clear in the lower panel (negative P waves first becoming progessively positive). Note concomitant QRS changes.*

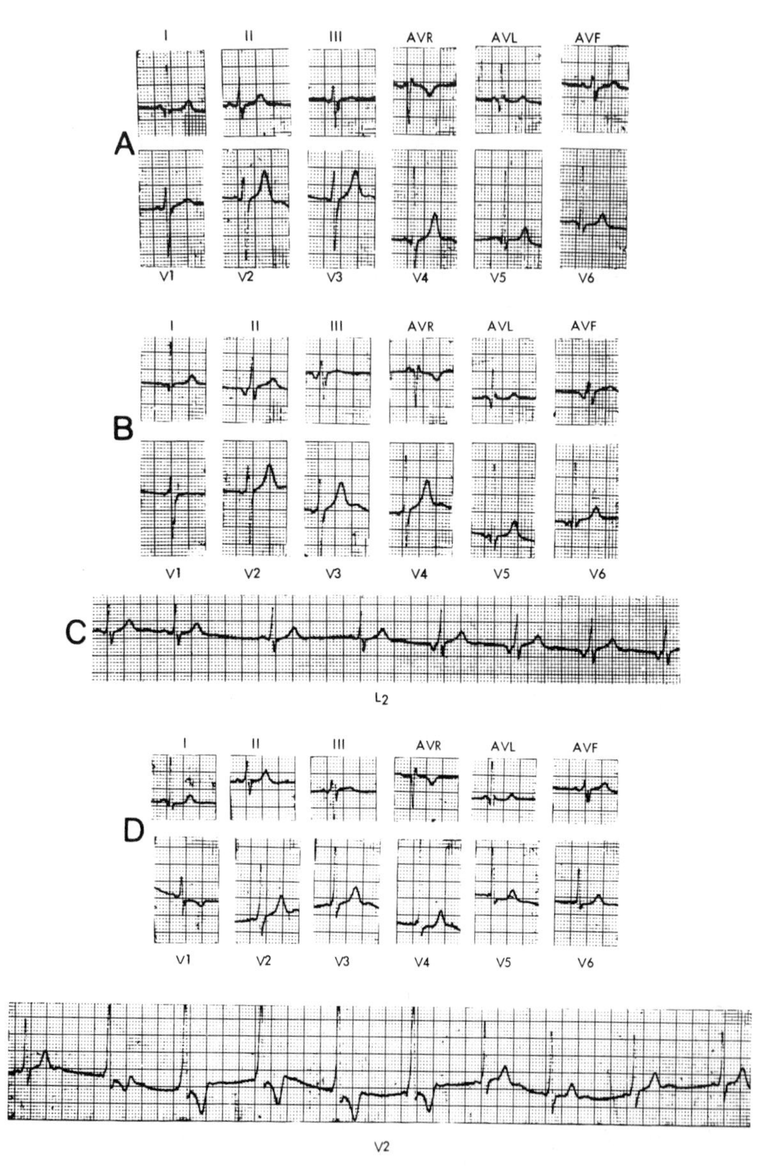

Figure 17. *P wave changes during normal and pre-excitation conduction in a 17-year-old man.* **a.** *Normal AV conduction.* **b.** *Negative P waves in leads II, III and aVF, positive in aVR (retrograde atrial activation).* **c.** *The same, long strip lead II (direct evidence).* **d.** *Pre-excitation conduction. Note clear delta waves in leads $V_{2,3}$.* **e.** *Direct evidence of nodal rhythm replacing sinus rhythm during pre-excitation (beats 2, 3, 4, 5, 6). Note concomitant changes in QRS and T configurations.*

nent (delta wave) followed by a slender, fast-conducting component. The complex is characterized by a prolonged duration in most cases, a wide variation in pattern among patients and occasional sequential variations in form within the same patient.

Characteristic Patterns

The presence of the slow conduction at the beginning of the QRS complex, the delta wave, is the essential part of the pre-excitation syndrome;[74] all other features may vary. It is generally agreed that this delta wave represents the slow spread of the depolarization through working, nonspecialized myocardial tissue. There is disagreement, however, as to the electrical background of the slender, fast-conducting later component of the QRS, although most authors agree that it represents a very fast conduction through specialized myocardial cells (Purkinje's fibers). Many investigators feel it represents the conduction of the sinus stimulus

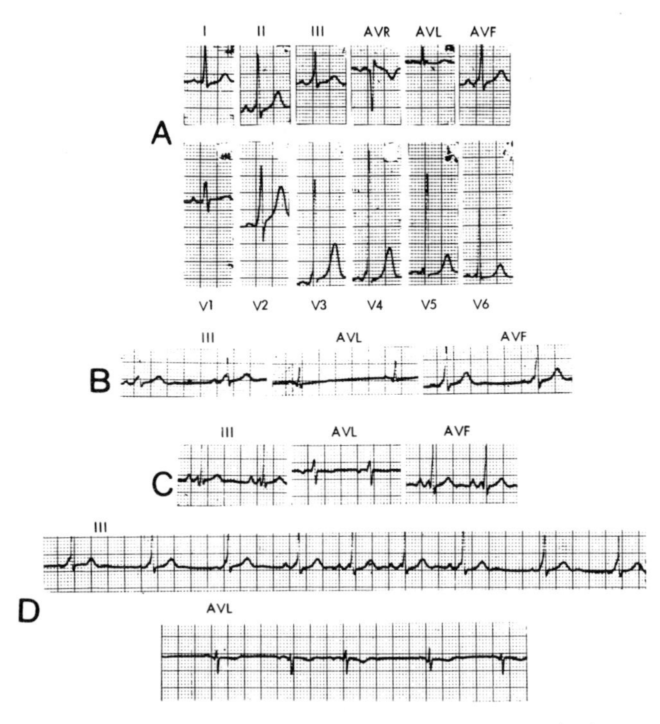

Figure 18. *P wave changes in a 27-year-old man.* **a.** *Pre-excitation conduction.* **b.** *Leads III, aVL and aVF during sinus bradycardia, positive P waves.* **c.** *The same during normal rate, negative P waves in aVL.* **d.** *Leads III and aVL. P wave changing with rate (nonrespiratory sinus arrhythmia).*

TABLE XI
Cases with Changes in Both P Wave Form and Shape of the QRS Complex

Author	Case #	Page #	Figure	Lead	Beats
Direct Evidence					
Fisch et al[51]	7	1007	3	V_1,V_4	2-3,3-4
Averill et al[11]	—	125	23	V_2,V_5,V_6	1-2,1-2
Samet et al[52]	7 group A	441	4a	T	2-3
Samet et al[52]	6 group B	443	6a	T	2-3
Sanghvi et al[53]	1	342	3	II,II	2-3
Ohnell[19]	31		E.F.		2-3
Sodi-Pallares[54]	—	603	482	III	3-4,10-11
Averill[11]	—	125	24	II,III	3-4,2-3
Littman et al[32]	9	112	11c	II,III	2-3
Rosenbaum[55]	10	311	15b	III	5-6
Wilson[38]	1	1015	12	III	3-4,8-9
Palatucci[41]	1	60	1a	II,III	3-4,2-3
Fosmoe[56]	—	90	6	II	5-6
Rodstein[57]	1	790	2c	CF3	3-4
Segers[58]	—	715	3	I	4-5,1-2
Bix[59]	2	41	4	I,V_3,V_5	5-6,4-5,2-3
Scherf et al[60]	—	388	153d	aVR	every sec
Swiderski et al[9]	39	564	3	?	1-3
Wolff[30]	5	18	6 (middle)	A,C	3-4
Harnischfeger[40]	1	31	4	I	3-4
Soffer[61]	1	87	1	aVF	1-2
Pick[62]	—	707	5c	I	2-3
Fox[63]	JPH	380	26	II	2-3
Scherf[64]	1	179	2	V_1	2-46
Chung[65]	1	217	38	II	3,4,9, *etc*
Grolleau[66]	1	15	3	I,II,III	2
Indirect Evidence					
Akesson[67]	—	4-5	2-5	I,II,III	—
Swiderski[9]	42	564	4b,c,e	aVL	—
Swiderski[9]	24	563	2a,b	aVL	—
Burchell[68]		392,394	1,4	II	—
Hunter[35]	1		Ib,c	I,II	—
Lepeschkin[26]	—	11-91	11-39b	V_1,V_6	—
Levine[69]	1	403	6-7-39 1-5-40	II,III	
Ramachandran[70]	1	531	2	III,V_1	—
Duthie[71]	1	97	2	II-III	—
Schiebler[28]	6	175	3	III,V_6	—
Boineau[72]	?	39	8	II,III	—
Katz[73]	—	704	412a,b	II,III	—

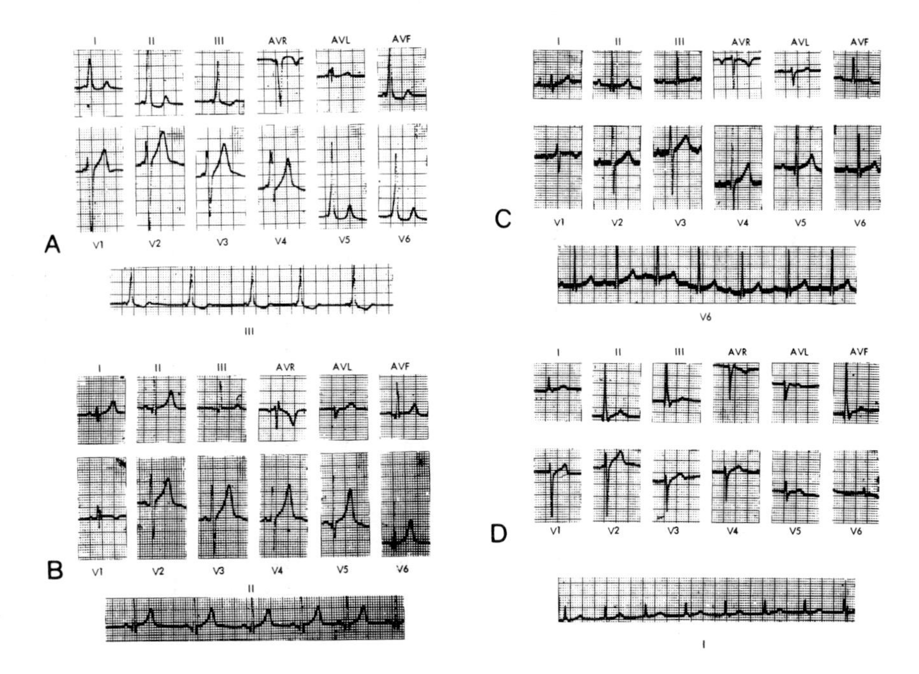

Figure 19. *Arrhythmias in pre-excitation in a 15-year-old boy.* **a.** *Pre-excitation conduction.* **b.** *Normal AV conduction, normal ECG tracing.* **c.** *Normal AV conduction. First degree AV block (P-R = 0.24 second).* **d.** *Junctional rhythm with retrograde atrial activation (P waves following QRS complexes). Note aberrant QRS complexes, unlike both normal AV conduction and pre-excitation.*

through the normal AV node–His bundle axis forming, together with the slow-conducting delta wave, the so-called "fusion beat" of pre-excitation.[75] Others believe that the later fast depolarization results from the unhindered advance of the slowly moving delta wave when it reaches some part of the empty channels of the ventricular conduction system; these channels are still in a nonrefractory stage because the delta wave reaches them before the normal stimulus descending through the regular AV conduction pathway.[24] This latter view is supported by the finding of a very narrow angle between the spatial axis of the delta vector and the mean vector of the remaining QRS. This has led some investigators to conclude that the direction of the delta wave actually "dictates" the vectorial direction of the whole QRS complex.[76,77] (See Chapter V.)

Duration of the QRS

The duration of the QRS during pre-excitation is usually longer than the normal 0.08-0.11 second. As shown in Table XII, the QRS

TABLE XII
Duration of the QRS Complex in Pre-excitation

Studies	0.06-0.08 sec	0.09-0.10 sec	0.11-0.12 sec	0.13-0.15 sec	0.16 sec	Total
Swiderski[9]	7	8	8	4	1	28*
Swiderski[9]	3	7	6	1	3	20**
Schiebler[10]	—	—	4	6	2	12*
Schiebler[10]	—	3	1	5	7	16**
Nasser[12]	—	3	18	7	1	29
Rosenkrantz[15]	—	1	7	8	4	20
Lombardi[13]	3	9	15	13	2	42
Hejtmancik[14]	4	25	31	17	3	80
Longhini[16]	—	2	18	4	—	24
Flensted-Jensen[3]	—	3	16	23	4	46
Averill[11]	—	6	31	27	2	66
Tel Hashomer	10	46	119	29	11	215
Total	27 4.5%	113 19%	274 45.8%	144 24%	40 6.7%	598

*without heart disease
**with heart disease

duration was 0.11-0.12 second in 274 of 598 cases—almost half; it was longer than 0.13 second in 184 cases, and 0.10 second or less in 140 cases. It is noteworthy that in the group with the shortest QRS duration (0.06-0.08 second, 27 cases), more than half were children where the QRS is generally very short during normal conduction.[7] The data from Table XII demonstrate that a broad QRS is not essential for a diagnosis of pre-excitation (Figure 8b and c, 20, 21). In our own 215 cases, more than half (119 cases) showed a QRS duration of 0.11-0.12 second, reflecting the proportion encountered in the total 598 cases.

Classification of QRS Patterns

Various attempts have been made to classify the great variety of unusual QRS patterns in pre-excitation.

Rosenbaum,[55] using the form of QRS in the right precordial leads (leads V_1, V_2 and V_E), defined two groups: 1) group A in which the R is the sole or the largest deflection in all these leads; and 2) group B in which an S or QS is the main QRS deflection in at least one of the leads. The author did not consider the QRS in the limb leads, and only stated that "left axis deviation is very common in atrioventricular excitation and the electrocardiogram in half our series of ten cases exhibited it."

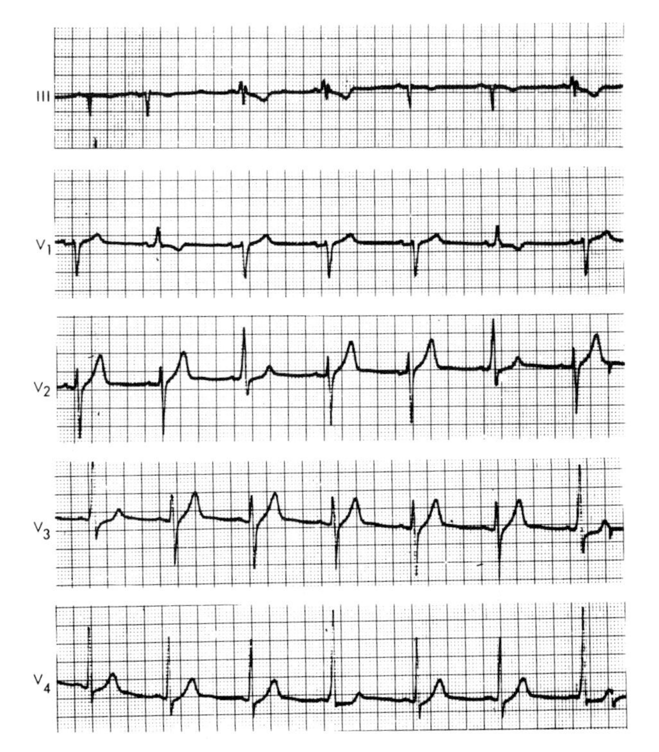

Figure 20. *Intermittent pre-excitation in a 50-year-old man. Relatively narrow QRS duration during pre-excitation beats (0.10 second). No concomitant P wave changes.*

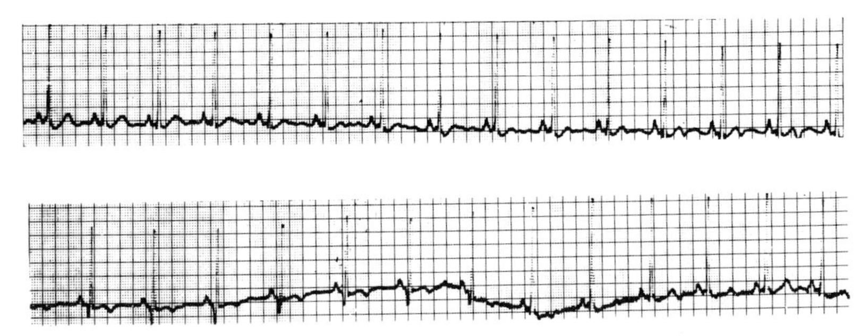

Figure 21. *Concertina phenomenon during pre-excitation in a 53-year-old man. Ambulatory ECG monitoring. Note T wave changes with disappearance of delta waves. Compare first complex in upper and lower panel.*

He added that no correlation was found between the inclination of the mean electrical axis of the QRS in the limb leads and from the right side of the precordium. Accordingly, left axis deviation occurs in cases that belong to group A as well as to group B.

Ueda et al[78] in 1966 introduced some minor changes in Rosenbaum's original classification by dividing group B into two parts. Type A is reserved for those cases where the QRS complex in V_1 is characterized by R or RS pattern, Type B is for those which show an rS form in V_1, while Type C is for cases with a QS or W pattern in V_1.

Burch and DePasquale[79] presented a four-group classification. Type A contains Rosenbaum's A and B, but the delta and QRS vectors are always directed superiorly, with a "QRS pattern" in lead III simulating an infarction of the diaphragmatic surface of the heart. In type B the vector is directed inferiorly and the QRS resembles the pattern of left bundle branch block. Type C appears similar to right bundle branch block, and in type D the QRS complex is minimally deformed and only slightly, if at all, prolonged. The important feature of this classification is the emphasis placed on the superior or inferior direction of the delta wave and of the whole QRS complex; Rosenbaum and Ueda stressed the anterior or posterior directions only of the deformed QRS complex. Burch and DePasquale's view was confirmed by some observations of Grant.[24]

We also noted in our cases the variety of spatial delta vectors with the remainder of the QRS complex usually following the direction of the delta wave. From a purely descriptive standpoint, the direction of the

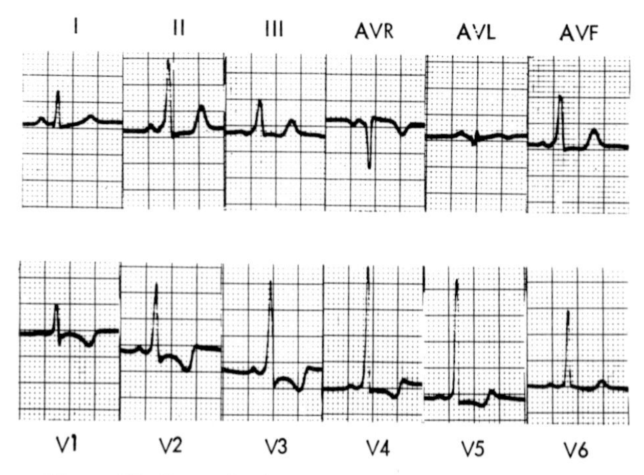

Figure 22. *Pre-excitation type AI in a 22-year-old man.*

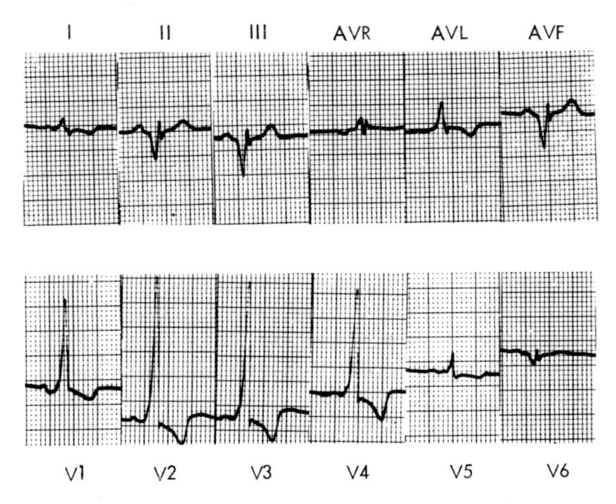

Figure 23. *Pre-excitation type AS in a 33-year-old woman.*

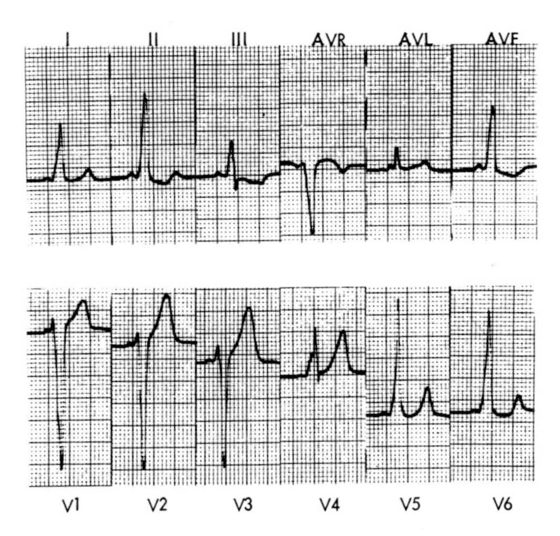

Figure 24. *Pre-excitation type BI in a 19-year-old man.*

vectors in the limb leads (frontal plane) seems to be as important as in the chest leads (horizontal plane).[80] While any combination of spatial directions is possible, the great majority of cases falls into four major groups: AI (anterior and inferior, Figure 22), AS (anterior and superior, Figure 23), BI (left posterior and inferior, Figure 24) and BS (left posterior and superior, Figure 25)(Diagram I). Our criteria for the anterior-posterior axis were the same as Rosenbaum's, with the exception that we replaced his V_E lead by V_4R. A superior vector was defined as one

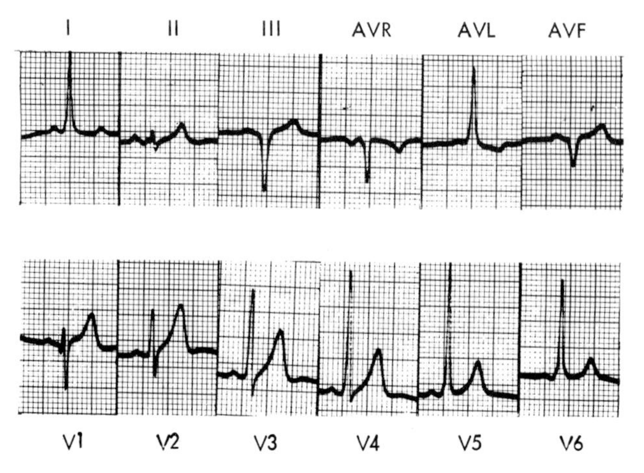

Figure 25. *Pre-excitation type BS in a 26-year-old man.*

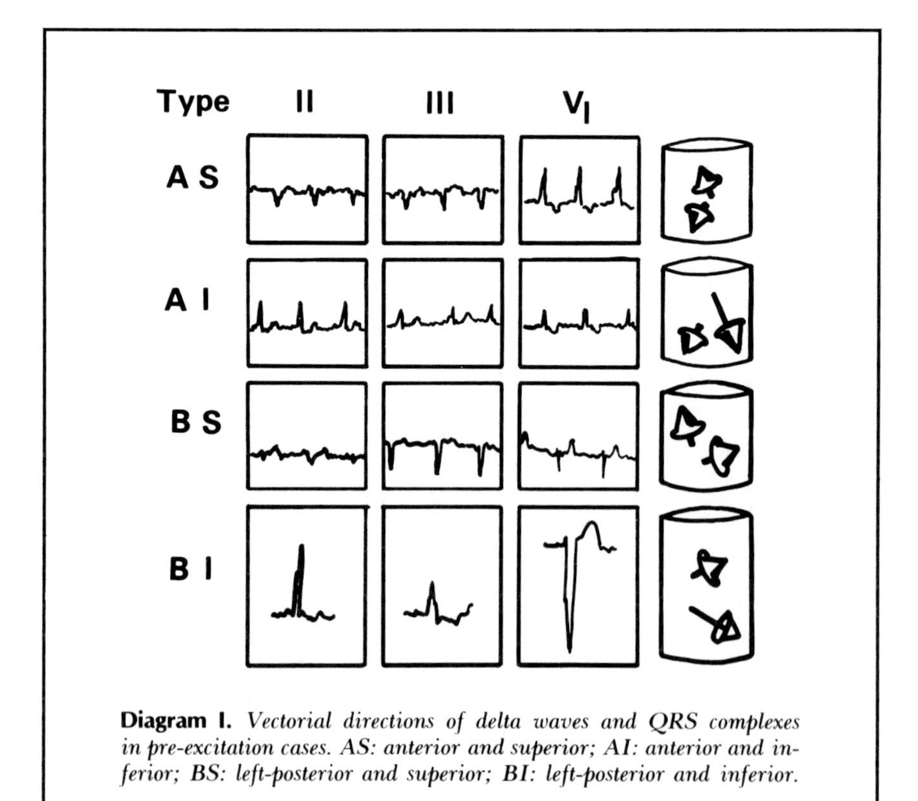

Diagram I. *Vectorial directions of delta waves and QRS complexes in pre-excitation cases. AS: anterior and superior; AI: anterior and inferior; BS: left-posterior and superior; BI: left-posterior and inferior.*

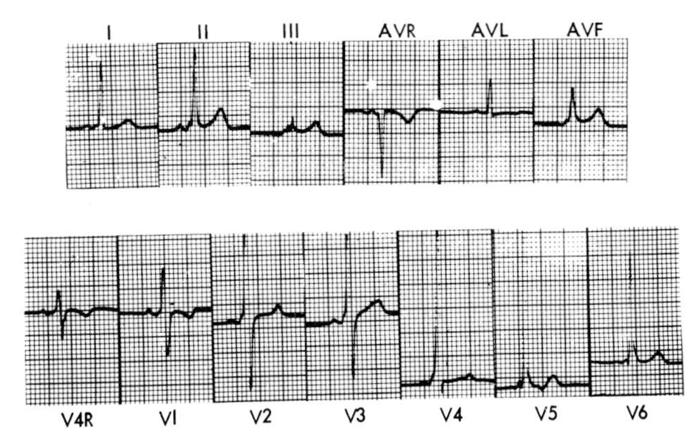

Figure 26. *Pre-excitation type ABI in an 18-year-old man.*

which shows a left axis deviation of 30° or more or a deep Q wave in leads II, III or aVF (usually in 2 of the 3 leads). A vector of +30° or more was considered an inferior vector. It was difficult in a few cases to fit the general vectors of a delta and a QRS complex into this relatively narrow schematic frame, and additional groups were required: vectors in the anterior-posterior axis located midway between the two extremities were labeled AB (Figure 26); and a horizontal vector in the frontal plane (between −30° and +30°) was indicated by M (middle direction, Figure 27).

Table XIII summarizes the findings in cases from our series using this classification. There were 80 instances where the vectors could be clearly located in groups AS and AI, and 113 cases in groups BS and BI

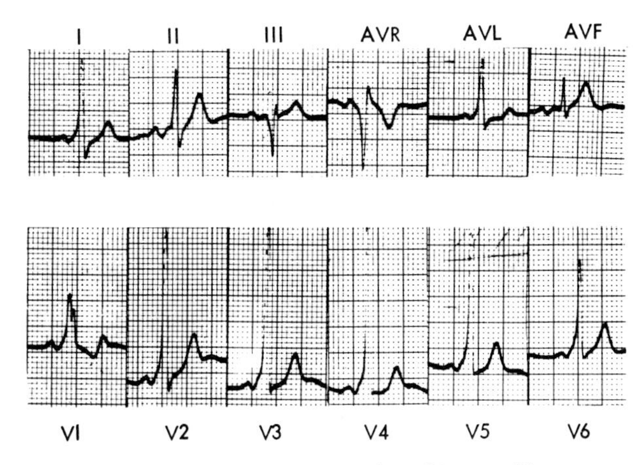

Figure 27. *Pre-excitation type AM in a 20-year-old woman.*

TABLE XIII
Proposed Descriptive Classification of QRS Patterns

Group*	# Cases
AS	32
AI	48
AM	3
BS	64
BI	49
BM	9
ABS	1
ABI	5
ABM	2
A + B	2

*See text for details

(a slight preponderance of cases in the B groups). Twenty cases (9%) did not fit into any of these four groups, and in 2 cases an A and B type were observed on different occasions in the same patient (Figure 7a and b, 15a and b). The advantage of this approach to the different directions of the delta and QRS vectors is that the form of the QRS can be predicted and a misdiagnosis such as "diaphragmatic infarction," avoided (AS and BS: Figure 4, 8b and c, 12a, b, c and d).

The classification systems of the various QRS complexes in pre-excitation have to date been descriptive in nature. Boineau et al[72] recently classified QRS patterns into five types, using epicardial mappings from 7 cases of pre-excitation. Their work was based on the assumption of a single accessory AV connection and no coexisting congenital or acquired cardiac abnormality. This condition fulfilled, they claimed to be able to define, from a simple 12-lead ECG tracing, the location of the anomalous muscular bridge. They stated that their system has practical applications for surgery as well as for the study of anomalous tracts in pathological material. This claim did not withstand the actual findings during surgery. Gallagher et al,[81] summarizing the findings in 30 patients who underwent surgery, concluded that "the electrocardiogram was useful as a first approximation of the site of pre-excitation but did not allow discrimination of free wall accessory pathways (18 patients) from septal accessory pathways (11 patients). Additional studies are thus required to predict the site of the accessory pathway prior to surgical intervention." Adding to this the qualifications introduced by Boineau et al,[72]—no additional heart diseases (found in 20-30% of cases with pre-excitation) and no more than one single anomalous connection (a fact which is usually unknown prior to surgery),[82]—the assumption that the topographic site of accessory bundles can be predicted from a single ECG tracing seems quite improbable. A similar but simplified approach (to divide all the forms of pre-excitation only in a left, right and septal type) was recently proposed,[83] but it seems that at present, factors such as etiology, pathological mechanism or topographical locations of anomalous AV connections are not possible to determine from simple ECG tracings, and therefore a descriptive classification seems preferable to us.

Variations in QRS Form in the Same Patient

Variations in QRS form are seen in the same individual. They may be obvious or subtle, and may appear in a single tracing or even in a single lead or in two or more different tracings. In most cases only a single change in pattern is detected, but sometimes as many as ten different forms of QRS complexes have been described in one patient.[21] The variations can be divided into two major groups, without and with P wave changes.

Without Concomitant P Wave Changes. *Intermittency.* Simple intermittent pre-excitation[84-88] refers to a sudden spontaneous normalization of the QRS complex (Figure 20). Only a few reports in the literature indicate the number of such cases: 16 of 45 cases in one series,[89] 7 of 29 in another,[12] and 57 of 215 in our series. The actual percentage of spontaneous intermittency must be much higher, however, since it is only a matter of chance that the phenomenon is recorded at all.

Concertina Effect. First described and labeled by Ohnell,[19] this consists of gradual normalization of the QRS accompanied by a progressive change in the P-R interval. The P-R interval is shortened in a step-by-step manner concomitant with a broadening of the QRS complexes ("pushing in"), or the P-R interval is gradually prolonged and accompanied by a multiphase normalization of the deformed QRS complex ("pulling out").[90,91] We have seen this effect only once with the help of Holter ambulatory monitoring (Figure 21).

Two or More Forms of Pre-excitation.[92-96] An alternation of two or more typical pre-excitation QRS forms may appear[18,97] (Figure 14a and b). This includes the "pseudo-normalization" forms which resemble normal conduction but retain some degree of delta wave and aberration;

they can be a sudden transition from one form to another or demonstrate a concertina-like effect.

Type A and Type B Pre-excitation in One Patient. This combination is rare, but was observed by Rosenbaum,[55] Ramachandran,[70] Matter,[98] Matsuda,[99] and in two of our cases (Figure 7a and b, 15a and b).

With Concomitant P Wave Changes. *Intermittency.*[100,101] Complete normalization of the QRS accompanied by P wave changes was observed in a single tracing (direct evidence) (Figure 11d) and in different tracings (indirect evidence) (Figure 3a and b, 8a and b, 10a and b). The P wave changed in form only or in polarity as well. When the new rhythm was faster or slower than the previous one, it was assumed that the pacemaker of the heart had been translocated to an ectopic focus in the atria.[48] When the rate remained the same, the change in the P wave vector[47] was limited to intra-atrial conduction disturbances.

Two Forms of Pre-excitation. The QRS complexes are all of the pre-excitation type, but different in form, and are always accompanied by P wave changes (Figure 3b and c, 17e, 18a, b and c).[21,45,66,72,102]

Type A and Type B Patterns of Pre-excitation. These are seen in the same patient, but preceded by P waves of different morphologies.[55,103,104]

Aberrant Conduction Pattern. This occurs together with a displacement of the main pacemaker of the heart into a supraventricular ectopic location. The QRS changes are neither typical of the patient's pre-excitation pattern nor resemble his normal QRS form (Figure 19a, b, c and d).

REFERENCES

1. **Wolff L, Parkinson J, White PD:** Bundle branch block with short P-R interval in healthy young people prone to paroxysmal tachycardia. *Am Heart J* 5:685, 1930.
2. **Wolff L:** Syndrome of short P-R interval with abnormal QRS complexes and paroxysmal tachycardia (Wolff-Parkinson-White syndrome). *Circulation* 10:282, 1954.
3. **Flensted-Jensen E:** Wolff-Parkinson-White syndrome. A long-term follow-up of 47 cases. *Acta Med Scand* 186:65, 1969.
4. **Coumel P, Waynberger M, Garnier JC, et al:** Ventricular pre-excitation syndrome associating short P-R and delta wave, without QRS widening (3 cases of WPW with narrow complexes). *Arch Mal Coeur* 64:1234, 1971.
5. **Dreifus LS, Arriaga J, Watanabe J, et al:** Recurrent Wolff-Parkinson-White tachycardia in an infant. Successful treatment by a radio-frequency pacemaker. *Am J Cardiol* 28:586, 1971.
6. **Mannheimer E:** Paroxysmal tachycardia in infants. *Acta Paediatr Scand* 33:383, 1945.
7. **Ziegler RF:** *Electrocardiographic Studies in Normal Infants and Children,* Springfield, Illinois: Charles C Thomas, 1951.
8. **Ritter O, Fattorusso V:** *Atlas der Elektrokardiographie,* Basel: Karger, 1951.
9. **Swiderski J, Lees MH, Nadas AS:** The Wolff-Parkinson-White syndrome in infancy and childhood. *Br Heart J* 24:561, 1962.
10. **Shiebler GL, Adams P Jr, Anderson RC:** The Wolff-Parkinson-White syndrome in infants and children. A review and a report of 28 cases. *Pediatrics* 24:585, 1959.

11. **Averill KH, Fosmoe RJ, Lamb LE:** Electrocardiographic findings in 67,375 asymptomatic subjects. IV. Wolff-Parkinson-White syndrome. *Am J Cardiol* 6:108, 1960.
12. **Nasser WK, Mishkin ME, Tavel ME, et al:** Occurrence or organic heart disease in association with the Wolff-Parkinson-White syndrome. Analysis of 29 cases. *J Indiana State Med Assoc* 64:111, 1971.
13. **Lombardi M, Masini G:** Electrocardiographic observations in Wolff-Parkinson-White syndrome. Observations on 43 cases. (Ital) *Mal Cardiovasc* 7:387, 1966.
14. **Hejtmancik MR, Herrman GR:** The electrocardiographic syndrome of short P-R interval and broad QRS complex. A clinical study of 80 cases. *Am Heart J* 54:708, 1957.
15. **Rosenkranz KA:** On the differential diagnosis of the electrocardiographic picture of the WPW syndrome. (Ger) *Z Kreislaufforsch* 54:1168, 1965.
16. **Longhini C, Pinelli G, Martelli A, et al:** Polycardiographic study of ventricular preexcitation. *Minerva Cardioangiol* 21:89, 1973.
17. **Doumer E, Dumez L:** Syndrome de Wolff-Parkinson-White familial. Syndrome de WPW avec PR apparemment normal sur certaines dérivations. *Arch Mal Coeur* 44:1134, 1951.
18. **Glushien AS, Goldblum HL:** Aberrant atrioventricular conduction with normal P-R interval and prolonged QRS complex simulating bundle branch block. *Am Heart J* 40:476, 1950.
19. **Ohnell RF:** Pre-excitation, a cardiac abnormality. *Acta Med Scand* (suppl 152), 1944.
20. **Wolff L, White PD:** Syndrome of short P-R interval with abnormal QRS complexes and paroxysmal tachycardia. *Arch Intern Med* 82:446, 1948.
21. **Lamb LE:** Multiple variations of WPW conduction in one subject. Intermittent normal conduction and a false-positive exercise tolerance test. *Am J Cardiol* 4: 346, 1959.
22. **Pick A, Katz LN:** Disturbances of impulse formation and conduction in the preexcitation (WPW) syndrome. Their bearing on its mechanism. *Am J Med* 19:759, 1955.
23. **Pick A, Fisch C:** Ventricular pre-excitation (WPW) in the presence of bundle branch block. *Am Heart J* 55:504, 1958.
24. **Grant RP, Tomlinson FB, Van Buren JK:** Ventricular activation in pre-excitation syndrome (Wolff-Parkinson-White). *Circulation* 18:355, 1958.
25. **James TN, Puech P:** De subitaneis mortibus. IX. Type A Wolff-Parkinson-White syndrome. *Circulation* 50:1264, 1974.
26. **Lepeschkin E:** Significance of pre-excitation of intraventricular and atrioventricular conduction disturbances. *J Electrocardiol* 2:185, 1969.
27. **McHenry PL, Knoebel SB, Fisch C:** The Wolff-Parkinson-White syndrome with supernormal conduction through the anomalous bypass. *Circulation* 34:734, 1966.
28. **Schiebler GL, Adams P Jr, Anderson RC, et al:** Clinical study of twenty-three cases of Ebstein's anomaly of the tricuspid valve. *Circulation* 19:165, 1959.
29. **Friedberg HD, Schamroth L:** Three atrioventricular pathways: reciprocating tachycardia with alternation of conduction times. *J Electrocardiol* 6:159, 1973.
30. **Wolff L:** Anomalous atrioventricular excitation (Wolff-Parkinson-White syndrome). *Circulation* 19:14, 1959.
31. **Wolferth CC, Wood FC:** Further observations on the mechanism of production of a short P-R interval in association with prolongation of the QRS complex. *Am Heart J* 22:450, 1941.
32. **Littmann D, Tarnower H:** Wolff-Parkinson-White syndrome. A clinical study with report of nine cases. *Am Heart J* 32:100, 1946.
33. **Dunaway MC, King SB, Hatcher CR, et al:** Disabling supraventricular tachycardia of Wolff-Parkinson-White syndrome type A, controlled by surgical A-V block and a demand pacemaker after epicardial mapping studies. *Circulation* 45:522, 1972.
34. **Haft JI, Gomes JAC:** The Wolff-Parkinson-White syndrome: the value of His bundle electrogram. *Cathet Cardiovasc Diagn* 2:113, 1976.
35. **Hunter A, Papp C, Parkinson J:** The syndrome of short P-R interval, apparent bundle branch block and associated paroxysmal tachycardia. *Br Heart J* 2:107, 1940.

36. **Hamburger WW:** Buncle branch block. Four cases of intraventricular block showing some interesting and unusual clinical features. *Med Clin North Am* 13:343, 1929.
37. **Wedd AM:** Paroxysmal tachycardia with reference to nomotopic tachycardia and the role of the extrinsic cardiac nerves. *Arch Intern Med* 27:571, 1921.
38. **Wilson FN:** A case in which the vagus influenced the form of the ventricular complex of the electrocardiogram. *Arch Intern Med* 16:1008, 1915.
39. **Holzmann M, Scherf D:** Über Elektrokardiogramme mit verkürtzter Vorhof-Kammer-Distanz und positiven P-Zacken. *Z Klin Med* 121:404, 1932.
40. **Harnischfeger WW:** Heredity occurrence of the pre-excitation (Wolff-Parkinson-White) syndrome with re-entry mechanism and concealed conduction. *Circulation* 19:28, 1959.
41. **Palatucci OA, Knighton JE:** Short P-R interval associated with prolongation of QRS complex; a clinical study demonstrating interesting variations. *Ann Intern Med* 21:58, 1944.
42. **Sanghvi LM, Misra SN, Baverjee K, et al:** Wolff-Parkinson-White syndrome. Report of a case with several types of P waves, varying QRS contour and A-V nodal rhythm with dissociation. *Am J Cardiol* 4:341, 1959.
43. **Bermudez GA, Childers RW:** Anomalous atrioventricular excitation (Wolff-Parkinson-White syndrome) persisting with ectopic impulse formation. *Chest* 58:405, 1970.
44. **Rőna G, Rusznak M:** Unusual rhythm disorder in the Wolff-Parkinson-White syndrome. (Ger) *Dtsch Gesundh* 22:2468, 1967.
45. **Mustakallio KK, Saikkonen JI:** Persistence of characteristic QRS pattern of Wolff-Parkinson-White syndrome in middle nodal rhythm. *Am Heart J* 46:607, 1953.
46. **Brody DA, Woolsey D, Arzbaecher RC:** Application of computer technique to the detection and analyzing of spontaneous P wave variations. *Circulation* 36:359, 1967.
47. **James TN, Sherf L:** Specialized tissues and preferential conduction in the atria of the heart. *Am J Cardiol* 28:414, 1971.
48. **James TN, Sherf L:** "P waves, atrial depolarization, and pacemaking site," in *Advances in Electrocardiography* (Schlant RC, Hurst JW, Eds), p 37, New York: Grune & Stratton, 1972.
49. **Sherf L, James TN:** A new electrocardiographic concept: synchronized sinoventricular conduction. *Dis Chest* 55:127, 1969.
50. **Waldo AL, Vitikainen KJ, Kaiser GA, et al:** The P wave and P-R interval. Effects of the site of origin of atrial depolarization. *Circulation* 42:653, 1970.
51. **Fisch C, Pinsky ST, Shields JP:** Wolff-Parkinson-White syndrome. Report of a case associated with wandering pacemaker, atrial tachycardia, atrial fibrillation, and incomplete A-V dissociation with interference. *Circulation* 16:1004, 1957.
52. **Samet P, Mednik H, Schwedel JB:** Electrokymographic studies of the relation between the electrical and mechanical events of the cardiac cycle in the Wolff-Parkinson-White syndrome. *Am Heart J* 40:430, 1950.
53. **Sanghvi LM, Misra SN:** Electrocardiograms with short P-R interval and aberrant QRS complex. *Br Heart J* 20:357, 1958.
54. **Sodi-Pallares D, Galder RM:** *New Bases of Electrocardiography,* St. Louis: Mosby, 1956.
55. **Rosenbaum FF, Hecht HH, Wilson FN, et al:** The potential variations of the thorax and esophagus in anomolous atrioventricular excitation (Wolff-Parkinson-White syndrome). *Am Heart J* 29:281, 1945.
56. **Fosmoe RJ, Averill KH, Lamb LE:** Electrocardiographic findings in 67,375 asymptomatic subjects. II. Supraventricular arrhythmias. *Am J Cardiol* 6:84, 1960.
57. **Rodstein M:** A case of anomalous auriculoventricular conduction with auriculoventricular block and a history of rheumatic fever. *NY State J Med* 51:789, 1951.
58. **Segers M:** La synchronisation auriculoventriculaire et le syndrome de Wolff-Parkinson-White. *Arch Mal Coeur* 44:712, 1951.
59. **Bix HH:** The electrocardiographic pattern of initial stimulation in the left auricle. A study with a report of unusual arrhythmias originating in the left auricle. *Sinai Hosp J* (Baltimore) 2:37, 1953.
60. **Scherf D, Cohen J:** *The Atrioventricular Node and Selected Cardiac Arrhythmias,* pp

373-447, New York-London: Grune & Stratton, 1964.

61. **Soffer A:** Impulse formation within the accessory conduction tissue in Wolff-Parkinson-White syndrome. *Dis Chest* 42:87, 1962.
62. **Pick A:** Aberrant ventricular conduction of escaped beats. *Circulation* 13:702, 1956.
63. **Fox T:** On the morphology of the delta wave in the Wolff-Parkinson-White syndrome. *Cardiologia* 42:377, 1963.
64. **Scherf D, Bornemann C:** Two cases of the pre-excitation syndrome. *J Electrocardiol* 2:177, 1969.
65. **Chung KY, Walsh TJ, Massie E:** Wolff-Parkinson-White syndrome. *Am Heart J* 69:116, 1965.
66. **Grolleau R, Puech P, Cabasson J, et al:** Particularités de la conduction auriculoventriculaire dans un syndrome de Wolff-Parkinson-White. *Arch Mal Coeur* 67:13, 1974.
67. **Akesson S:** Pre-excitation and auricular fibrillation. *Acta Med Scand,* 120:1, 1945.
68. **Burchell HB:** Atrioventricular nodal (reciprocating) rhythm. Report of a case. *Am Heart J* 67:391, 1964.
69. **Massumi RA, Vera Z:** Patterns and mechanisms of QRS normalization in patients with Wolff-Parkinson-White syndrome. *Am J Cardiol* 28:541, 1971.
70. **Ramachandran S:** Wolff-Parkinson-White syndrome: conversion of type A to type B. Electrocardiographic changes. *Circulation* 45:529, 1972.
71. **Duthie RJ:** Mechanism of the Wolff-Parkinson-White syndrome. *Br Heart J* 8:96, 1946.
72. **Boineau JP, Moore EN, Spear JF, et al:** Basis on static and dynamic electrocardiographic variations in Wolff-Parkinson-White syndrome: anatomic and electrophysiologic observations in right and left ventricular preexcitation. *Am J Cardiol* 32:32, 1973.
73. **Katz LN, Pick A:** *Clinical Electrocardiography: The Arrhythmias,* pp 100-104, Philadelphia: Lea & Febiger, 1956.
74. **Quaglia GB:** The "Delta Complex" in the Wolf-Parkinson-White syndrome. Interpretation of intermediate notching. *Minerva Med* 6:275, 1964.
75. **Rogel S, Kaplinsky E:** Electrocardiographic features in clinical and experimental ventricular pre-excitation. *Am Heart J* 66:453, 1963.
76. **Bleifer S, Kahn M, Grishman A, et al:** Wolff-Parkinson-White syndrome. A vectorcardiographic, electrocardiographic and clinical study. *Am J Cardiol* 4:321, 1959.
77. **Zao ZZ, Herrman CR, Hejtmancik MR:** A vector study of the delta wave in "nondelayed" conduction. *Am Heart J* 56:920, 1958.
78. **Ueda H, Harumi K, Shimomura K, et al:** A vectorcardiographic study of WPW syndrome. *Jpn Heart J* 7:255, 1966.
79. **Burch GE, DePasquale NP:** Electrocardiographic and vectorcardiographic detection of heart disease in the presence of the pre-excitation syndrome (Wolff-Parkinson-White syndrome). *Ann Intern Med* 54:387, 1961.
80. **Schamroth L:** Observations on the QRS complex in the Wolff-Parkinson-White syndrome. *Adv Cardiol* 14:210, 1975.
81. **Gallagher JJ, Gilbert M, Svenson RH, et al:** Wolff-Parkinson-White syndrome. The problem, evaluation, and surgical correction. *Circulation* 51:767, 1975.
82. **Sealy WC, Gallagher JJ, Wallace AG:** The surgical treatment of Wolff-Parkinson-White syndrome: evolution of improved methods for identification and interruption of the Kent bundle. *Ann Thor Surg* 22:443, 1976.
83. **Burch GE:** Of simplifying classification of WPW syndrome (left, right, and septal types of WPW syndrome). *Am Heart J* 90:807, 1975.
84. **Bazika V, Bazikowa K, Slaby A, et al:** Undulant Wolff-Parkinson-White syndrome. (Cz) *Sb Lek* 71:275, 1969.
85. **Blaszczakiewicz M, Kwietniewski W, Podgórski J:** Spontaneous regression in a case of Wolff-Parkinson-White syndrome. (Pol) *Wiad Lek* 19:571, 1966.
86. **Kovaliv IuM, Mironenko VN:** Transient course of the Wolff-Parkinson-White syndrome. (Rus) *Vrach Delo* 5:41, 1966.

87. **Moulopoulos SD, Plassaras GC, Sideris DA:** Heart rate and intermittent Wolff-Parkinson-White syndrome. *Br Heart J* 33:513, 1971.
88. **Sherf L, James TN:** A new look at some old questions in clinical electrocardiography. *Henry Ford Hosp Med Bull* 14:265, 1966.
89. **Mortensen V, Nielsen AL, Eskildsen P:** Wolff-Parkinson and White's syndrome. *Acta Med Scand* 118:506, 1944.
90. **De Petra V, Cecchetti E:** On a rare case of "concertina" Wolff-Parkinson-White syndrome during acute coronary insufficiency. *Cardiol Prat* 20:315, 1969.
91. **Rafalowicz A, Zieliński J:** The accordion sign in the WPW syndrome. (Pol) *Kardiol Pol* 6:201, 1963.
92. **Adamska-Dyniewska H:** Electric alternation in Wolff-Parkinson-White Syndrome, (Pol) *Wiad Lek* 22:2199, 1969.
93. **Bugoslavskaia TV, Krivolutskaia OI:** Several variants of the Wolff-Parkinson-White syndrome. (Rus) *Kardiologiia* 9:125, 1969.
94. **Cotti L, Mergon G:** Sulla variabilitá e sul polimorfismo del quadro elettrocardiografico di Wolff-Parkinson-White. *Minerva Med* 44:271, 1953.
95. **Heikkilä J, Jounela A:** Intra-A-type variation of WPW syndrome. *Br Heart J* 37:767, 1975.
96. **Sarkas A, Sideries DA, Valianos G:** Multiform Wolff-Parkinson-White syndrome. *J Electrocardiol* 7:87, 1974.
97. **Chung EK:** Type A Wolff-Parkinson-White syndrome with dual anomalous conduction. *Postgr Med* 51:266, 1972.
98. **Matter WJ, Hayes WL:** Wolff-Parkinson-White syndrome: report of a case with both type A and type B pre-excitation. *Am J Cardiol* 13:284, 1964.
99. **Matsuda A, Harumi K, Murao S, et al:** Case of WPW syndrome with various arrhythmias and alternating Rosenbaum's A and B patterns. (Jap) *Naika* 20:353, 1967.
100. **Amat-y-Leon F, Dhingra RC, Rosen M:** Ectopic atrial rhythm with pre-excitation. *Chest* 69:538, 1976.
101. **Gould L, Reddy CV, Gomprecht RF:** Anomalous conduction in Wolff-Parkinson-White syndrome. *NY State J Med* 75:906, 1975.
102. **Segers M, Lequime J, Denolin H:** L'activation ventriculaire précoce de certains coeurs hyperexcitables. Etude de l'onde de l'electrocardiogramme. *Cardiologia* 8:113, 1944.
103. **Bruyneel KJ:** Wolff-Parkinson-White syndrome. *Circulation* 47:433, 1973.
104. **Josephson ME, Caracta AR, Lau SH:** Alternating type A and type B Wolff-Parkinson-White syndrome. *Am Heart J* 87:363, 1974.

III

The Electrocardiogram in the Pre-excitation Syndrome: Disturbances of Rate and Rhythm

PREMATURE BEATS

Premature beats in pre-excitation[1-7] are either supraventricular or ventricular. The supraventricular form appears in two major QRS patterns —one mimicking the typical pre-excitation complex and the other presenting the configuration of ordinary supraventricular premature beats —and a number of less commonly seen patterns. These include: atrial premature beats with normally shaped QRS patterns (Moia and Inchauspe[8]); del Zar et al's case[9] where a pre-excitation beat was conducted in a retrograde manner to the atria and the impulse was again followed by another extrasystole (an "echo beat"); bigeminy, where one tracing showed a classic pattern of pre-excitation and another exhibited normal sinus rhythm and pre-excitation-like premature beats (Robinson and Talmage[10]); Chung, Walsh and Massie's case[11] where the supraventricular extrasystole originated from an atrial parasystolic center; and others (Scherf and Schoenbruner[12]).

Ventricular premature beats are more rare in this syndrome.[1,13] Katz and Pick,[14] Littman and Tarnover,[15] and Cloetens and De Mey[16] all reported the phenomenon, and Rosenbaum et al[17] found it in one patient followed by inverted P waves in lead I and positive P waves in lead III. Gaspary,[18] Katz and Pick[14] and Chung et al[11] described cases of pre-excitation in which a ventricular parasystolic center was active. Sometimes an alternation of normally and abnormally conducted beats is

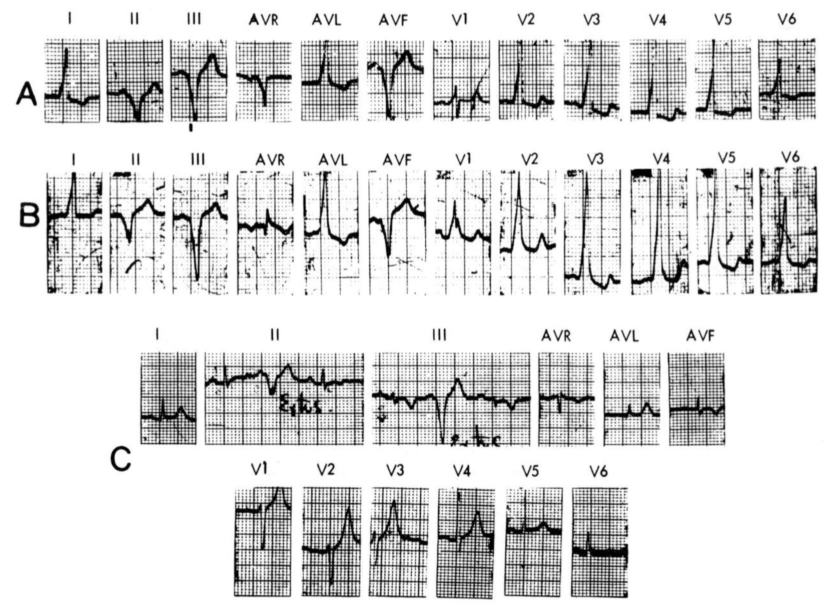

Figure 28. *Two forms of pre-excitation in a 45-year-old man.* **a.** *Pre-excitation type BS.* **b.** *Note the appearance of Rr in lead V_1 (RBBB).* **c.** *Pre-excitation beats during intermittency (leads II and III) misdiagnosed as ventricular premature beats (see Chapter IX).*

misinterpreted as atrial or ventricular bigeminy, and the pre-excitation beat is taken for an extrasystole occurring late in diastole after each normal beat (two of our cases, Figure 11d, 28). It must be stressed that some apparently ventricular premature beats may be of atrial origin, as demonstrated by Fleischman[19] with the help of esophageal leads.

Although premature beats of both kinds appear to be a common feature in the pre-excitation syndrome, there are almost no data available on incidence. In our series, 41 (19%) of 215 patients exhibited premature beats: 7 had supraventricular beats, 5 had pre-excitation-like forms of extrasystoles (Figure 29a), 11 had ventricular premature beats only (Figure 11c), and 14 showed both supraventricular and ventricular premature beats (Figure 14f). Four patients had a history suggestive of extrasystoles.

Chung, Walsh and Massie[11] found premature beats in 7 (18%) of their 40 cases, supraventricular beats in 5 cases and ventricular beats in 2 cases. Recently Hindman et al,[20] using ambulatory monitoring (Holter apparatus), published findings in 27 pre-excitation patients monitored for 22-24 hours: 17 (63%) registered premature beats—4 ventricular, 8 supraventricular and 5 both forms. No information on pre-excitation–

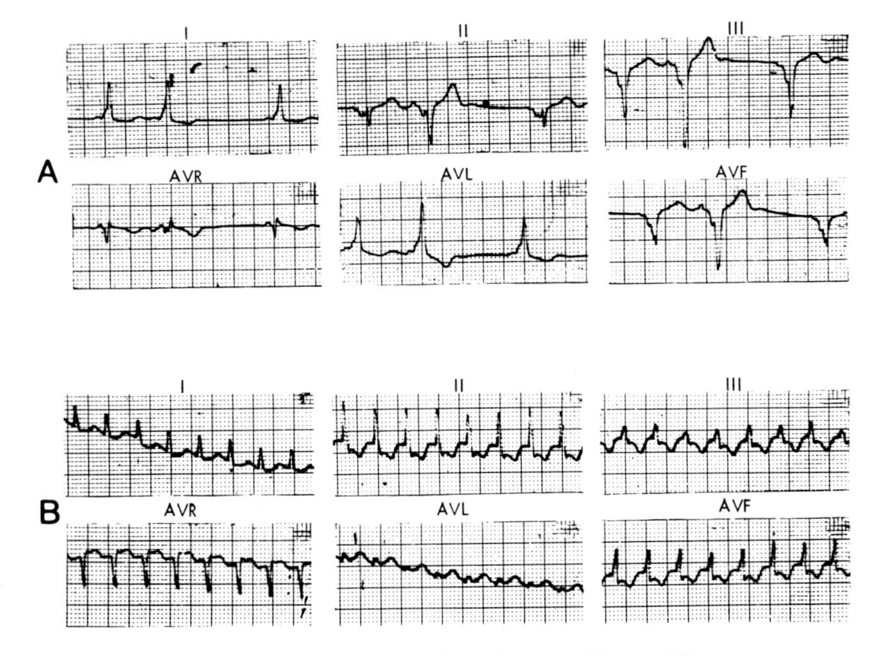

Figure 29. *Premature beats, tachycardia in a 60-year-old man.* **a.** *Supraventricular premature beats with pre-excitation configuration.* **b.** *Paroxysmal atrial tachycardia (same patient as in Figure 15).*

like premature beats was given.

Premature beats in the pre-excitation syndrome have received much attention recently because of the known role of extrasystoles in the initiation of tachycardias.[3,21-23] However, the important question of why there is such a frequent occurrence of premature beats in this syndrome has been generally ignored. The prevalence is even more remarkable when compared to a random series of healthy or sick people: Hiss and Lamb[24] found premature beats in only 4% of 122,043 healthy people, and Katz and Pick[14] found them in 14% of 50,000 hospitalized patients. The premature beats in both these series were usually of ventricular origin, while the majority of those among cases of pre-excitation were supraventricular. These facts point to the likelihood of an unusual activity of ectopic supraventricular pacemaking in pre-excitation.

TACHYCARDIAS

Wolff, Parkinson and White[25] first pointed out the frequent association of pre-excitation with paroxysmal tachycardia. Previously published cases,[26,27] which were later recognized as pre-excitation, also showed fast

TABLE XIV
Incidence of Tachycardia

Series	# Cases	# with Tachycardia	%
Flensted-Jensen[28]	47	29	62
Averill[29]	67	8	12
Swiderski[30]*	28	19	68
Swiderski[30]**	20	9	45
Schiebler[31]*	12	7	58
Schiebler[31]**	16	8	50
Hejtmancik[32]	80	45	56
Chung[11]	40	29	72
Rosenkranz[33]	20	10	50
Mortensen[34]	45	25	55.5
Longhini[35]	24	19	79
Hunter[4]	19	15	79
Nasser[36]	29	12	41
Ohnell[37]	70	44	63
Wellens[38]	82	67	82
Lowe[39]	45	26	58
Pfisterer[40]	27	19	70
Bleifer[41]	34	17	50
Giardina[42]	62	35	56
Tel Hashomer	215	109	50.6
Isaeff et al[22]	68	35	51
Hindman et al[20]	27	19	70
Total	1,077	605	56.17%

*Without heart disease
**With heart disease

heart rates. It has been estimated that approximately 5-10% of all patients suffering from attacks of paroxysmal tachycardia have pre-excitation.[4] On the other hand, some 50-60% of all cases of pre-excitation present a history of palpitations or have documented ECG tracings of tachycardias. An analysis of 19 series from the literature and our own group (Table XIV) revealed 605 cases (56.17%) with a history of palpitations out of a total of 1,077 patients. Twelve of the 20 series showed an incidence of tachycardia ranging between 40 and 60%. Wellens[38] found the highest number (67 out of 82 patients, 82%), and Averill[29] the lowest (8 out of 67, 12%). This large discrepancy can be explained by the fact that numerous cases with tachycardia were referred to Wellens' group because of the sophisticated electronic studies performed in their institute prior to surgery. Averill's series[29] consisted of military personnel

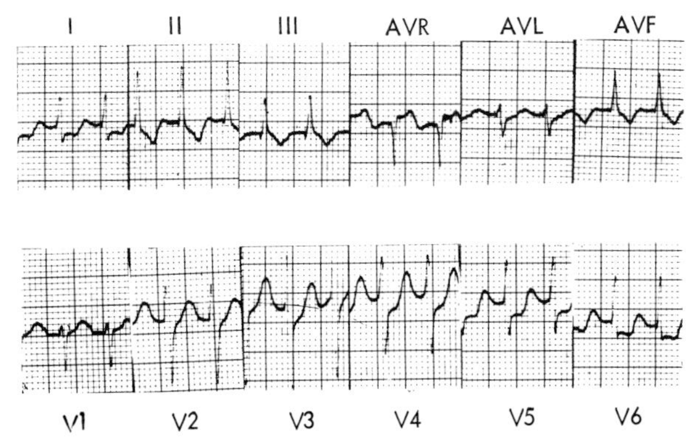

Figure 30. *Paroxysmal atrial tachycardia (PAT) in a 33-year-old woman (same patient as in Figure 8).*

where those with a history of palpitations might already have been excluded. Children and mixed populations showed no differences in the incidence of tachycardia. Paradoxically, in one group of young patients,[30] children with associated heart disease had fewer complaints of palpitations compared to children with pre-excitation alone (45 and 68%, respectively). Among our own 215 cases, 109 gave a history of palpitations (50.6%).

Tachycardias in pre-excitation[43-58] can be divided roughly into three groups: supraventricular, pseudoventricular and true ventricular arrhythmias.

Supraventricular Tachycardias

These constitute the greatest number of documented cases with tachycardias. According to Katz and Pick,[14] three-fourths of the different tachycardias in pre-excitation are supraventricular in origin. The ventricular complexes are usually of normal configuration. These tachycardias can be subdivided into five subgroups:

Paroxysmal atrial tachycardia with a regular heart rate is the most common.[59] It was observed in 10 of 11 cases with tachycardias by Flensted-Jensen,[28] in 5 of 8 by Schiebler,[31] 29 of 41 by Hejtmancik,[32] 24 of 29 by Chung,[11] 5 of 7 by Hunter,[4] 7 of 10 by Mortensen[34] and 19 of 26 by Lowe.[39] In 16 of our patients it was the predominant documented tachycardia observed (Figure 9c, 10c, 30, 31b).

Paroxysmal atrial fibrillation has been observed in sporadic cases by almost all authors.[60-68] Hejtmancik[32] registered it in 6 of his 41 documented cases, Schiebler[31] in 2, Chung[11] in 4, and Lowe[39] in one. Four of our own cases showed paroxysmal atrial fibrillation with narrow QRS complexes (Figure 1, 14e). Katz and Pick[14] stated that this kind of tachy-

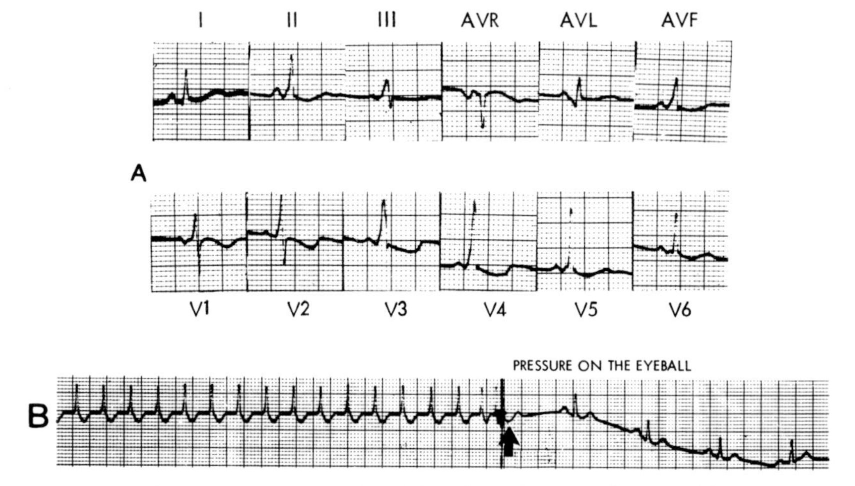

Figure 31. *Paroxysmal atrial tachycardia in a 25-year-old man.* **a.** *Pre-excitation type ABI.* **b.** *PAT, yielding after pressure on the eyeball (see Chapter II and VIII).*

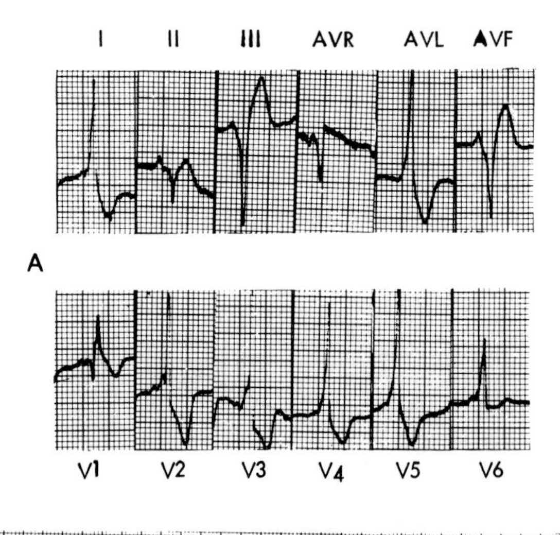

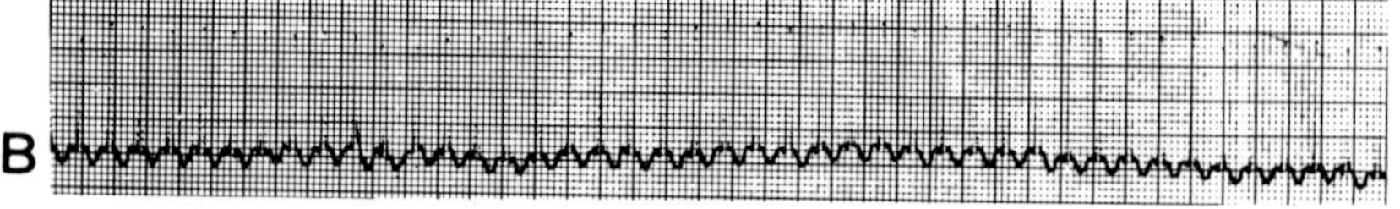

Figure 32. *Atrial flutter in a one-year-old male child.* **a.** *Pre-excitation type AS.* **b.** *Atrial flutter, rate 300, with 1:1 AV conduction.*

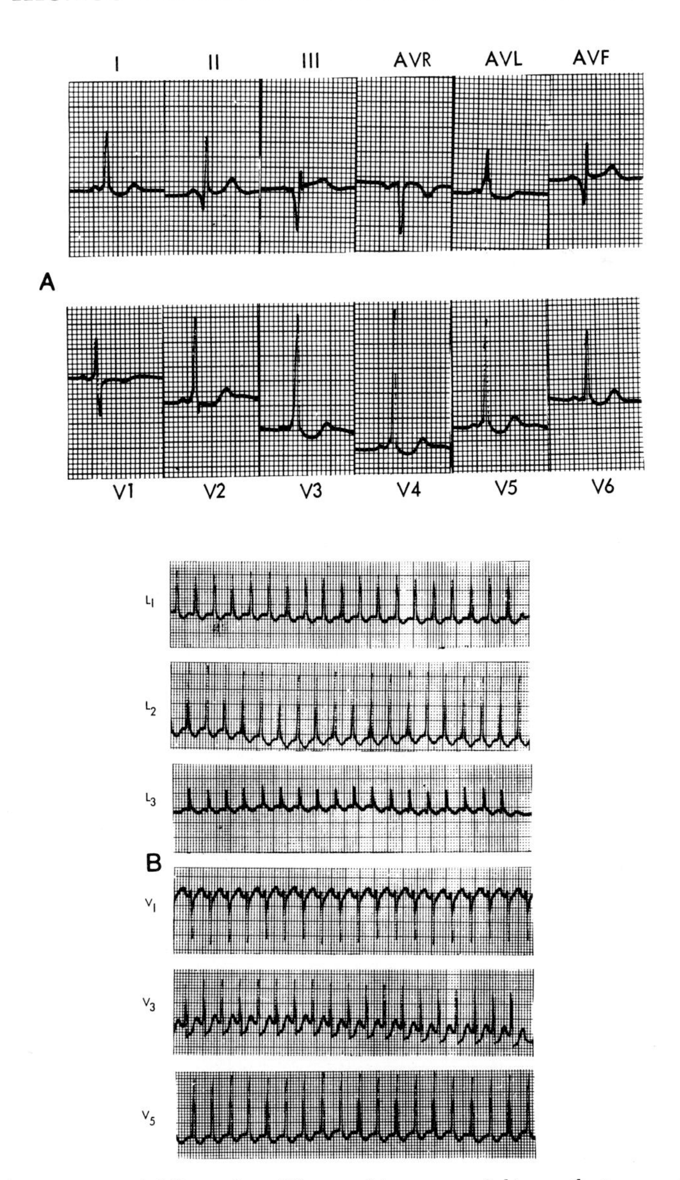

Figure 33. *Atrial flutter in a 25-year-old woman, eight months pregnant.* **a.** *Pre-excitation type ABS.* **b.** *Atrial flutter, 1:1 AV conduction.*

cardia is seen more in patients also suffering from rheumatic heart disease. We did not see it in any of our 9 cases with rheumatic heart disease; tachycardias were recorded in 4 of them, and all were of the paroxysmal atrial tachycardia type (Figure 9c, 10c).

Cases of *atrial flutter* are less common.[69-71] They usually show a very fast heart rate and a 1:1 ventricular response.[72] Two cases each were reported by Wolff and White (quoted by Scherf and Cohen[73]), Schiebler,[31] Hejtmancik[32] and Lowe.[39] Flensted-Jensen[28] and Chung[11] each reported one case. We observed attacks of atrial flutter in two of our patients; both had heart rates close to 300 beats per minute (Figure 32b, 33b).

AV junctional tachycardia was reported by Ferrer et al,[74] Giraud et al,[75] Puech[76] and Harnischfeger.[77] Scherf and Cohen,[73] however, suspected that most of these were instances of atrial tachycardia in which the P waves were buried in the T or QRS deflections. Burchell[78] reported a case of suspected AV nodal tachycardia, and we saw one such episode with negative P waves following each QRS complex (Figure 34).

Sinus tachycardia. In Hindman et al's[20] 27 cases with Holter ambulatory monitoring, 3 cases with bursts of non-exertional sinus tachycardia during complaints of palpitations were found. All three occurred during rapid heart rates, one as high as 180 beats per minute. No comparable findings have been reported in the literature. Six of our cases exhibited

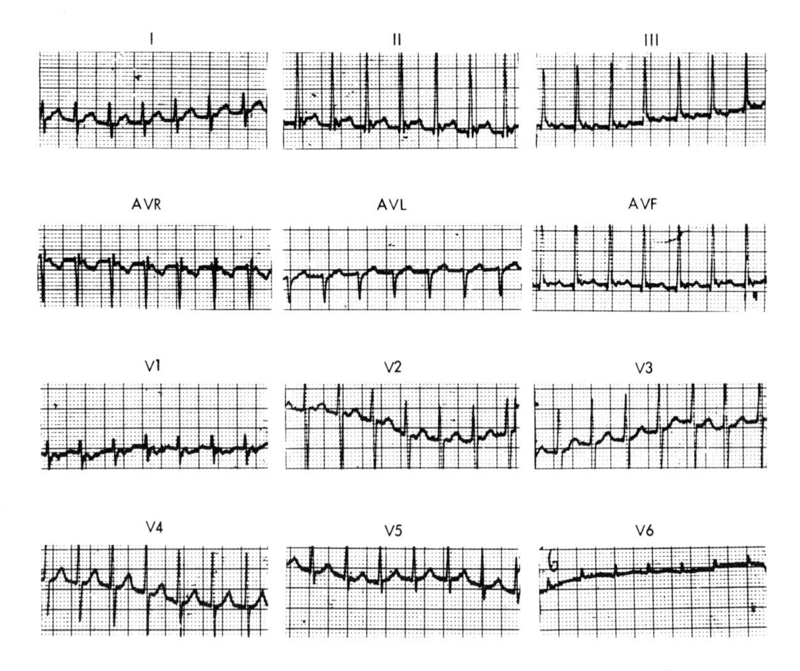

Figure 34. *Junctional tachycardia in a 15-year-old boy. Supraventricular tachycardia with retrograde atrial activation (negative P waves after QRS complexes in leads II, III and aVF, positive in aVR). This is same patient as in Figure 19.*

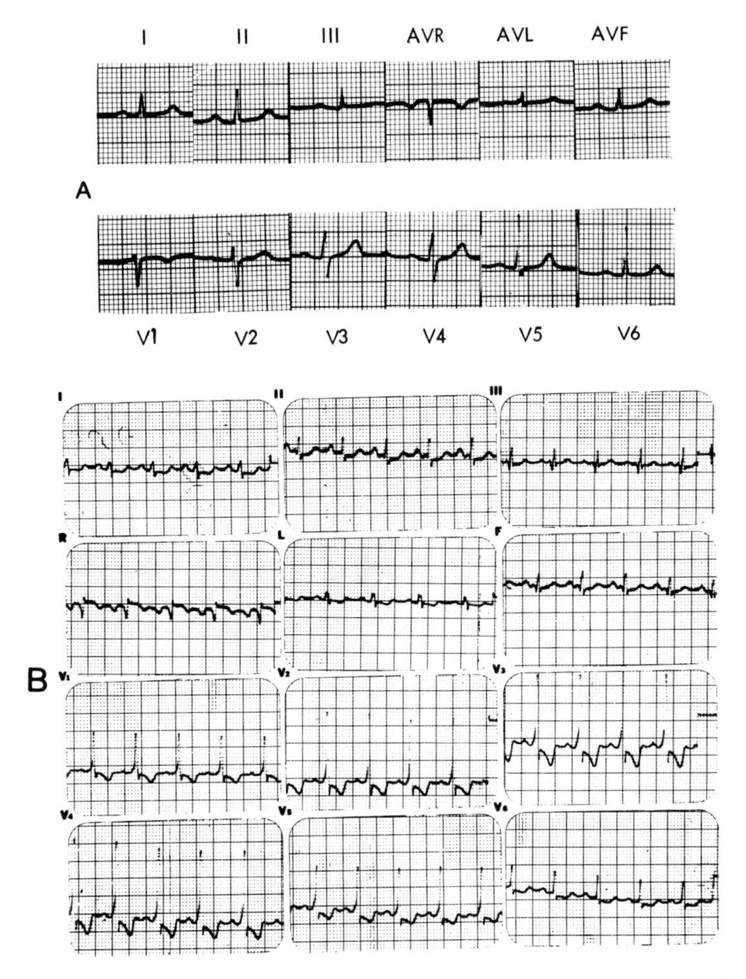

Figure 35. *Paroxysmal sinus tachycardia in a 32-year-old man.* **a.** *Normal AV conduction, normal ECG tracing.* **b.** *Paroxysmal sinus tachycardia, rate 120, with typical pre-excitation QRS pattern (type AS).*

sinus tachycardia (Figure 35b), also during palpitations but not at such fast rates as reported by Hindman.

Pseudoventricular Tachycardias

Cases of pre-excitation showing episodes of ventricular tachycardia were reported for about a decade after Wolff, Parkinson and White's paper: Gilchrist,[79] Arana and Cossio,[80] Levine and Beeson,[81] Cooke and White,[82] Franke and Vetter,[83] Palatuci and Knighton,[84] Missal et al,[85]

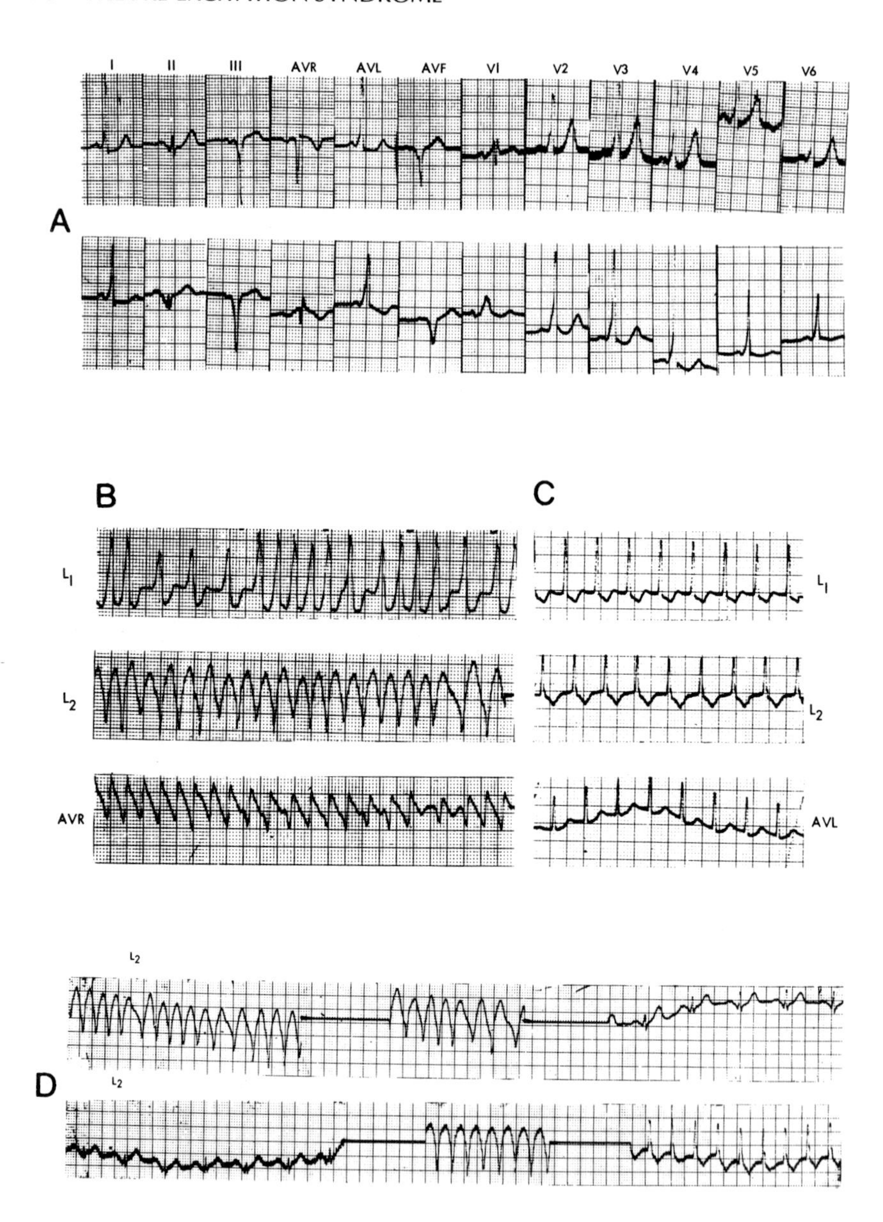

Figure 36. *Mixed tachycardias in a 50-year-old man who experienced sudden death outside the hospital.* **a.** *Two different forms of pre-excitation.* **b.** *Pseudoventricular tachycardia, fast atrial fibrillation with broad QRS complexes.* **c.** *Paroxysmal atrial tachycardia (PAT).* **d.** *Different forms of spontaneous tachycardia and normal sinus rhythm during 15 minutes of ECG tracing.*

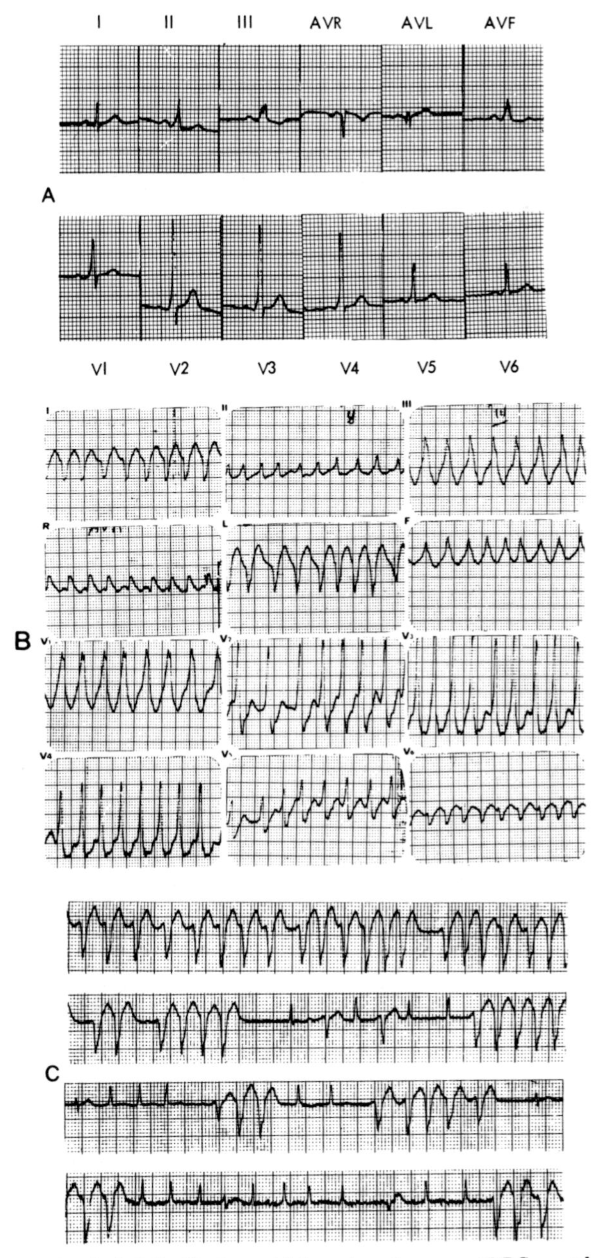

Figure 37. *Atrial fibrillation with broad and narrow QRS complexes in a 58-year-old man.* **a.** *Pre-excitation type AI.* **b.** *Pseudoventricular tachycardia (atrial fibrillation with broad QRS complexes).* **c.** *Atrial fibrillation with broad and narrow QRS complexes.*

Klainer and Joffe,[86] Stein,[87] Cain,[88] Beers and de la Chapelle,[89] Giraud et al,[90] Jordan and Canuteson,[91] Williams and Ellis,[92] Armbrust and Levine,[21] Bruce et al,[93] Broustet et al,[94] Fleishman,[95] Cordeiro,[96] Sears and Manning,[97] Dunn et al.[98] Then Gouaux and Ashman[99] in 1947 and Langendorf et al[100] in 1952 came to the conclusion that episodes of tachycardia labeled "ventricular tachycardias" were mostly cases of atrial fibrillations, or sometimes other types of supraventricular tachycardia (atrial flutter,[101] PAT), either with associated ventricular aberration or retaining the pre-excitation pattern during the tachycardias.[102-106] Eleven of our cases had patterns of "pseudoventricular tachycardia" (Figure 36b and d, 37b). It is our opinion as well that most, if not all, tachycardias encountered in the pre-excitation syndrome are supraventricular tachycardias.

True Ventricular Tachycardias

Although it is not definite that there are episodes of ventricular tachycardia, in some instances such a possibility cannot be excluded. In 1964 Scherf and Cohen[73] quoted some papers which may have included cases of true ventricular tachycardia: Gaspary,[18] Cooke and White,[82] Katz and Pick,[14] Burack and Scherf,[107] Hoffman et al,[108] Quaglia,[109] Suarez,[110] and Tsonchev.[111]

The presence of ventricular fibrillation in the pre-excitation syndrome is questionable, although sporadic cases of true ventricular fibrillation have been observed.[112] Dreifus et al[113] recently found 6 such cases in a review of the literature (Fox et al,[114] Touche et al,[115] Okel,[116] Wojtasik et al,[117] Kaplan et al[118]), and added one of their own. Two additional cases were reported by Gallagher et al[119] in 1975. It is noteworthy that in 5 of these 9 cases, the ventricular fibrillation followed episodes of rapid atrial fibrillation with pre-excitation patterns (pseudoventricular tachycardias).

OTHER ARRHYTHMIAS

There are other arrhythmias in the pre-excitation syndrome which are apparently more rare than tachycardias and premature beats, or perhaps are simply observed less frequently.[120-122] Most of them are related to sinoatrial or atrioventricular conduction anomalies.

Sinoatrial Conduction Anomalies

Sinus Bradycardia. Wolff[123] first noted this phenomenon in 1960. He stated that "when the anomalous mechanism prevails, the heart rate is usually slow." This observation was not picked up, however, until 1973

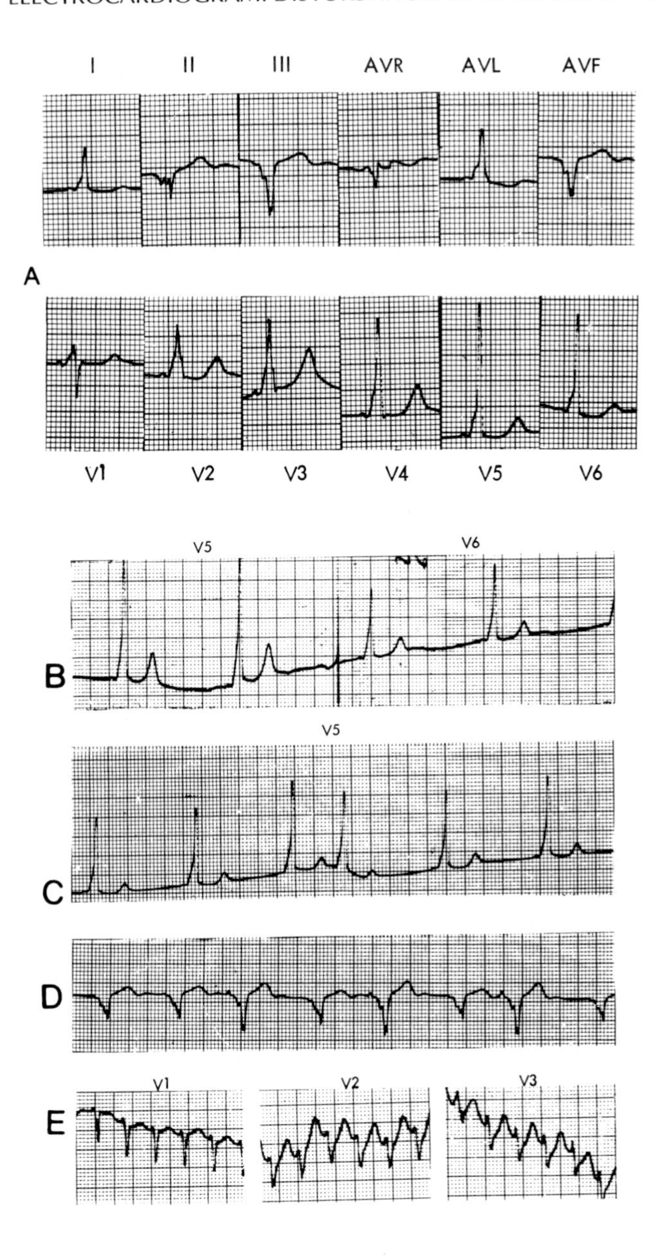

Figure 38. *Pre-excitation and sick sinus syndrome in a 23-year-old man.* **a.** *Pre-excitation type BS.* **b.** *Sinus bradycardia, rate 45-48 beats per minute ($V_{5,6}$).* **c.** *Supraventricular premature beat with pre-excitation (V_5).* **d.** *Concomitant P and QRS changes.* **e.** *PAT.*

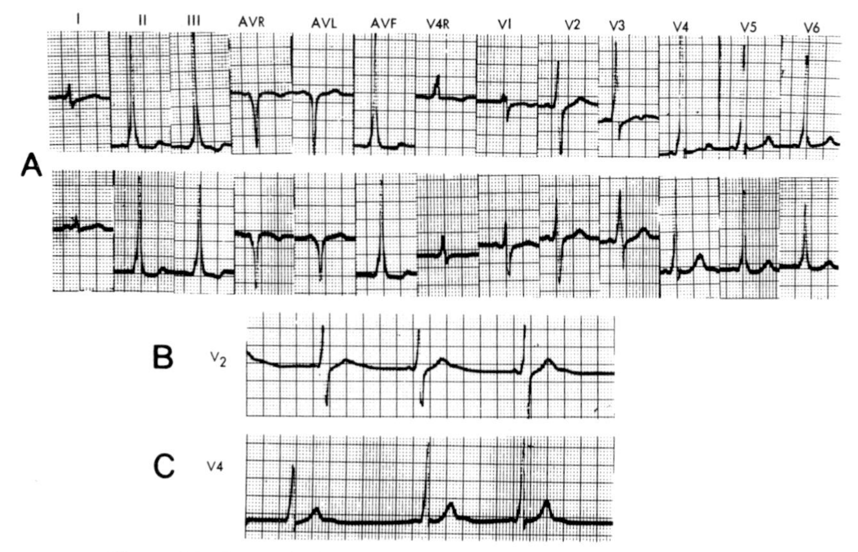

Figure 39. *Pre-excitation.* **a.** *Two different forms of pre-excitation in a 20-year-old woman.* **b.** *Sinus bradycardia ±50 beats per minute (V_2).* **c.** *Escape beat with pre-excitation conduction (V_4).*

when Dreifus et al[124] presented a paper at the 46th Scientific Session of the American Heart Association, stressing the coincidence of sinus bradycardia in the pre-excitation syndrome with a higher incidence of atrial fibrillation. Since the bradycardia-tachycardia syndromes are also often associated with atrial fibrillation, the presence of bradycardia was examined in 19 of their patients with pre-excitation and recurrent atrial fibrillation: they found sinus rates of 40-50 beats per minute in 5 cases, and below 40 beats per minute in 3 cases. Unfortunately, no mention was made of the incidence of bradycardias in the whole series, and the authors concluded that sinus bradycardia is frequently asssociated with intermittent atrial fibrillation in the pre-excitation syndrome and could be a manifestation of a diseased sinus node. These same authors recently added another 14 such patients and came to the same conclusions.[125] Hindman et al,[20] in their 27 cases with portable monitoring over a 24-hour period, found 6 cases with bradycardia (22%). They did not indicate the heart rate or the incidence of tachycardias in these 6 cases as compared to the whole series. These authors and others also concluded that there is a rather high frequency of sinus bradycardia in the pre-excitation syndrome.[20,126]

In the Tel Hashomer series of 215 cases, 39 presented sinus bradycardia (18%), 30 with sinus rates between 51-60 and 9 with 50 or less.

Twenty-four of these 39 patients also had attacks of palpitations and tachycardias (61.5%) compared with 50.7% in the whole series (Figure 38a and b).

It is not known whether sinus bradycardia is more frequent in cases with pre-excitation than in the general population because of lack of comparative data.

Sinus arrhythmias, on the other hand, are strikingly more common among pre-excitation patients than in the general population. Hindman et al[20] found 13 (48%) in their 27 cases with ambulatory monitoring, and we found 25 (11.6%) in 215 cases. In contrast, Greybiel et al[127] found only 5.6% in a series of 1,000 young healthy pilots (and sinus arrhythmias are more common in the young), and Hiss et al[24] found it in 3.4% of 57,942 normal adults. In our cases the sinus arrhythmias were often accompanied by bradycardia, 64% of which had a history of tachycardia.

Sinus Asystole and Arrest. Cases of sinus asystole or arrest have appeared sporadically in the literature. Burchell[78] reported a 44-year-old woman with pre-excitation who showed periods of asystole and tachycardias. Kaplan[128] published a case with prolonged asystole after supraventricular tachycardia. In James' case[129] the sinus rhythm disappeared in a young woman prone to attacks of tachycardia and pre-excitation, and was replaced by a slow AV junctional rhythm (without pre-excitation). Of special interest was Hindman et al's[20] finding of 3 cases of sinus arrest in their 27 patients with ambulatory monitoring.

Junctional Escape Beats. The three dysfunctions of the sinus node described above predispose the appearance of junctional escape beats or rhythm. They appeared in 3 instances in Hindman's cases and in 3 of our cases (Figure 39, 40) where their configurations were similar to the QRS complex seen during pre-excitation.

Summarizing all these sinoatrial conduction anomalies, Hindman concluded that "the high frequency of sinus bradycardia and sinus arrhythmia, and the occurrence of AV dissociation, AV nodal escapes and sinus arrest is a significant finding, implying the possible existence of conduction system abnormalities in addition to anatomical or functional bypass of the AV node."

AV Dissociation, Wandering Pacemaker. AV dissociation due to one or a few escape beats is a passive and physiological depolarization of a subordinate ectopic pacemaker which occurs in cases with sinus bradycardias or arrest. We have also observed cases in which a junctional pacemaker shows enhanced activity where, by a rate of depolarization faster than that of the sinus node, it produces another type of AV dissociation. In one case (Figure 17c and e), a lower and faster pacemaker took over the pacing of the ventricles, once during normal conduction

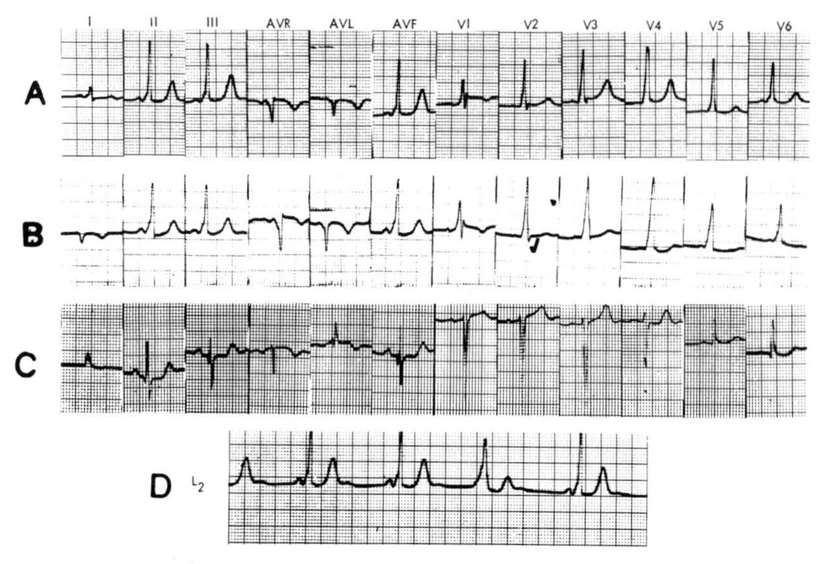

Figure 40. *Pre-excitation and sinus bradycardia.* **a, b.** *Two forms of pre-excitation.* **c.** *Normal conduction.* **d.** *Lead II, Sinus bradycardia, junctional escape beat with pre-excitation configuration.*

without any change in the QRS configuration, and once during pre-excitation producing a changed pattern of anomalous conduction. In both instances the rate of the active junctional pacemaker was identical. In another case (Figure 41), an AV dissociation was observed (isorhythmic AV dissociation) with aberrant ventricular conduction during the dissociation. In this same case (Figure 19b and d), a continuous junctional rhythm with aberration was also seen. In a newborn infant suffering from myocarditis (Figure 42), a third instance of AV dissociation with slight ventricular aberration (best seen in V_4) was registered. In this last case, as in the first one, the pattern of pre-excitation did not disappear during the dissociation.

Cases of active AV dissociation are also reported in the literature. Olesch and Belz[5] described a 25-year-old man with attacks of palpitation which started at the age of 3; in addition to a typical pre-excitation pattern and supraventricular premature beats in the ECG tracing, there were also periods of AV dissociation with normalization of the QRS pattern during the junctional conduction and reappearance of the pre-excitation pattern upon the return of normally conducted sinus beats. Chung et al[11] noted a case of AV dissociation in their series of 40 cases with pre-excitation. Pick and Katz[130] mentioned a case with an AV disso-

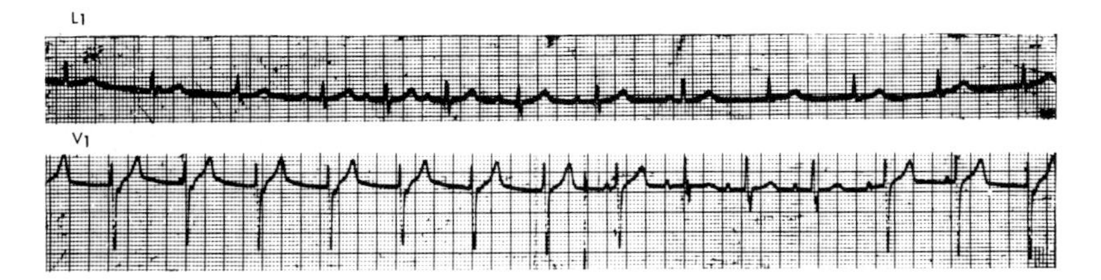

Figure 41. *AV dissociation with aberration and first degree AV block, in a 15-year-old boy. Note appearance of an rS pattern with AV dissociation, while during sinus conduction an Rs pattern is prevailing in lead V_1 (see also Figure 19 and 34 of same patient).*

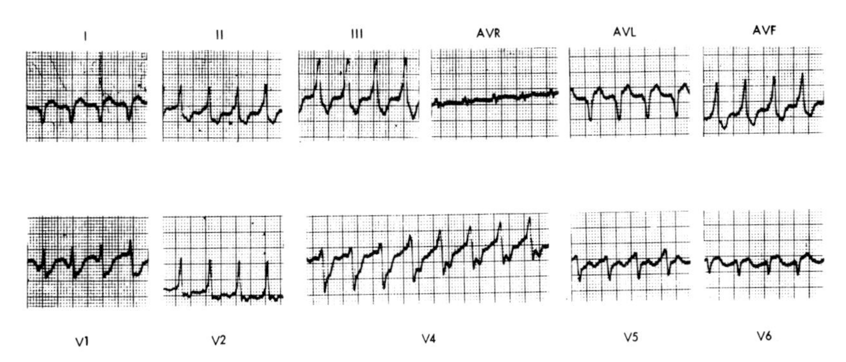

Figure 42. *Pre-excitation in acute myocarditis in a one-week-old male infant. Note AV dissociation (best seen in lead V_4) and QRS aberration in addition to pre-excitation.*

ciation which began with retrograde conducted premature beats accompanied by a change (but not disappearance) in the pre-excitation pattern, followed by the lower ectopic center taking over all the driving of the ventricles, with retrograde atrial conduction and pre-excitation QRS patterns. The authors[130] postulated that the active low ectopic pacemakers may be located in the assumed accessory AV connection. Fisch et al[131] reported a 58-year-old man with an AV dissociation and pre-excitation who also had other arrhythmias, including atrial standstill, concurrent P and QRS changes and a wandering pacemaker. We saw such a wandering pacemaker twice—once in a case of Ebstein's anomaly of the tricuspid valve during Holter portable monitoring (Figure 16) and in another case not included in this series (James and Sherf[132,Fig 14]).

Characteristic of this last group of arrhythmias is an enhanced pacemaking activity in the atria, which expresses itself in active junctional rhythms or wandering pacemaker. This, together with the P wave changes and a high incidence of supraventricular premature beats typical of other arrhythmias, suggest a functional unrest in the atria in pre-excitation.

Atrioventricular Conduction Anomalies

The following different degrees of atrioventricular conduction anomalies were described in patients with pre-excitation.[133-140]

First Degree AV Block. A large number of classic cases of pre-excitation were reported which showed a prolonged P-R interval (first degree AV block) during normal conduction.[29,31,79,129,130,141-146] In most of these cases the duration of the pre-excitation P-R interval was normal, 0.12 second and more (see also Table X). Two of our patients exhibited this (Figure 11a, b, and d, 14d).

In Ebstein's anomaly of the tricuspid valve a prolonged P-R interval (first degree AV block) can often be detected.[30,31] When such cases also show pre-excitation, a normal P-R interval (0.12-0.20 second) is sometimes seen during the anomalous conduction. In one particularly interesting case[146] there was a direct correlation between a prolonged P-R interval during regular conduction (0.24 second) and a normal interval during pre-excitation (0.14 second). It is noteworthy that both P-R intervals were observed in the same patient during intermittency.

Another interesting observation concerns two cases (Grant et al[147] and Friedberg and Schamrot[141]), in which the P-R interval during normal conduction was so prolonged (first degree AV block) that the whole P-QRS of the pre-excitation complex was registered before the normal conducted (intermittent) QRS pattern of the same patient started. Both authors concluded that the entire depolarization of the conduction in the ventricles during pre-excitation was anomalous, and stimuli descending through the normal AV node–His axis could not possibly participate in formation of pre-excitation pattern (no fusion beat!).

Although first degree AV block during normal conduction is not a rare occurrence in cases with pre-excitation, there are to date only 2 published cases which show a prolongation of the P-R of more than 0.20 second, *ie,* a first degree AV block during pre-excitation conduction. The phenomenon in the case reported by Pick and Katz[130] was intermittent and there is no doubt about the diagnosis of both pre-excitation and first degree AV block. In an additional case by Varriale et al,[148] the P-R was 0.22 second, but the normal rhythm showed a left bundle branch block during intermittency.

Second Degree AV Block. The simultaneous occurrence of the pre-excitation syndrome and second degree AV block is rare. While cases with first degree AV block have a prolonged P-R (0.20 second or more) during normal conduction (with the exception of the cases described by Pick and Katz[130] and Varriale et al[148]), a second degree AV block presents the opposite picture. Most of the cases showed a 2:1 AV block (or a Wenckebach phenomenon) during pre-excitation, while only one case[149] showed a 2:1 block along with the normal sinus conduction during intermittency.

In 1959 Moret et al[150] collected 6 cases of different forms of second degree AV block from the literature and added one of their own. Two of the 7 cases represented the Wenckebach phenomenon type of second degree AV block together with the pre-excitation QRS pattern. Moret's own case (one of the two with Wenckebach) was an 86-year-old man with hypertension and acute myocardial infarction, whose ECG tracings showed a classic pre-excitation syndrome with consecutive P-R intervals of 0.08, 0.12 and 0.20 second, and then a dropped beat. Scherf et al[151] described a 70-year-old woman with hypertension and diaphragmatic myocardial infarction with a clear Wenckebach phenomenon and 3:4 conduction. (A 2:1 AV block was present as well.) Another possible case of Wenckebach phenomenon was cited by De Rango,[152] but without details.

The other form of second degree AV block, namely 2:1 conduction with a pre-excitation QRS pattern, was reported in 5 cases. One was Scherf et al's[151] patient with fresh diaphragmatic infarction. The second case (Coelho[153]) was a 62-year-old woman also suffering from hypertension and diaphragmatic myocardial infarction. The ECG showed a 2:1 AV block, a short P-R interval and the characteristic abnormal QRS complex; later complete AV block appeared and the pre-excitation forms of the QRS complex disappeared. Fox et al[114] reported a similar case of a 70-year-old woman with an anginal syndrome, whose ECG showed typical pre-excitation complexes. Following 30 mg procainamide, a typical 2:1 block appeared without any change in the QRS pattern. An additional 200 mg of procainamide resulted in a complete AV block and disappearance of the pre-excitation complexes. De Rango et al[152] described a 60-year-old man with a history of frequent attacks of paroxysmal tachycardia. During hospitalization a classic 2:1 second degree AV block developed with continuation of the pre-excitation pattern. In this case the transition of the 2:1 block to a complete AV block, recorded in one tracing, was associated with a concomitant normalization of the QRS patterns. When sinus conduction reappeared (ventricular captures), the QRS pattern again showed pre-excitation conduction. Masini and Milazzotto[154] added another case of a 66-year-old man suffering from diabetes, angina pectoris, palpitations and fainting spells. He showed right bundle branch block alternating with the pre-excitation syndrome type B and periods of second (or high 4:1) degree AV block as well as complete AV block. The authors felt that during the complete AV block the idioventricular rhythm was left bundle branch block–like (in our opinion, a pre-excitation pattern of the QRS complexes cannot be excluded; see *Fig. 2B in Masini and Milazzotto's paper*). Two cases were reported in which the 2:1 block appeared only during supra-ventricular tachycardias. One, a 64-year-old man reported by Segers et al,[155] had an atrial tachycardia, 2:1 AV block and pre-excitation pattern of the QRS complexes. Pressure on the eyeballs produced a lengthening of the P-R interval. The second case, described by Scherf et al,[151] was a 10-year-old girl. Here too, a 2:1, 3:1 Wenckebach type of second degree AV block appeared during an attack of paroxysmal atrial tachycardia. The authors noted particularly the QRS patterns which showed that the AV block was present during both normal AV and pre-excitation conduction.

Complete AV Block. Pre-excitation cases with complete AV block fall into three groups: 1) complete AV block and a QRS pattern without any characteristics of pre-excitation;[154] 2) incomplete AV block and general absence of pre-excitation QRS pattern,[156] but occasional ventricular captures showing the previously diagnosed QRS pre-excitation pattern; and 3) complete AV block with an idioventricular rhythm resembling pre-excitation QRS complexes.[157]

Masini and Milazzotto's case[154] belongs to the first group, where the complete AV block appeared after a 3:1 and a 4:1 conduction with pre-excitation complexes. According to the authors (only one lead V_1 is presented in the paper), a left bundle branch block–like idioventricular rhythm was observed during complete AV block. James'[129] unusual case, the etiology of which remained obscure even after the postmortem examination and which displayed in one phase of the disease a clear picture of pre-excitation, can be included in this group; a complete atrial standstill appeared later, during which a slow junctional rhythm without pre-excitation drove the ventricles. Coelho's[153] case also developed into a complete AV block with an idioventricular non–pre-excitation rhythm. Lev et al's[149] case also falls in this group.

The second group contains, among others, Levine and Burge's[158] case of pre-excitation and complete AV block in a 62-year-old man just after an acute myocardial infarction. His ECG showed complete heart block with occasional conduction of atrial impulses to the ventricles (ventricular captures); whenever an impulse was conducted, the complexes had the short P-R interval and the wide QRS suggestive of pre-excitation. The importance of this case lies in the postmortem finding of a fresh diaphragmatic myocardial infarction and, for the first time documented, an anomalous bundle bridging the AV groove at the right posterolateral aspect of the heart. The phenomenon of ventricular capture beats with a pre-excitation pattern during complete AV block was also reported by Addarii and Protolani.[159] This group also claims De Rango's[152] patient, in whom a 2:1 AV block with pre-excitation developed into a complete AV block with narrow normal-looking QRS complexes (idiojunctional rhythm) which, from time to time, exhibited ventricular capture beats with a QRS pre-excitation pattern.

Of special importance in this group is Timmis et al's[160] case of a 10-year-old boy in whom cardiac investigation revealed a membranous ventricular septal defect of the type resulting in a dominant left ventricular–right atrial shunt. Preoperative ventricular excitation was normal. The defect was surgically closed, and the immediate postoperative course was complicated by complete heart block which persisted for 18 days, after which conducted sinus rhythm reappeared and a type B pre-excitation pattern was identified. However, during the next 9 months increasingly severe AV block developed with occasional conduction of atrial complexes (ventricular captures). These ventricular complexes retained the same configuration of type B pre-excitation conduction noted in the early postoperative period. Sectioning of the normal AV conduction system (His bundle), which most probably occurred by accident in this case at time of surgery, is now used by surgeons as a method for treating pre-excitation cases suffering from intractable tachycardias (iatrogenic complete AV block).[161]

Finally, one of our cases makes up the third group of complete AV block and pre-excitation. The patient was a one-year-old child with endocardial sclerosis who developed complete AV block, without showing pre-excitation drove the ventricles. Coelho's[153] case also developed into a during the complete AV block was identical to a classic pre-excitation form seen in regular cases, with the exception of the relation to the P waves. The absence of this relationship prompted us to call these cases "pseudo–pre-excitation."[157]

REFERENCES

1. **Anghelescu F:** Ventricular extrasystolic arrhythmia in the Wolff-Parkinson-White syndrome. Considerations on two clinical cases. (Rum) *Med Intern* (Bucur) 22:241, 1970.
2. **Bodrogi G, Bereczky A, Boda E, et al:** WPW syndrome with multiple extra-systoles. *Acta Cardiol* (Bruxelles) 18:449, 1963.
3. **Durrer D, Schoo L, Schuilenburg RM, et al:** The role of premature beats in the initiation and the termination of supraventricular tachycardia in the Wolff-Parkinson-White syndrome. *Circulation* 36:644, 1967.
4. **Hunter A, Papp C, Parkinson J:** The syndrome of short P-R interval, apparent bundle branch block and associated paroxysmal tachycardia. *Br Heart J* 2:107, 1940.

5. **Olesch K, Belz GG:** Arrhythmia in a patient with pre-excitation WPW syndrome. (Ger) (Eng Abstr) *Z Kreislaufforsch* 59:585, 1970.
6. **Scherf D, Schott A:** *Extrasystoles and Allied Arrhythmias,* London: Wm Heinemann Med Book Ltd, and New York: Grune & Stratton, 1953.
7. **Wolferth CC, Wood FC:** Further observations on the mechanism of production of a short P-R interval in association with prolongation of the QRS complex. *Am Heart J* 22:450, 1941.
8. **Moia B, Incauspe LH:** Sobre un caso de P-R corto con QRS ancho y mellado presentando asicronismo ventricular. *Rev Argent Cardiol* 5:114, 1938.
9. **del Zar LE, Batlle FF, Bronstein J:** Sindrome de Wolff-Parkinson-White y cardiopatía reumática. *Rev Argent Cardiol* 16:241, 1949.
10. **Robinson RW, Talmage WG:** Wolff-Parkinson-White syndrome. Report of five cases. *Am Heart J* 29:569, 1945.
11. **Chung KY, Walsh TJ, Massie E:** Wolff-Parkinson-White syndrome. *Am Heart J* 69:116, 1965.
12. **Scherf D, Schoenbrunner E:** Beitrage zum Problem der verkuerzten Hofkammerleitung. *Z Klin Med* 128:750, 1935.
13. **Berkun MA, Kesselman RH, Donoso RH, et al:** The spatial ventricular gradient: intermittent Wolff-Parkinson-White syndrome, intermittent left bundle branch block and ventricular premature contractions. *Circulation* 13:562, 1956.
14. **Katz LN, Pick A:** *Clinical Electrocardiography: The Arrhythmias,* pp 100-104, Philadelphia: Lea & Febiger, 1956.
15. **Littmann D, Tarnower H:** Wolff-Parkinson-White syndrome. A clinical study with report of nine cases. *Am Heart J* 32:100, 1946.
16. **Cloetens W, De Mey D:** Evolution particuliere d'un syndrome de Wolff-Parkinson-White. Influence pharmacologique de la procaine-amide. *Acta Cardiol* 8:632, 1953.
17. **Rosenbaum FF, Hecht HH, Wilson FN, et al:** The potential variations of the thorax and esophagus in anomalous atrioventricular excitation (Wolff-Parkinson-White syndrome). *Am Heart J* 29:281, 1945.
18. **Gaspary F:** Sindrome de Wolff-Parkinson-White per interferencia de ina parasistolia ventricular con el regimen sinusal. *Rev Argent Cardiol* 17:259, 1950.
19. **Fleischmann P:** Interpolation of atrial premature beats of intra-atrial origin due to concealed A-S conduction. *Am Heart J* 66:309, 1963.
20. **Hindman MC, Last JH, Rosen KM:** Wolff-Parkinson-White syndrome observed by portable monitoring. *Ann Intern Med* 79:654, 1973.
21. **Armbrust CA Jr, Levine SA:** Paroxysmal ventricular tachycardia: a study of one hundred and seven cases. *Circulation* 1:28, 1950.
22. **Isaeff DM, Harper Gaston J, Harrison DC:** Wolff-Parkinson-White syndrome. Long-term monitoring for arrhythmias. *JAMA* 222:449, 1972.
23. **Scherf D, Cohen J, Parangi A, et al:** Paroxysmal tachycardia precipitated by atrial or ventricular extrasystoles. *Am J Cardiol* 11:757, 1963.
24. **Hiss RG, Lamb LE:** Electrocardiographic findings in 122,043 individuals. *Circulation* 25:947, 1962.
25. **Wolff L, Parkinson J, White PD:** Bundle branch block with short P-R interval in healthy young people prone to paroxysmal tachycardia. *Am Heart J* 5:685, 1930.
26. **Hamburger WW:** Bundle branch block. Four cases of intraventricular block showing some interesting and unusual clinical features. *Med Clin North Am* 13:343, 1929.
27. **Wedd AM:** Paroxysmal tachycardia with reference to nomotopic tachycardia and the role of the extrinsic cardiac nerves. *Arch Intern Med* 27:571, 1921.
28. **Flensted-Jensen E:** Wolff-Parkinson-White syndrome. A long-term follow-up of 47 cases. *Acta Med Scand* 186:65, 1969.
29. **Averill KH, Fosmoe RJ, Lamb LE:** Electrocardiographic findings in 67,375 asymptomatic subjects. IV. Wolff-Parkinson-White syndrome. *Am J Cardiol* 6:108, 1960.
30. **Swiderski J, Lees MH, Nadas AS:** The Wolff-Parkinson-White syndrome in infancy and childhood. *Br Heart J* 24:561, 1962.
31. **Schiebler GL, Adams P Jr. Anderson RC:** The Wolff-Parkinson-White syndrome in

infants and children. A review and a report of 28 cases. *Pediatrics* 24:585, 1959.

32. **Hejtmancik MR, Herrman GR:** The electrocardiographic syndrome of short P-R interval and broad QRS complex. A clinical study of 80 cases. *Am Heart J* 54:708, 1957.
33. **Rosenkranz KA:** On the differential diagnosis of the electrocardiographic picture of the WPW syndrome. (Ger) *Z Kreislaufforsch* 54:1168, 1965.
34. **Mortensen V, Nielsen AL, Eskildsen P:** Wolff-Parkinson and White's syndrome. *Acta Med Scand* 118:506, 1944.
35. **Longhini C, Pinelli G, Martelli A, et al:** Polycardiographic study of ventricular pre-excitation. *Minerva Cardioangiol* 21:89, 1973.
36. **Nasser WK, Mishkin ME, Tavel ME, et al:** Occurrence of organic heart disease in association with the Wolff-Parkinson-White syndrome. Analysis of 29 cases. *J Indiana State Med Assoc* 64:111, 1971.
37. **Ohnell RF:** Pre-excitation, a cardiac abnormality. *Acta Med Scand* (suppl 152), 1944.
38. **Wellens HJ, Durrer D:** Wolff-Parkinson-White syndrome and atrial fibrillation. Relation between refractory period of accessory pathway and ventricular rate during atrial fibrillation. *Am J Cardiol* 34:777, 1974.
39. **Lowe KG, Esmile-Smith D, Ward C, et al:** Classification of ventricular pre-excitation. Vectorcardiographic study. *Br Heart J* 37:9, 1975.
40. **Pfisterer M:** Wolff-Parkinson-White syndrom-Langzeit Ekg. *Schweiz Med Wochenschr* 105:61, 1975.
41. **Bleifer S, Kahn M, Grishman A, et al:** Wolff-Parkinson-White syndrome. A vectorcardiographic, electrocardiographic and clinical study. *Am J Cardiol* 4:321, 1959.
42. **Giardina AC, Ehlers KH, Engle MA:** Wolff-Parkinson-White syndrome in infants and children. A long-term follow-up study. *Br Heart J* 34:839, 1972.
43. **Brokhes LI:** Paroxysmal cardia rhythm disorders in the Wolff-Parkinson-White syndrome. (Rus) *Kardiologiia* 12:110, 1972.
44. **Czerwinski H:** Recurrent electrocardiographic syndrome after tachycardia in a case of Wolff-Parkinson-White syndrome. (Pol) *Wiad Lek* 20:1723, 1967.
45. **Faxen N:** Paroxysmal tachycardia and bundle branch block in a boy of 11. *Acta Paediatr* 18:491, 1936.
46. **Fontaine G, Frank R, Coutte R, et al:** Antidromic reciprocal rhythm in a type A Wolff-Parkinson-White syndrome. *Ann Cardiol Angeiol* (Paris) 24:59, 1975.
47. **Fosmoe RJ, Averill KH, Lamb LE:** Electrocardiographic findings in 67,375 asymptomatic subjects. II. Supraventricular arrhythmias. *Am J Cardiol* 6:84, 1960.
48. **González Maqueda I, Martín Jadraque L, Liste Jiménez D, et al:** Arrhythmias and the pre-excitation syndrome. (Sp) *Rev Esp Cardiol* 27:269, 1974.
49. **Grolleau R, Dufoix R, Puech P, et al:** Tachycardia due to reciprocal rhythm in Wolff-Parkinson-White syndrome. *Arch Mal Coeur* 63:74, 1970.
50. **Holzmann M, Volkert M:** Paroxysmal tachyarrhythmia in Wolff-Parkinson-White syndrome. (Fr) *Atti Soc Ital Cardiol* 2:143, 1967.
51. **Lüderitz B, Steinbeck G:** Arrhythmias in the Wolff-Parkinson-White syndrome. (Ger) *Munch Med Wochenschr* 118:377, 1976.
52. **Mark AL, Basta LL:** Paroxysmal tachycardia with atrioventricular dissociation in a patient with a variant of pre-excitation syndrome. *J Electrocardiol* 7:355, 1974.
53. **Martin-Noel P, Grunwald D, Denis B:** Rhythm disorders during Wolff-Parkinson-White syndrome. (Fr) *Lyon Med* 221:317, 1969.
54. **Motté G, Bellanger P, Vogel M, et al:** Disappearance of a bundle branch block with the acceleration of reciprocal tachycardia in Wolff-Parkinson-White syndrome. *Ann Cardiol Angeiol* (Paris) 22:343, 1973.
55. **Pautrat J, Vincent P, Pochet P, et al:** Tachycardia crises due to reciprocal rhythm. Diagnostic components apropos of a case observed in the course of Wolff-Parkinson-White syndrome. *Ann Cardiol Angeiol* (Paris) 21:153, 1972.
56. **Pawluk W:** Wolff-Parkinson-White syndrome and arrhythmias. *Kardiol Pol* 9:261, 1966.
57. **Trifunovic S:** The problem of arrhythmia in Wolff-Parkinson-White syndrome. (Ser)

Med Pregl 18:23, 1965.
58. **Zakopoulos KS, Tsatas AT, Liokis TE:** Type B Wolff-Parkinson-White syndrome associated with right bundle branch block. *Dis Chest* 46:346, 1964.
59. **Luria MH, Gordon Hale C:** Wolff-Parkinson-White syndrome in association with atrial reciprocal rhythm and reciprocating tachycardia. *Br Heart J* 32:134, 1970.
60. **Cárdenas M, Urina-Daza M, Sanchez A, et al:** Auricular fibrillation and Wolff-Parkinson-White syndrome. (Sp) *Arch Inst Cardiol Mex* 37:38, 1967.
61. **Chia BL, Yap MH, Lee YK:** Atrial fibrillation in the Wolff-Parkinson-White syndrome, *Singapore Med J* 14:494, 1973.
62. **Colonna L:** On a case of ventricular pre-excitation with atrial fibrillation crises and fetal outcome in a patient with mitral valve defect. (Ital) *Cuore e Circul* 51:155, 1967.
63. **Curry PV, Krikler DM:** Proceedings: atrial fibrillation in the Wolff-Parkinson-White syndrome. *Br Heart J* 37:779, 1975.
64. **DiMichele A, Pellegrino L, Prencipe D, et al:** Case of Wolff-Parkinson-White syndrome with attacks of atrial fibrillation in a young non-cardiopathic subject. *Minerva Med* 64:64, 1973.
65. **Laham J:** Paroxysmal auricular fibrillation and Wolff-Parkinson-White syndrome. *Coeur Med Interne* 11:97, 1972.
66. **Kerin N, Davies B, Kopelson M, et al:** Paroxysmal atrial fibrillation in Wolff-Parkinson-White syndrome type B. *Cardiology* 58:251, 1973.
67. **Nicita-Mauro V, Munafo G:** Atrial fibrillation in a case of Wolff-Parkinson-White syndrome. (Ital) *Boll Soc Ital Cardiol* 11:733, 1966.
68. **Pellegrino L, DiMichele A, Principe D, et al:** Wolff-Parkinson-White syndrome with atrial fibrillation in a young non-cardiopathic individual. *Boll Soc Ital Cardiol* 17:611, 1972.
69. **Giraud G, Puech P, Latour H, et al:** Auricular flutter and Wolff-Parkinson-White syndrome. (Fr) *Mal Cardiovasc* 4:967, 1963.
70. **Picart N, Lanoy M, Courtoy P:** 2-1 and 1-1 auricular flutter attack and Wolff-Parkinson-White syndrome. (Fr) *Acta Cardiol* (Bruxelles) 25:77, 1970.
71. **Poveda Sierra J, Pajaron A, de Michelli A:** Atrial fibrillation and flutter in pre-excitation syndrome (an analysis of 28 cases). *Acta Cardiol* (Bruxelles) 29:455, 1974.
72. **Laham J, Brechenmacher C, Bensaidi J, et al:** Paroxysmal auricular flutter and ventricular pre-excitation syndromes. *Ann Cardiol Angeiol* (Paris) 24:41, 1975.
73. **Scherf D, Cohen J:** *The Atrioventricular Node and Selected Cardiac Arrhythmias,* pp 373-447, New York-London: Grune & Stratton, 1964.
74. **Ferrer MI, Harvey RM, Weiner MH, et al:** Hemodynamic studies in two cases of Wolff-Parkinson-White syndrome with paroxysmal AV nodal tachycardia. *Am J Med* 6:725, 1949.
75. **Giraud G, Latour H, Puech P, et al:** Les troubles du rhythme du syndrome de Wolff-Parkinson-White. Analyse électrocardiographique endocavitaire. *Arch Mal Coeur* 49:101, 1956.
76. **Puech P:** *L'Activité Électrique Auriculaiare Normale et Pathologique,* Paris: Masson et Cie, 1956.
77. **Harnischfeger WW:** Heredity occurrence of the pre-excitation (Wolff-Parkinson-White) syndrome with re-entry mechanism and concealed conduction. *Circulation* 19:28, 1959.
78. **Burchell HB:** Atrioventricular nodal (reciprocating) rhythm. Report of a case. *Am Heart J* 67:391, 1964.
79. **Gilchrist AR:** Paroxysmal ventricular tachycardia. A report of five cases. *Am Heart J* 1:547, 1926.
80. **Arana R, Cossio P:** Fibrilación auricular y taquicardia ventricular como eventualidad posible en el P-R corto con QRS ancho y mellado. *Rev Argent Cardiol* 5:43, 1938.
81. **Levine SA, Beeson PB:** The Wolff-Parkinson-White syndrome with paroxysms of ventricular tachycardia. *Am Heart J* 22:401, 1941.
82. **Cooke WT, White PD:** Paroxysmal ventricular tachycardia. *Br Heart J* 5:33, 1943.
83. **Franke H, Vetter R:** Gibt es auf Grund der Analyse des Ekg-Syndroms nach Wolff,

Parkinson und White beim Menschen ein zweites Reizleitungssystem? *Z Klin Med* 144:21, 1944.
84. **Palatucci OA, Knighton JE:** Short P-R interval associated with prolongation of QRS complex; a clinical study demonstrating interesting variations. *Ann Intern Med* 21:58, 1944.
85. **Missal ME, Wood DJ, Leo SD:** Paroxysmal ventricular tachycardia associated with short P-R intervals and prolonged QRS complexes. *Ann Intern Med* 24:911, 1946.
86. **Klainer MJ, Joffe HH:** A case of short PR interval and prolonged QRS complex with a paroxysm of ventricular tachycardia. *Ann Intern Med* 24:902, 1946.
87. **Stein I:** Short P-R interval, prolonged QRS complex (Wolff-Parkinson-White syndrome), report of fourteen cases and a review of the literature. *Ann Intern Med* 24:60, 1946.
88. **Cain EF:** Wolff-Parkinson-White syndrome presenting unusual features. *Am Heart J* 33:523, 1947.
89. **Beers SD, de la Chapelle CE:** An unusual case of ventricular tachycardia. *Ann Intern Med* 27:441, 1947.
90. **Giraud G, Cazal P, Parati H, et al:** Syndrome permanent de Wolff-Parkinson-White avec paroxysmes tachycardiques dimorphes et signes d'hyperlixtation ventriculaire. Fait clinique. *Sud Med Chirurg* 81:1258, 1949.
91. **Jordan RA, Canuteson RI:** Wolff-Parkinson-White syndrome. Report of 2 cases. *Lancet* 69:38, 1949.
92. **Williams C:** Ventricular tachycardia. An analysis of thirty-six cases. *Arch Intern Med* 71:137, 1943.
93. **Bruce RA, Yu PNG, Lovejoy FW Jr, et al:** Ventricular tachycardia during cardiac catheterization of patient with Wolff-Parkinson-White syndrome. Report of a case showing effects of atropine sulfate. *Circulation* 2:245, 1950.
94. **Broustet P, Capsec-Laporterie J, Gazeau J:** Les formes graves du syndrome de Wolff-Parkinson-White. *Arch Mal Coeur* 44:901, 1951.
95. **Fleishman SJ:** A case of Wolff-Parkinson-White syndrome with paroxysmal ventricular tachycardia. *Am Heart J* 44:897, 1952.
96. **Cordeiro Arsenio:** Syndrome WPW d'origine ventriculaire. *Acta Cardiol* 9:473, 1954.
97. **Sears GA, Manning GW:** The Wolff-Parkinson-White pattern in routine electrocardiography. *Can Med Assoc J* 87:1213, 1962.
98. **Dunn JJ, Sarrell W, Franklin RB:** The Wolff-Parkinson-White syndrome associated with paroxysmal ventricular tachycardia. *Am Heart J* 47:462, 1954.
99. **Gouax JL, Ashman R:** Auricular fibrillation with aberration simulating ventricular paroxysmal tachycardia. *Am Heart J* 34:366, 1947.
100. **Langendorf R, Lev M, Pick A:** Auricular fibrillation with anomalous A-V excitation (WPW syndrome) imitating ventricular paroxysmal tachycardia. A case report with clinical and autopsy findings and critical review of the literature. *Acta Cardiol* 7:241, 1952.
101. **Chung KY, Thomas J:** Paroxysmal atrial flutter in Wolff-Parkinson-White syndrome. 1 to 1 atrioventricular conduction resembling ventricular tachycardia. *J Lancet* 86:214, 1966.
102. **Dominguez P:** Pseudoventricular tachycardia in Wolff-Parkinson-White syndrome. (Sp) (Eng Abstr) *Arq Bras Cardiol* 23:65, 1970.
103. **Hawker RE:** Rapid atrial fibrillation simulating ventricular tachycardia in the Wolff-Parkinson-White syndrome. *Med J Aust* 2:621, 1971.
104. **Hellman E, Altchek MR:** Paroxysmal pseudoventricular tachycardia with ventricular rate of 290 in a patient with accelerated A-V conduction. *NY State J Med* 58:2427, 1958.
105. **Herrmann GR, Oates JR, Runge TM, et al:** Paroxysmal pseudoventricular tachycardia and pseudoventricular fibrillation in patients with accelerated A-V conduction. *Am Heart J* 53:254, 1957.
106. **Pericoli Ridolfini F, Pagliari M:** Pseudo-ventricular tachycardia in the Wolff-Parkinson-White syndrome. Description of a case. (Ital) *Policlinico* (Med) 72:415, 1965.

107. **Burak M, Scherf D:** Angina Pectoris und paroxysmale Tachycardia. *Wien Arch Inn Med* 23:475, 1933.
108. **Hoffman I, Morris MH, Friedfeld L, et al:** Aberrant beats of Wolff-Parkinson-White configuration in arteriosclerotic heart disease. *Br Heart J* 18:301, 1956.
109. **Quaglia GB:** Ventricular paroxysm in the Wolff-Parkinson-White syndrome. Clinical case and treatment. (Ital) *Gazz Med Ital* 124:181, 1965.
110. **Suarez LD, Sciandro EE, Macchi RJ, et al:** Sindrome de Wolff-Parkinson-White. III. Taquicardia ventricular paroxistica. *Prensa Med Argent* 47:1850, 1960.
111. **Tsonchev I, Andonova S:** Paroxysmal ventricular tachycardia in WPW syndrome. *Vutr Boles* 12:139, 1973.
112. **Lim CH, Toh CC, Chia BL:** Ventricular fibrillation in type B Wolff-Parkinson-White syndrome. *Aust NZ J Med* 4:515, 1974.
113. **Dreifus LS, Haiat R, Watanabe Y, et al:** Ventricular fibrillation. A possible mechanism of sudden death in patients and Wolf-Parkinson-White syndrome. *Circulation* 43:520, 1971.
114. **Fox TT, Weaver J, March HW:** On the mechanism of the arrhythmias in aberrant atrioventricular conduction (Wolff-Parkinson-White). *Am Heart J* 43:507, 1952.
115. **Touche M, Jouvet M, Touche S:** Ventricular fibrillation in Wolff-Parkinson-White syndrome. Reduction by external electric shock. *Arch Mal Coeur* 59:1122, 1966.
116. **Okel BB:** The Wolff-Parkinson-White syndrome, report of a case with fatal arrhythmia and autopsy findings of myocarditis, interatrial lipomatous hypertrophy, and prominent right moderator band. *Am Heart J* 75:673, 1968.
117. **Wojtasik W, Hakalo Z, Gawryś A:** Resuscitation in a case of ventricular flutter and fibrillation in a patient with WPW syndrome. (Pol) *Wiad Lek* 23:823, 1969.
118. **Kaplan MA, Cohen KL:** Ventricular fibrillation in the Wolff-Parkinson-White syndrome. *Am J Cardiol* 24:259, 1969.
119. **Gallagher JJ, Gilbert M, Svenson RH, et al:** Wolff-Parkinson-White syndrome. The problem, evaluation, and surgical correction. *Circulation* 51:767, 1975.
120. **Bleecker ER, Engel BT:** Learned control of cardiac rate and cardiac conduction in the Wolff-Parkinson-White syndrome. *NEJM* 288:560, 1973.
121. **Chignon JC, Distel R, Stephan H, et al:** A case of Wolff-Parkinson-White syndrome. Problems raised by the interpretation of the electrical diagrams. (Fr) *Presse Med* 77:801, 1969.
122. **Schamroth L, Coskey RL:** Reciprocal rhythm, the Wolff-Parkinson-White syndrome. *Br Heart J* 31:616, 1969.
123. **Wolff L:** Wolff-Parkinson-White syndrome: historical and clinical features. *Prog Cardiovasc Dis* 2:677, 1960.
124. **Dreifus LS, Kimbiris D, Wellens HJ:** Sinus bradycardia and atrial fibrillation associated with Wolff-Parkinson-White syndrome. *Circulation* 48 (suppl IV): 17, 1973.
125. **Dreifus LS, Wellens HJ, Watanabe Y, et al:** Sinus bradycardia and atrial fibrillation associated with the Wolff-Parkinson-White syndrome. *Am J Cardiol* 38:149, 1976.
126. **Santucci F, Lisi B, Palumbo R, et al:** An unusual "natural history" and other uncommon aspects of the Wolff-Parkinson-White syndrome. (Ital) *Boll Soc Ital Cardiol* 13:10, 1968.
127. **Graybiel A, McFarland RA, Gates DC, et al:** Analysis of the electrocardiograms obtained from 1000 young healthy aviators. *Am Heart J* 27:524, 1944.
128. **Kaplan G, Cohn DT:** Syndrome of auriculoventricular accessory pathway. *Ann Intern Med* 21:824, 1944.
129. **James TN, Puech P:** De subitaneis mortibus. IX. Type A Wolff-Parkinson-White syndrome. *Circulation* 50:1264, 1974.
130. **Pick A, Katz LN:** Disturbances of impulse formation and conduction in the preexcitation (WPW) syndrome. Their bearing on its mechanism. *Am J Med* 19:759, 1955.
131. **Fisch C, Pinsky ST, Shields JP:** Wolff-Parkinson-White syndrome. Report of a case associated with wandering pacemaker, atrial tachycardia, atrial fibrillation, and incomplete A-V dissociation with interference. *Circulation* 16:1004, 1957.

132. **James TN, Sherf L:** Specialized tissues and preferential conduction in the atria of the heart. *Am J Cardiol* 28:414, 1971.
133. **Ballarino M, Rumolo R, Folli G:** Syndrome of ventricular pre-excitation (WPW) associated with partial sino-atrial block and partial atrio-ventricular block (Ital) *Cardiol Prat* 17:357, 1966.
134. **Giegler I, Rübesamen M:** WPW syndrome particular A-V block. *Cor Vasa* 7:73, 1965.
135. **Jazienicki B, Zimmermann-Górska I:** Wolff-Parkinson-White syndrome developing into atrioventricular block. (Pol) *Kardiol Pol* 11:317, 1968.
136. **Mielczarek H, Sapinski A:** Periodical appearance of the WPW syndrome in a case with atrioventricular block and paroxysmal tachycardia due to retrograde conduction. (Pol) *Wiad Lek* 21:307, 1968.
137. **Parameswaran R, Ohe T, Nakhjavan FK, et al:** Spontaneous variations in atrioventricular conduction in pre-excitation. *Br Heart J* 38:427, 1976.
138. **Pretolani EG:** On a rare case of complete transitory A-V block with re-establishment of normal A-V conduction through a phase of the Wolff-Parkinson-White syndrome. (Ital) *Clin Med* 47:404, 1966.
139. **Storti D:** Ventricular pre-excitation syndrome alternating with intra-ventricular disorder of the conduction associated with alternating 1st grade atrio-ventricular block. (Ital) *Folia Cardiol* (Milano) 23:73, 1964.
140. **Yahini JH, Neuman A, Nathan D, et al:** Wolff-Parkinson-White syndrome in the presence of intermittent atrioventricular block. *Isr J Med Sci* 4:258, 1968.
141. **Friedberg HD, Schamroth L:** Three atrioventricular pathways: reciprocating tachycardia with alternation of conduction times. *J Electrocardiol* 6:159, 1973.
142. **Lepeschkin E:** Significance of pre-excitation of intraventricular and atrioventricular conduction disturbances. *J Electrocardiol* 2:185, 1969.
143. **Lombardi M, Masini G:** Electrocardiographic observations in Wolff-Parkinson-White syndrome. Observations on 43 cases. (Ital) *Mal Cardiovasc* 7:387, 1966.
144. **Mihalkovics T:** Vom Elektrokardiogramm des Paladino Kent-Typus. *Z Kreislaufforsch* 35:388, 1943.
145. **Rodstein M:** A case of anomalous auriculoventricular conduction with auriculoventricular block and a history of rheumatic fever. *NY State J Med* 51:789, 1951.
146. **Schiebler GL, Adams P Jr, Anderson RC, et al:** Clinical study of twenty-three cases of Ebstein's anomaly of the tricuspid valve. *Circulation* 19:165, 1959.
147. **Grant RP, Tomlinson FB, Van Buren JK:** Ventricular activation in pre-excitation syndrome (Wolff-Parkinson-White). *Circulation* 18:355, 1958.
148. **Varriale P, Alfenito J, Kennedy RJ:** The simultaneous occurrence of ventricular pre-excitation, left bundle branch block, and delayed A-V conduction. *Am Heart J* 71:803, 1966.
149. **Lev M, Leffler WB, Langendorf R, et al:** Anatomic findings in a case of ventricular pre-exictation (WPW) terminating in complete atrioventricular block. *Circulation* 34:718, 1966.
150. **Moret PR, Schwartz ML, White TJ:** The Wolff-Parkinson-White syndrome and second degree heart block. Report of a case and a discussion of the significance of the association of these phenomena. *Cardiologia* 34:43, 1959.
151. **Scherf D, Blumenfeld S, Mueller P:** A-V conduction disturbance in the presence of the pre-excitation syndrome. *Am Heart J* 43:829, 1952.
152. **De Rango F, Amoroso A, Pozzi LL:** Disturbances of A-V conduction in Wolff-Parkinson-White syndrome. (Ital) *Mal Cardiovasc* 7:293, 1966.
153. **Coelho E, Fonseca JM, Borges AS, et al:** Etude des dérivations intra-cavitaires du syndrome de Wolff-Parkinson-White et son déclenchement au moyen de l'excitation de la cloison inter-ventriculaire. *Sem Hop Paris* 27:8, 1951.
154. **Masini V, Milazzotto F:** Wolff-Parkinson-White syndrome of type B with right branch block and atrio-ventricular block. *G Ital Cardiol* 1:585, 1971.
155. **Segers M, Leguime J, Denolin H:** L'activation ventriculaire precoce de certains coeurs hyperexcitables. Etude de l'onde de l'electrocardiogramme. *Cardiologia* 8:113, 1944.

156. **Seipel L, Both A, Breithardt G, et al:** His bundle recording in a case of complete atrioventricular block combined with pre-excitation syndrome. *Am Heart J* 92:623, 1976.
157. **Sherf L, James TN:** A new look at some old questions in clinical electrocardiography. *Henry Ford Hosp Med Bull* 14:265, 1966.
158. **Levine HD, Burge JC:** Septal infarction with complete heart block and intermittent anomalous atrioventricular excitation (Wolff-Parkinson-White syndrome): histologic demonstration of a right lateral bundle. *Am Heart J* 36:431, 1948.
159. **Addarii F, Pretolani E:** Microelectrocardiographic study of a case of complete atrioventricular block associated with unstable Wolff-Parkinson-White syndrome. (Ital) *Cardiol Prat* 20:451, 1969.
160. **Timmis GC, Henke J, Gordon S, et al:** The Wolff-Parkinson-White syndrome in advanced atrioventricular block. *Am J Cardiol* 28:592, 1971.
161. **Coumel P, Gourdon R, Slama R, et al:** Conduction auriculo-ventriculaire par des fibres de pré-excitation, associée à un bloc complet de la voie nodo-hissienne. Etude electrocardiographique de quatre cas. *Arch Mal Coeur* 66:285, 1973.

IV

Pathology

Holzman and Scherf,[1] and Wolferth and Wood[2] suggested at about the same time that the pre-excitation syndrome might be due to an anomalous accessory direct atrioventricular connection, outside the regular conduction system. Their suggestion, based on the questionable anatomic studies of Kent,[3,4] launched a search for a Kent bundle in the hearts of pre-excitation cases; this search continues today in pathology institutes the world over.[5-11] Wood et al[12] were the first to find such an accessory connection. In 1943 they described a case with three such bundles connecting the right atrium and the right ventricle (Table XV, # 1). In 1975 Lev et al[13] reported that they were unable to find such direct muscular accessory bundles in carefully investigated cases (Table XV, # 46). During the 32 years between these two papers, a large number of pathological studies were published. The literature is difficult to summarize due to variations in methods of sectioning the hearts, and in the quantity of material collected for histological examination. Advances in technology also add to the complexities of comparing studies ranging over three decades. It was decided to include all studies in which the purpose of the heart autopsy was to determine the anatomicopathological background of the pre-excitation syndrome, even when only a partial dissection was done. The wealth of material was classified into five groups, summarized in Table XV. In Group A only the atrioventricular groove was studied, completely or partially (# 1-7). In Group B only the conduction system of the heart was investigated, completely or partially (# 8-17). Group C includes cases from the literature with very little information but where the author's intention was to find the anatomical background of the disorder (# 18-23). Group D cases are complete studies of the atrioventricular groove and the conduction system (# 24-52). This last group is emphasized in the ensuing discussion. Group E consists of one additional case (# 53) where it was not clear from the text what kind of histological investigation was performed.

The location of accessory bundles, their dimensions and cell composition are quoted, insofar as possible, in the author's own words to avoid misinterpretation. The findings of two animal experiments are also listed. Recently two more series and one individual case were reported briefly (abstracts), but are not included in Table XV.[14-16] The pathological findings were similar to those in the Table and do not change the conclusions.

Nomenclature

Since Holzman and Scherf[1] and Wolferth and Wood[2] based their theory on the anatomic studies of Stanley Kent,[4] the term "Kent bundle" came to be used to indicate the atrioventricular connection in pre-excitation. This covered both cases in which neither pathological nor surgical proof for such an additional muscle bundle was sought, as well as those in which a Kent bundle was not seen even in carefully performed histological studies.[13] In the latter cases "accessory bundles" were added to account for the electrical atrioventricular shortcut, including Mahaim's paraspecific fibers[17] and "James' AV nodal bypass fibers." These terms are misused, since Mahaim fibers are seen mostly in newborns and only rarely in adults, and AV nodal bypass fibers are present in every human heart and are part of the posterior internodal tract described by James. In Rossi's words, this use of anatomical bundles is "oversimplified schemes of atrioventricular (AV) anomalous pathways, currently taken for granted in anatomofunctional interpretation of WPW, thus often dangerously superseding histology."[18] A more accurate, up-to-date and all-inclusive nomenclature is needed, covering all the possible normal and abnormal atrioventricular bypass tracts in the heart. An attempt to develop such a nomenclature was presented by Anderson et al in 1975.[19] They divided the possible anatomical background of pre-excitation into six groups: 1) accessory atrioventricular muscle bundles; 2) accessory nodoventricular muscle bundles; 3) atriofascicular bypass tracts; 4) fasciculoventricular fibers; 5) intranodal bypass tracts; and 6) nodal malformations. As one of their groups (nodal malformations) is, in their own opinion, only hypothetical and another (intranodal bypass tracts) still disputed, we prefer the less complicated scheme and classification postulated by Rossi and published the same year.[18] We used this classification with the omission of terms like Kent bundles and Mahaim fibers, replacing them with abbreviations of anatomically descriptive terms (Table XVI). Findings such as inflammation, fibrosis, *etc,* often discovered as the only pathological finding with all kinds of normal or abnormal atrioventricular connections, were also abbreviated and used.

Postmortem Findings in 53 Pre-excitation Cases (Table XV)

Direct AV Bypass Outside the Conduction System. In 21 of the 53 cases no pathological formations of any kind were found (# 2, 3, 6, 8,

9, 11-15, 18-23, 28, 33, 35-37). In 23 cases direct muscular AV junctions were identified: 7 on the right side only (Table XV, # 1, 5, 25, 27, 29, 38, 39), 9 on the left side (# 7, 24, 42, 43, 44, 49, 50, 51, 52) and 4 in the septal area (# 10, 26, 31, 34). In 3 cases fibers were found in more than one location: right and left AV connecting fibers in # 4, right and septal fibers in # 40, and right, left and septal fibers in # 41. More than 31 AV accessory bundles were found in these 23 cases due to some cases of multiple fibers.

In 9 cases the bundles were located in the subendocardial region, in 3 the subepicardial, and in one the bundle started deep in the myocardium and ended in the subepicardial zone.

In 17 muscle bundles the cells comprising the accessory bypass were of the regular working myocardial type. In four cases unusual cells were identified: one showed cells larger than ordinary myocardial cells but not typical Purkinje's cells (# 29); the second showed what was termed by the author "cells of the conducting tissue" (# 38), and in two others the accessory bundle was composed mostly of small P cells (# 44), resembling the ones seen in the sinus node in one, and slender transitional cells resembling the AV node in another (# 50).

Mediate Bypass Through the Conduction System. *Inlet connections* (Table XV) were present in 10 cases: in one an AN connection was described (and interpreted by us as a James' bundle) (# 16), in a second AN + AH bundles were found (# 47), and in three cases AH connections only were identified (# 17, 32 and 53). In these 5 cases the connections were mentioned as single and isolated findings, without any other AV junctions. In # 29, 43 and 50, the inlet connections were identified in association with accessory bundles outside the conduction system, and in # 30 and 46 they were found together with outlet connections (Mahaim fibers in Rossi's classification).

Outlet connections were present in 11 cases. In 2 cases these fibers were described as the only AV junction (# 45, 48). In 7 cases the connection accompanied accessory bundles outside the AV conduction system (# 25, 27, 29, 40, 41, 49, 51), and in 2 cases (# 30, 46) it was found together with inlet connections only.

Pathological Findings In or Around the Conduction System. Pathological findings were present in different parts of the conduction system in addition to abnormal or normal (AN) anatomical fibers, bundles or AV junctions in 19 cases. In almost all of them more than one abnormality was observed histologically, the most common combination being cell infiltration and fibrosis. There were 17 instances of fibrosis, 4 of cell infiltration, 3 of cell degeneration, 3 of fat infiltration and one of calcification.

TABLE XV
Accessory AV Connections and Other Anatomical and Anatomopathological Findings in Cases with Pre-excitation Syndrome

Author, Case #, Year of Publication	Localization of Accessory or Other Unusual AV Connections and Cell Composition	Basic Disease, ECG Findings	Summary of Pathological Findings (abbreviations in Table XVI)	Remarks
Group A: AV groove only (complete or partial)				
1. Wood et al,[12] 1943	Three connections were found between the right atrium and ventricle at approximately the right lateral border of the heart. They consisted of a bridge of tissue across the ventricular cavity just below the attachment of the tricuspid valve. "A muscle band."	Debility? Premature beats.	3-RAB-Endocardial (en)	Sudden death; 897 sections. Investigation was partial since work was stopped after the bundles were found.
2. Ohnell,[29] 1944	—	? Myocarditis? (4 years before death); tachycardia.	I, F	Sudden death; many round cell foci, moderate quantity of connective tissue.
3. Ohnell,[29] 1944	—	Syphilis.	I-RBB I-LBB	—

4. Kimball & Burch,[30] 1947	Two aberrant bundles of muscle tissue were found, one connecting the right atrium with the right ventricle and another connecting the left atrium with the left ventricle.	Tachycardia.	1-RAB 1-LAB	The type of autopsy is not clear. A promised paper with details (Deerhake et al) never appeared.
5. Levine & Burge,[31] 1948	An accessory bundle across the atrioventricular groove at the right posterolateral aspect of the heart. Healthy cardiac muscle.	Acute myocardial infarction; tachycardia; complete AV block (CAVB) with few conducted beats simulating short P–R (0.10 sec) and pre-excitation.	1-RAB	Only one block from the AV groove was taken; "further studies of the AV groove to demonstrate the possible presence of additional muscular connections between atrium and ventricle were not undertaken."
6. Sondergaard,[32] 1953 Case 2	—	Atrial septal defect; ventricular septal defect; patent ductus arteriosus. ECG: P–R 0.8 sec, QRS: 0.10 sec.	—	The tissue containing the AV groove was freed by incisions 1 cm below the groove. The tissue was divided into 30 blocks, and one section was cut from each. Monozygotic twin sister healthy.
7. Villeneuve,[33] 1958	Folds running between left atrium and ventricle made up by aberrant myocardial bundles connecting A with V.	Newborn. Open foramen ovale; ventricular septal defect; bicuspid aortic valve. ECG: paroxysmal tachycardia. Died during an attack.	1-LAB	Quoted by Rossi. The heart was not investigated for possible additional bundles in other parts.

Author/case	Localization of Connections	ECG Findings	Pathology	Remarks
Group B: Conduction system only (complete or partial)				
8. Yater and Shapiro,[34] 1938	—	Ebstein's anomaly. ECG: direct evidence of QRS changes and of T changes, P–R 0.16 sec. Left axis deviation −30°.	—	Very meticulous anatomic dissection; 4,500 sections taken from the AV node–His bundle.
9. Mahaim and Bogdanovic,[35] 1974	Inflammation of the left bundle branch (LBB); no Mahaim fibers.	Myocarditis. ECG: P–R 0.08 sec. QRS: 0.10-0.11 sec. Tachycardia.	I-LBB I-RBB	Dissection of the His bundle (3,400 sections) AVN, bifurcation of His bundle and bundle branches (BB).
10. Segers et al,[36] 1947	A bundle connecting the right auricle directly to the right ventricle. This was a supernumerary pathway, located in the septal zone immediately under the node. Ordinary myocardial tissue. Normal composition.	Ventricular septal defect. ECG: no tachy-cardia, pre-excitation appearing only during pregnancy, effort and during nodal rhythm.	1-SAB	—
11. Prinzmetal et al,[37] 1952 Case 1	Increased connective and fat tissue in the AVN (collagen).	Acute myocardial infarction. Lymphosarcoma. ECG: P–R 0.10 sec, only leads I, II, III registered.	—	—

12. Prinzmetal et al,[37] 1952 Case 8	The AVN was not positively identified. Marked fibrosis in this area.	Cerebral vascular accident. ECG: P–R 0.09 sec, one single tracing, no tachycardia.	F-AVN	60 sections.
13. Prinzmetal et al,[37] 1952 Case 9	The AVN and surroundings were markedly pathologic. The central fibrous body and adjacent base of tricuspid and mitral valve were markedly calcified, which compressed the AVN. The fibers of the AVN were narrow, granular and contained an increased amount of connective tissue.	Rheumatic heart disease.	C-AVN F-AVN	This case was also published by Langendorf et al.[38] In addition to the conduction system, all of the right AV orifice was investigated (12,437 sections). The branching portion of His bundle and bundle branches were not available for investigation.
14. Plavsic et al,[39] 1956	—	Acute myocardial infarction. Syphilis. ECG: intermittent pre-excitation, P–R 0.16 sec, type B, ST elevation in V_1 and V_2 during infarction.	—	2,165 sections.
15. Okel,[40] 1968	Fatty infiltration of the His bundles. An RV band producing stenosis in the RV. Inflammatory infiltrate of mononuclear leukocytes in the intraventricular septum. Lipoma of the interatrial septum.	Myocarditis. ECG: type B, tachycardia, atrial fibrillation (AF), ventricular-like tachycardia (VLT).	F-His bundle (HB) I-atrial septum and ventricular septum (AS+VS)	Not clear if the AV groove was systematically sectioned, but may have been since the authors state: "No accessory AV muscle connections could be identified" and "the apparent absence of a Bundle of Kent . . ."

Author/case	Localization of Connections	ECG Findings	Pathology	Remarks
16. Cole et al,[41] 1970	One aberrant pathway was defined histologically in the atrial septum 3-5 mm anterior to the AV node and His bundle. The cells were diagnosed as P cells (electron microscopy).	? ECG: type B, AF; paroxysmal atrial tachycardia.	AN	This is a case of post-operative death in pre-excitation. The entire conduction tissue was not investigated. It may be suggested that the abnormal pathway was the posterior internodal tract (James' bypass fibers) and the P cells were Purkinje cells.
17. Brechenmacher et al,[42] 1974 Case 2	*Auricular-Hisian fibers:* an abnormal finding. Degeneration and fibrosis of the sinus node.	Arteriosclerotic cardiovascular disease (ASCVD). Abnormally low insertion of the septal tricuspid valve. ECG: Lown-Ganong-Levine syndrome, tachycardia, AF, VLT.	AH D-SN F-SN	It is not clear that this is a case of Lown-Ganong-Levine syndrome. The tachycardia resembles that of pre-excitation (broad QRS complexes, pseudo-ventricular tachycardia).
Group C: Insufficient data given				
18. Holzman,[43] 1939	—	Intercurrent infection.	—	In serial sections in the AV border area a so-called "Kent bundle" could not be detected. The conduction system was not included.

19. Soedenstroem,[44] 1946	—	—	—	—
20. Prinzmetal et al,[45] 1974	—	—	—	—
21. Prinzmetal et al,[45] 1974	—	—	—	—
22. Robertson,[46] 1953	—	—	—	Quoted but findings were questioned by Scherf and Cohen.[47]
23. Schiebler et al,[48] 1959	—	—	—	11
Group D: AV groove and conduction system				
24. Ohnell,[29] 1944	A bundle was found in the left half of the heart muscle. It lay dorsal to the mitral ostium, 4 cm from the septum and peripheral to the annulus fibrosis. Surrounded by fat tissue; from inside the atrial myocardium to the epicardium of the ventricle: 6 mm long, 0.3 mm diameter. Normal myocardium cells.	Normal His bundle. Familial cardiomyopathy; familial paroxysmal tachycardia. ECG: tachycardia.	LAB-myocardium to epicardium (MEp)	17,000 sections. No mention of SN or AVN.

Author/case	Localization of Connections	ECG Findings	Pathology	Remarks
25. Lev et al,[49] 1955	"In the connective tissue of the AV groove an intermediary muscle bundle was noted. This arose from the connective tissue of the right AV ring on its parietal aspect a short distance from the central fibrous body. It coursed through the AV ring along the parietal wall adjacent to the endocardium and at the level of the right circumflex artery. This intermediary bundle gave off numerous fasciculi to the right atrial appendage and the parietal wall of the right ventricle." No cell descriptions. Other findings: a small communication of the right branch and the right side of the ventricular septum; encasement of the right branch in dense, fibroelastic tissue; normal SN, AVN, HB.	Ebstein's anomaly. ECG: type B, P.-R 0.08 sec. Supraventricular tachycardia.	RAB-en HV (R) BV (L) BV F-RBBB	12,083 sections.
26. Truex et al,[50] 1960	An accessory atrioventricular communication outside the conduction system.	Chronic myocarditis, endocardial	SAB F-SN F-AVN	11,500 sections.

	Other findings: chronic inflammation with fibrosis of the SN, AVN and LBB.	fibroelastosis. ECG: atrial flutter, atrial fibrillation, intermittent RBBB.	F-LBB	
27. Lev et al,[51] 1961 Case 1	"Numerous small and large muscular communications extended over a distance of 0.52 cm between the endocardial portions of the musculature of the right atrium and ventricle in the lateral wall." Ordinary cardiac musculature surrounded by masses of lymphocytes. Other findings: the approaches to the AVN showed marked infiltration of lymphoid cells with degenerative changes of fat tissue. The center of AVN and His bundle were moderately infiltrated; degenerative changes in the His bundle.	Chagas disease. ECG: type B.	RAB-en (L) BV I-AVN I-HB D-HB	9,405 sections. The SN was not investigated.
28. Lev et al,[51] 1961 Case 2	SN—occasional lymphocyte, abnormal RBB with complete focal fibrosis, chronic granulomatous myocarditis with involvement of the conduction system.	Chagas disease. Congenital pulmonary bicuspid valve. ECG: type A (+RBBB?). Supraventricular tachycardia.	F-RBB I-conduction system	This paper contains a very good discussion of the pathological findings in pre-excitation, plus many references.

Author/case	Localization of Connections	ECG Findings	Pathology	Remarks
29. Lev et al,[52] 1963	Right AV junction: there was a distinct communication between A and V subendocardially in the superior part of the free walls, measuring 1.3 mm in width. This communication consisted of muscle cells which were not typical Purkinje cells. These cells showed degenerative changes with fibrosis. Other findings: all parts of the conduction system showed marked fatty infiltration of mononuclear cells; James' fibers—Mahaim fibers.	ASCVD. Alcoholism. ECG: type B. Supraventricular tachycardia.	RAB-en D-RAB AN HV I;L;F—conduction system	23,740 sections. The LBBB was found to correlate with atrophy (but not complete destruction) of the LBB. Alterations of cardiac nerve cell.
30. Lev et al,[53] 1966	The atrial septal musculature at the approaches of the AV node sent a large fasciculus into the central fibrous body. This tract (about $^{1}/_{10}$ of the cross-section of the AVN) made a fine connection with the beginning of the His bundle (where it gave off Mahaim fibers). This fascicle consisted of atrial muscle cells. Other findings: normal SN;	Cardiomyopathy? Small artery disease? There was fine infiltration of mononuclear cells and neutrophils. Some of the arterioles showed acute degeneration with narrowing. ECG: type B; 2:1 AV	AH-en (septal) NV HV —D;F; conduction system	14,662 sections.

	abnormal formation of the fibrous skeleton of the heart; severe degeneration and fibrosis of the conduction system; copious diffuse Mahaim communications from AVN and from His bundle; vacuolar degeneration with fibrosis of the approaches to the AVN, His bundle, Mahaim fibers and BB.	block with normal conduction. CAVB (supraventricular focus). CAVB (idioventricular focus).		
31. Niessen,[54] 1966	Septal bundle.	—	SAB	No details are available on this case; however the critical approach of the author on the methods of pathological investigation suggest that his histological work was thorough.
32. Rossi,[55] 1969 Case 78	On the whole no properly anomalous AV pathway was found, but there was evidence that several atrial fascicles bypassed the AVN to join the insertion of the tricuspid valve and the common bundle. Right atrial bundles. Other findings: low insertion of posterior part of medial tricuspid valve; chronic inflammation in upper ventricular septum.	Cancer of lung. ECG: type B, tachycardia, P–R: 0.10 sec. QRS: 0.12 sec.	AH?	This case was previously described by Rossi and Rovelli.

Author/case	Localization of Connections	ECG Findings	Pathology	Remarks
33. Rossi,[55] 1969 Case 79	Normal SN, AVN James' fibers, His bundle, RBB; atrophy of anterior and posterior fascicle of LBB.	Hypertension, acute myocardial infarction, pericarditis. ECG: type B, P–R: 0.09 sec. QRS: 0.13 sec.	A-LBB	—
34. Schumann et al,[56] 1970 Case 1	On the border of the anterior and middle third of the ventricular septum a bundle of muscle was found which connected the myocardium of the atrial septum with one of the ventricular septum. No connection between this bundle and the AVN or His bundle were detected. Thickness 6.3 mm. Nonspecific myocardial cells. Other findings: the AVN showed a discrete fibrosis, and in the region of the SN a low grade diffuse fibrosis of the myocardium was found.	Acute myocardial infarction, old rheumatic valvular disease? ECG: type B, tachycardia.	SAB-en F-SN F-AVN	18,465 sections. Lipomatosis and fibrosis of the atria but especially of the LV. Fibrosis of the tricuspid valve.
35. Schumann et al,[56] 1970 Case 2	All the conduction system normal.	Active pulmonary tuberculosis, heart failure, mitral	—	19,724 sections. Moderate fibrosis and hypertrophy of all the chambers. No

		stenosis? Post-mortem: heredodegenerative disease of the central nervous system. Friedreich's disease? Cardiomyopathy? ECG: type B, no tachycardia.		marked coronary disease. Moderate thickening of small coronary arteries.
36. Schumann,[57] 1970 Case 1	Extensive hypoplasia of the annulus fibrosus.	Chronic myocarditis. ECG: type B, tachycardia. P–R: 0.08 sec. QRS: 0.17 sec.	—	17,420 sections.
37. Schumann,[57] 1970 Case 2	Small gaps in the annulus fibrosus.	Mongolism (Trisomy 21); ventricular septal defect; ulcus duodeni; ECG: type A, P–R: 0.10 sec. QRS: 0.14 sec.	—	13,078 sections. This and case 1 do not give details on the histology of the conduction system, but the author's thorough method of investigation and his statement that "all the AV region was investigated" justifies classifying these cases in Group D rather than Group A.
38. Rosenberg et al,[58] 1971 Case 1	Through the fibrous ring, individual small muscle groups extended from the A to the V. The continuity of these fibers was not in the plane of section. Despite the gross appearance of extensive fusion of A and V myocardium at the tricuspid valve, evidence for communication was subtle. No details of cells. Conduction system was normal.	Ventricular septal defect; anomalous right papillary muscle; supravalvular mitral stenosis; pulmonary emboli. ECG: tachycardia.	RAB-en?	The left AV groove was not completely investigated.

Author/case	Localization of Connections	ECG Findings	Pathology	Remarks
39. Rosenberg et al,[58] 1971 Case 2	The right coronary artery occupied the epicardial fat with muscular trabeculae connecting the RA & RV, superficial to the coronary artery and vein. This trabecular pattern between RA & RV extended 2 cm about the valve circumference at the base of the anterior and posterior cusps. These small muscle fibers were of the same configuration as the adjacent atrial myocardium. Conduction normal.	Ebstein's anomaly. Persistent left superior cava present. ECG: tachycardia.	RAB-epicardial (ep)	The SN was not investigated. A Table with 11 previously dissected hearts with pre-excitation syndrome is presented.
40. Verduyn Lunel,[59] 1972 Case 1	About 1 mm ventral to the dorsal interventricular sulcus, fibers were located subendocardially connecting the RA with the RV wall over a distance of 3 mm, through holes in the annulus fibrosus; more ventrally, myocardial fibers 7 mm in width connected subendocardially the RA with the RV wall, through holes in the annulus fibrosus. At the dorsal junction	Mitral stenosis. Death during mitral commissurotomy. ECG: P–R: 0.16 sec. Ventricular fibrillation.	2 RAB-en SAB-en NV HV	—

	of the interatrial and interventricular septum the annulus fibrosus was much more discontinuous. Myocardial fibers connected the most dorsal parts of the interatrial and the interventricular septa through many holes. Fibers with features of conduction tissue (the first bundle described). Myocardial fibers (the septal and the second right accessory bundle). Other findings: Mahaim fibers.			
41. Verduyn Lunel,[59] 1972 Case 2	One mm ventral to the dorsal interventricular sulcus, myocardial fibers connected the RA with the RV subendocardially; between the LA and the LV, at 3 locations fibers were found also subendocardially, and others between the interatrial and interventricular septum. Ordinary myocardial fibers. Other findings: Mahaim fibers from the AVN, His bundle and LBB to the interventricular septum.	Sudden death. 4 years previous there had been syncope while playing football. ECG: type B, sinus bradycardia rate 50.	RAB 3 LAB M-SAB (multiple) NV HV (L)BV	—
42. Mann et al,[60] 1973	Two accessory bundles on the left side.	ECG: type A.	2-LAB	—

Author/case	Localization of Connections	ECG Findings	Pathology	Remarks
43. Brechenmacher et al,[42] 1974 Case 1	At the level of the posterior part of the left AV ring, 2.5 cm from the septum, 3 muscular fascicles were found connecting the left ventricular myocardium with the left atrium. They ran as individual fibers for only 2 mm. Each one of the 3 was accompanied by a vein. Nonspecific myocardial tissue. Other findings: James' fibers; the sinus node was the seat of degenerative cellular lesions, together with a rich fibrosis.	ECG: type AB, AF.	3-LAB AN D-SN F-SN	We consider this case to be one of the forms of sick sinus node syndrome.
44. James and Puech,[61] 1974	Beneath the left atrial appendage there was a fault in the mitral annulus, through which there was a direct AV connection. The atrial connection was to a Bachmann bundle. The ventricular was to working myocardium, midway between epicardium and endocardium (30-100 mm in cross section, 400 mm long). Most of this connection was composed of P cells and thus resembled the patient's own SN. Transitional	Degenerative disease of the atria: fibrosis, recent degenerative hemorrhage, recent necrosis of cells, fatty degeneration. Paper-like appearance of the atria. ECG: tachycardia,	LAB-Ep-M D inter-nodal tracts.	14,165 sections. Within the central fibrous body there were numerous lacunae and one lone cleft extending from atrial to ventricular septum, parallel to the AVN. Although cleft contained cells, there was no direct AV connection.

	cells and Purkinje cells were also encountered as in the case with P cell connections in the normal SN. Other findings: both the SN and the AVN arteries were narrowed by mural thickening. The SN was generally intact but atrionodal connections were degenerated. The AVN was similarly detached from its atrial connections, but was in direct continuity with the His bundle. The RBB was interrupted at its origin.	junctional rhythm.		
45. Okada et al,[62] 1974	A bypass tract from the AVN to the ventricles and 2 sets of Mahaim fibers, one from the proximal end and one from the distal AV bundle.	Idiopathic cardiomyopathy. ECG: a typical type B, slightly prolonged P–R.	NV 2-HV	
46. Lev et al,[13] 1975	AVN normal. SN—considerable hemorrhage in and on the periphery. Fatty infiltration and incomplete destruction of the AVN approaches. AVB—mild to moderate fibroelastosis (His bundle), normal James' fibers entered the terminal portion of AVN. Mahaim fibers (His bundle) tissue, Purkinje cells, from the AV bundle to both the right and left sides of the septum.	Carcinoma of the breast; metastases including pericardium. ECG: type undefined.	AN HV L-INT F-HB	18,600 sections.

Author/case	Localization of Connections	ECG Findings	Pathology	Remarks
47. Rossi,[18] 1975 Case 2	A fascicle of James' fibers seems to override the bulk of a rather small AVN and to interconnect the distal part of it with Paladino's bundle.	ECG: type B.	AN AH	Very scanty clinical information.
48. Rossi,[18] 1975 Case 4	An upper Mahaim fascicle was detected, piercing the central fibrous body to reach the myocardium of the ventricular septum.	ECG: pre-excitation complexes, leads V_1, V_2, type A, AF, LBBB.	NV	—
49. Rossi,[18] 1975 Case 5	Left posterolateral Kent bundle was seen to cross the AV ring on the subepicardial side, 20 mm long, 0.2 mm thick. Normal myocardial findings. Other findings: middle Mahaim fibers were found on the left side of the proximal tract of the His bundle piercing the pars membranacea septi atriorum to approach the ventricular septal myocardium.	ECG: type B.	LAB-ep HV	—
50. Brechenmacher et al,[63] 1977	Two unusual AV connections were found: (1) on the left AV ring near the obtuse margin of the left ventricle, 4.5 cm from the crux of the heart. It coursed from the left atrium deep to the coronary sinus and entered the epicardial margin of the left ventricular myocardium. Composed of working myocardial cells; (2) from the atrial septum to the junction	ECG: type A P–R 0.16 sec. PAT and VLT	1-LAB 1-AH F-SN F-AVN L-AVN	16,000 slides investigated

	slightly toward the left side. Composed of slender transitional cells. Additional findings: closely adjacent to the left AV connection, but not connected to it, embedded within the collagen of the mitral valve there were a few islands of slender transitional cells. More than normal fibrosis in the SN and INP. Focal fibrosis and abnormal deposits of fat in the AVN.			
51. Rossi et al,[64] 1975	A left lateral AV connection was found connecting directly the posterolateral atrial wall with the adjacent ventricle across a normal annulus fibrosus and insertion of mitral valve. Common myocardial cells. In addition Mahaim fibers were present between the His bundle and the left bundle and ventricular septum.	— Type A(?) SVT	1-LAB HV BV	— —
52. Dreifus et al,[65] 1976	Three accessory muscle bridges crossed the AV junction to connect the left atrium with the left ventricle. The first connected the posterior wall of the left atrium to the top of the interventricular septum and was 300 μ wide. The second and the third accessory tracts were located in an extreme lateral position near the lateral commissure of the mitral valve, and were respectively 380 and 330 μ wide. Atrial muscle cells. Additional findings: Most of the sinus nodal cells have been replaced by collagen and elastic fibers.	Carcinoma of the breast. ASHD SSS ECG: Type A P–R 0.12 sec. VLT., VF.	3-LAB F-SN	—

Author/case	Localization of Connections	ECG Findings	Pathology	Remarks
Group E: see comments				
53. Hammou et al.[66] 1971	An accessory pathway originates behind the Node of Tawara in the left auricle, then pierces the fibrous body at some distance from the AVN, penetrates the membranous septum, and joins the left border of the bundle of His before its division into left and right bundles.	IHSS (Obstructive myocardiopathy) —	1-AH	From the text, it is not clear if a general histological investigation of the heart was performed with the intention to find accessory bundles, or only a partial study was done. No ECG tracings.
ANIMAL STUDIES				
Boineau and Moore,[20] 1970	In a dog, a bundle of Kent on the posterior wall of the right ventricle close to the septum, 8 mm x 1.5 mm. Working myocardial cells? The velocity of 0.3 mm/sec is comparable to that of ventricle and would tend to rule out specialized conductive properties of this tissue.	ECG: type A	RAB	No histological studies available.
Boineau et al,[21] 1973	Right atrial to right ventricular connections: a pseudopod of atrial fibers approached the ventricle, a small interconnecting bundle, and became apparent in the ventricle. On the left side: a larger AV connection appeared to directly join the most proximal parts of the atrium and ventricle; ordinary working muscle.	ECG: spontaneous pre-excitation. Functional dissociation and block in bilateral pre-excitation.	RAB LAB	No histological studies available.

The AV node (AVN) was involved in 9 cases, the His bundle in 6, the bundle branches in 7, the internodal tracts (or approaches to the SN and/or AVN) in 4, and the sinus node (SN) in 10 cases with 2 additional not yet published reports (personal communication, James TN, March 1976). The pathology found in more than one-third of the 53 investigated hearts is discussed in Theories (see Chapter X).

Analysis of the 29 Most Thoroughly Studied Cases

In cases # 24-52, Group D, the entire conduction system and AV groove and septal areas were carefully investigated. In 5 of them no abnormal AV connection of any type could be found (# 28, 33, 35, 36, 37).

Direct AV bypass outside the conduction system was identified in 18 cases: 5 on the right AV border (# 25, 27, 29, 38, 39); 8 on the left (# 24, 42, 43, 44, 49, 50, 51, 52); 3 between the atrial and the ventricular septa (# 26, 31, 34); in one right and septal (# 40); and in another right, left and septal connections (# 41). In 7 cases, in addition to these direct, right, left, septal or multiple AV connections outside the conduction system, NV, HV and BV (upper, middle and lower Mahaim fibers, according to Rossi's classification) were described (# 25, 27, 29, 40, 41, 49, 51). Although the assumed AV connections (James' fibers according to Rossi) were present in all the cases, since they are an integral part of the normal posterior internodal tract they were particularly emphasized in one case (# 43).

Mediate bypass through the conduction system. In 6 cases inlet and outlet connections were present as the only connections between the atria and the ventricles: in 2 cases AH tracts were described (# 32, 47, in the latter an AN bundle also); and in 2 cases only NV and HV fibers were found (# 45, 48); and in 2 cases both inlet and outlet connections fibers were the only unusual pathological findings (# 30, 46).

Pathological findings in and around the conduction system of one kind or another were present in 11 cases: in 2 cases (# 28, 33) they were the only pathological findings; in another 2 (# 30, 46) they accompanied AV fibers mediated through the conduction system; and in 9 cases (# 25, 26, 27, 29, 34, 43, 44, 50, 52) they were associated with direct AV bypass tracts outside the conduction system. It is noteworthy that all the cases where pathological findings were described in the sinus node fell into Group D, with the exception of # 17.

Animal Studies

Two autopsies of hearts were described in animals whose ECG tracings showed signs considered to be compatible with pre-excitation.

TABLE XVI
Proposed Nomenclature for Normal and Abnormal Muscular Connections in the Pre-excitation Syndrome and Additional Pathological Findings

A. Direct AV Bypass Outside the Conduction System
*RAB: right accessory bundle
*LAB: left accessory bundle
*SAB: septal accessory bundle

B. Mediate Bypass Through the Conduction System
**a) Inlet connections
AN: atrial-AV nodal fibers (low AVN)
AH: atrial-His bundle fibers
AB: atrial-bundle branch fibers
***b) Outlet connections
NV: AV nodal-ventricular (septum) fibers
HV: His bundle-ventricular (septum) fibers
BV: bifurcation of His bundle or bundle branch fibers ventricular (septum)

C. Pathological Findings In or Around the Conduction System
A: Atrophy; C: Calcification;
D: Degeneration; F: Fibrosis;
I: Cell Infiltration; L: Fat Infiltration

In Rossi's scheme*: Kent bundles (K);** James' fibers(J);*** Upper, middle, lower and lower most Mahaim fibers (uM, mM, lM and lmM).

Boineau and Moore[20] described a bundle of Kent on the posterior wall of the right ventricle of a dog showing type A pre-excitation.[20] Boineau et al[21] found right and left AV connections in a squirrel monkey, but no convincing ECG tracings were published for this animal. The pre-excitation syndrome has also been described in cattle.[22-24]

Pathological Studies in Cases in which Pre-excitation was Not Found during Lifetime

The complete pathological investigation of human hearts performed to find an accessory AV bypass completely separated from or partially mediated through the conduction system is a difficult, time-consuming task, as evidenced by the thousands of histological sections required for such an undertaking. Most of the hearts so thoroughly investigated were from known pre-excitation cases. A control series of cases without pre-excitation is still not available for comparison. However, during the past decade some series and individual studies were undertaken in cases without pre-excitation to find pathological AV connections and provide comparative data.

Truex, Bishof and Hoffman,[25] in a study of the hearts of 15 newborns and fetuses, found a total of 17 accessory connections in 11, all on the right side of the heart. In children and adults, 3 instances of abnormal accessory AV bundles completely separated from the conducting system were demonstrated. Edwards[26,27] found a right AV bundle in a case with Ebstein's anomaly. Mahaim[28] described a left connection in a 32-year-old man who suffered from cor triloculare biatriatum with a complete agenesis of the bundle of His (this patient had normal sinus rhythm without pre-excitation). Rossi[18] extensively studied the hearts of 9 patients with and without pre-excitation, and found such an accessory fascicle of right atrial myocardium joining the underlying ventricular septum in the posteromedial part of the heart of a patient who suffered from recurrent tachycardias, but not pre-excitation. This same author found a number of AH, HV and BV fibers[18] in 5 hearts.

In summarizing this as yet small number of carefully studied hearts, it should be pointed out that due to the intermittent nature of pre-excitation, one cannot eliminate the possibility that in some of these cases the disorder was not documented by ECG tracings.

REFERENCES

1. **Holzmann M, Scherf D:** Über Elektrokardiogramme mit verkürtzter Vorhof-Kammer-Distanz und positiven P-Zacken. *Z Klin Med* 121:404, 1932.
2. **Wolferth CC, Wood FC:** Further observations on the mechanism of production of a short P-R interval in association with prolongation of the QRS complex. *Am Heart J* 22:450, 1941.
3. **James TN:** The Wolff-Parkinson-White syndrome. *Ann Intern Med* 71:399, 1969.
4. **James TN:** The Wolff-Parkinson-White syndrome: evolving concepts of its pathogenesis. *Prog Cardiovasc Dis* 13:159, 1970.
5. **Anderson RH, Thapar MK, Arnold R, et al:** Study of connecting tissue in a case of ventricular pre-excitation. *Br Heart J* 35:566, 1973.
6. **Downing DF:** Accessory atrioventricular muscle. II. Cardiac conduction system in a human specimen with Wolff-Parkinson-White syndrome. *Anat Rec* 137:417, 1960.
7. **Lev M, Sodi-Pallares M, Friedland M:** A histopathologic study of the atrioventricular communication in a case of Wolff-Parkinson-White with complete left bundle branch block. Presented at the 4th World Cardiol Congress, Mexico City, July 1962.
8. **Ohnell RF:** Postmortem examination and clinical report of a case of the short P-R interval and wide QRS wave syndrome (Wolff-Parkinson-White). *Cardiologia* 4:249, 1940.
9. **Rossi L, Rovelli F:** Anomalia del sistema atrioventricolare in un caso di Wolff-Parkinson-White (nota preventiva). *Riv Anat Pathol Oncol* 9:161, 1955.

10. **Segers M, Sanabria T, Lequine J, et al:** Le mécanisme del l'invasion ventriculaire precoce. Mise en évidence d'une connexion septale directe entre l'oreillette et le ventricule. *Comp Ren Soc Biol* 139:801, 1945.
11. **Soderstrom N:** Absence of accessory AV-muscle connections in a case with pre-excitation electrocardiogram. *Acta Med Scand* (suppl 170): 119, 1946.
12. **Wood FC, Wolferth CC, Geckeler GD:** Histologic demonstration of accessory muscular connections between auricle and ventricle in a case of short P-R interval and prolonged QRS complexes. *Am Heart J* 25:454, 1943.
13. **Lev M, Fox SM, Greenfield JC, et al:** Mahaim and James' fibers as a basis for a unique variety of ventricular pre-excitation. *Am J Cardiol* 36:880, 1975.
14. **Becker AE, Anderson RH, Durrer D, et al:** The anatomy of ventricular pre-excitation. (Abstr) *Circulation* 54 (suppl II): 186, 1976.
15. **Ogata K, Okada R:** A study on a case of Wolff-Parkinson-White syndrome type A with a bypass between the right atrium and right ventricle. Presented at the VIth Asian-Pacific Congress of Cardiology, Oct. 3-8, 1976.
16. **Okada R:** Pathology of the pre-excitation syndrome. Presented at the VIth Asian-Pacific Congress of Cardiology, Oct. 3-8, 1976.
17. **Ferrer MI:** New concepts relating to the pre-excitation syndrome. *JAMA* 201:1938, 1967.
18. **Rossi L:** A histological survey of pre-excitation syndrome and related arrhythmias. *G Ital Cardiol* 5:817, 1975.
19. **Anderson RH, Becker AE, Brechenmacher C, et al:** Ventricular pre-excitation. A proposed nomenclature. *Eur J Cardiol* 3:27, 1975.
20. **Boineau JP, Moore EN:** Evidence for propagation of activation across an accessory atrioventricular connection in types A and B pre-excitation. *Circulation* 41:375, 1970.
21. **Boineau JP, Moore EN, Spear JF, et al:** Basis on static and dynamic electrocardiographic variations in Wolff-Parkinson-White syndrome: anatomic and electrophysiologic observations in right and left ventricular pre-excitation. *Am J Cardiol* 32:32, 1973.
22. **Sugeno H, Murao S, Ueda H:** WPW syndrome on bovine: a case report. *Jpn Heart J* 5:140, 1964.
23. **Van Arsdel WC III:** The WPW syndrome in cattle. *Am Heart J* 70:146, 1965.
24. **Van Arsdel WC III, Bogart R:** The Wolff-Parkinson-White syndrome in a Hereford cow. *Zentralbl Veterinaermed* 11:57, 1964.
25. **Truex RC, Bishof JK, Hoffman EL:** Accessory atrioventricular muscle bundles. II. Cardiac conduction system in a human specimen with Wolff-Parkinson-White syndrome. *Anat Rec* 131:45, 1958.
26. **Edwards JE:** Symposium on cardiac catheterization. III. Pathologic features of Ebstein's malformation of tricuspid valve. *Proc Staff Meet Mayo Clin* 28:89, 1953.
27. **Edwards JE:** Pathologic feature of Ebstein's malformation of the tricuspid valve. *Proc Staff Meet Mayo Clin* 28:94, 1953.
28. **Mahaim I:** Cor triloculare biatriatum. Agénésie du faisceau de His-Tawara et des deux branches. Electrocardiogrammes examens histologiques. *Arch Inst Cardiol Mex* 18:42, 1948.
29. **Ohnell RF:** Pre-excitation, a cardiac abnormality. *Acta Med Scand* (suppl 152), 1944.
30. **Kimball JL, Burch G:** The prognosis of the Wolff-Parkinson-White syndrome. *Ann Intern Med* 27:239, 1947.
31. **Levine HD, Burge JC:** Septal infarction with complete heart block and intermittent anomalous atrioventricular excitation (Wolff-Parkinson-White syndrome): histologic demonstration of a right lateral bundle. *Am Heart J* 36:431, 1948.
32. **Sondergaard G:** The Wolff-Parkinson-White syndrome in infants. *Acta Med Scand* 145:386, 1953.
33. **de Villeneuve, Schornagel HE:** Voorkomen van een Accessoire Atrio-ventriculaire Spierbundel bij een zuigeling met Paroxysmale Tachycardia en het Wolff-Parkinson-White Syndrom. *Maandschr Kindergeneesk* 26:23, 1958.
34. **Yater WM, Shapiro MJ:** Congenital displacement of the tricuspid valve (Ebstein's

disease): review and report of a case with electrocardiographic abnormalities and detailed histologic study of the conduction system. *Ann Intern Med* 11:1043, 1938.

35. **Mahaim I, Bogdanovic P:** Un cas mortel de syndrome de Wolff-Parkinson-White. Examen histologique de faisceau de His-Tawara. *Folia Cardiol* 7:626, 1948.
36. **Segers M, Sanabria T, Lequime J, et al:** Le syndrome de Wolff-Parkinson-White. Mise en evidence d'une connexion A-V septale directe. *Acta Cardiol* 2:21, 1947.
37. **Prinzmetal M, Kennamer R, Corday E, et al:** *Accelerated Conduction. The Wolff-Parkinson-White Syndrome and Related Conditions,* New York: Grune & Stratton, 1952.
38. **Langendorf R, Lev M, Pick A:** Auricular fibrillation with anomalous A-V excitation (WPW syndrome) imitating ventricular paroxysmal tachycardia. A case report with clinical and autopsy findings and critical review of the literature. *Acta Cardiol* 7:241, 1952.
39. **Plavsic C, Maric D, Zimolo A:** Infarctus du myocarde et syndrome de Wolff-Parkinson-White. *Acta Cardiol* 11:190, 1956.
40. **Okel BB:** The Wolff-Parkinson-White syndrome, report of a case with fatal arrhythmia and autopsy findings of myocarditis, interatrial lipomatous hypertrophy, and prominent right moderator band. *Am Heart J* 75:673, 1968.
41. **Cole JS, Wills RE, Winterscheid LC, et al:** The Wolff-Parkinson-White syndrome: problems in evaluation and surgical therapy. *Circulation* 42:111, 1970.
42. **Brechenmacher C, Laham J, Iris L, et al:** Etude histologique des voies abnormales de la conduction dans un syndrome de Wolff-Parkinson-White et dans un syndrome de Lown-Ganong-Levine. *Arch Mal Coeur* 67:507, 1974.
43. **Holzmann M:** Bemerkungen zum Ekg mit verkuerzter P-Q Distanz und Schenkelblockbild des Kammerkomplexes. *Verh Dtsch Ges Kreislaufforsch* 12:101, 1939.
44. **Soederstroem N:** Absence of accessory A-V muscle connections in a case with pre-excitation electrocardiogram. Report of clinical and post-mortem findings. *Acta Med Scand* (suppl 170): 119, 1946.
45. **Prinzmetal M, Kennamer R:** Anomalous atrioventricular excitation. Panel discussion. *Ann NY Acad Sci* 65:852, 1957.
46. **Robertson D:** Wolff-Parkinson-White syndrome with paroxysmal ventricular tachycardia. *Br Heart J* 15:466, 1953.
47. **Scherf D, Cohen J:** *The Atrioventricular Node and Selected Cardiac Arrhythmias,* pp. 373-447, New York-London: Grune & Stratton, 1964.
48. **Schiebler GL, Adams P Jr, Anderson RC:** The Wolff-Parkinson-White syndrome in infants and children. A review and a report of 28 cases. *Pediatrics* 24:585, 1959.
49. **Lev M, Gibson S, Miller RA:** Ebstein's disease with Wolff-Parkinson-White syndrome: report of a case with a histopathologic study of possible conduction pathways. *Am Heart J* 49:724, 1955.
50. **Truex RC, Bishof JK, Downing DF:** Accessory atrioventricular muscle bundles. II. Cardiac conduction system in a human specimen with Wolff-Parkinson-White syndrome. *Anat Rec* 137:417, 1960.
51. **Lev M, Kennamer R, Prinzmetal M, et al:** A histopathologic study of the atrioventricular communication in two hearts with the Wolff-Parkinson-White syndrome. *Circulation* 24:41, 1961.
52. **Lev M, Sodi-Pallares D, Friedland C:** A histopathologic study of the atrioventricular communications in a case of WPW with incomplete left bundle branch block. *Am Heart J* 66:399, 1963.
53. **Lev M, Leffler WB, Langendorf R, et al:** Anatomic findings in a case of ventricular pre-excitation (WPW) terminating in complete atrioventricular block. *Circulation* 34:718, 1966.
54. **Niessen KH:** Das Syndrom von Wolff-Parkinson-White. Klinische und histologische Untersuchung eines Falles. *Inaugural dissertation at U of Heidelberg,* 1966.
55. **Rossi L:** *Histopathologic Features of Cardiac Arrhythmias,* p 240, Milano: Casa Edit Ambrosiana, 1969.
56. **Schumann G, Jansen HH, Anschütz F:** On the pathogenesis of the WPW syndrome. (Ger) *Virchows Arch (Pathol Anat)* 349:48, 1970.

57. **Schumann G:** Acquired WPW syndrome. *Z Kreislaufforsch* 59:1081, 1970.
58. **Rosenberg HS, Klima T, McNamara DG, et al:** Atrioventricular communication in the Wolff-Parkinson-White syndrome. *Am J Clin Pathol* 56:79, 1971.
59. **Verduyn Lunel AA:** Significance of annulus fibrosus of heart in relation to AV conduction and ventricular activation in cases of Wolff-Parkinson-White syndrome. *Br Heart J* 34:1263, 1972.
60. **Mann RB, Fisher RS, Scherlis S, et al:** Accessory left atrio-ventricular connection in type A Wolff-Parkinson-White syndrome. *Johns Hopkins Med J* 132:242, 1973.
61. **James TN, Puech P:** De subitaneis mortibus. IX. Type A Wolff-Parkinson-White syndrome. *Circulation* 50:1264, 1974.
62. **Okada R, Mizutani T, Mochizuki S:** A morphologic study of a case of idiopathic cardiomyopathy which showed the electro-cardiographic features of a WPW syndrome. *Heart* 6:630, 1974.
63. **Brechenmacher C, Coumel P, Fauchier JP, et al:** De subitaneis mortibus. XXII. Intractable paroxysmal tachycardias which proved fatal in type A Wolff-Parkinson-White syndrome. *Circulation* 55:408, 1977.
64. **Rossi L, Knippel M, Raccardi B:** Histological findings, His bundle recordings and body-surface potential mappings in a case of Wolff-Parkinson-White syndrome; an anatomoclinical comparison. *Cardiology* 60:265, 1975.
65. **Dreifus LS, Wellens HJ, Watanabe Y, et al:** Sinus bradycardia and atrial fibrillation associated with the Wolff-Parkinson-White syndrome. *Am J Cardiol* 38:149, 1976.
66. **Hammou JC, Grosgogeat Y, el-Hachimi A, et al:** Obstructive myocardiopathy and Wolff-Parkinson-White syndrome. Anatomical study. (Fr) *Arch Anat Pathol* (Paris) 20:65, 1972.

V

Noninvasive Methods of Investigation

VECTORCARDIOGRAPHY

Many papers on spatial vectorcardiography have appeared in the literature since 1959, when Donzelot et al[1] first used this procedure in pre-excitation cases.[2-36] It is employed mainly for in-depth study of the unusual ECG findings, including the order of depolarization of the ventricles during pre-excitation and the presence of bundle branch block (BBB) or other pathological conditions. Providing three-dimensional information, it is also used to help determine the location of the accessory bypass from the atria to the ventricles (structural or functional bridges).

The Tel Hashomer Series (Table XVII)

Vectorcardiograms (VCG) were obtained in 25 of our 215 patients: 14 type A and 11 type B pre-excitation cases, according to Rosenbaum et al's ECG classification.[37] The VCG tracings were classified according to the orientation of the mean delta and QRS vector in the horizontal plane, as set down by Bleifer[10]: Group A had a forward orientation of the delta vector (+120° to +30°), and Group B had a backward or leftward orientation of the delta vector (−60° to +30°). The delta waves were plainly seen in all the patients and in all 3 planes, and their onset and termination were easily distinguishable. The waves appeared as straight lines in some of the recordings and as a slight curve in most (Figure 43, 44, 45, 46). The rest of the QRS loop continued initially in the direction of the terminal portion of the delta wave, later changing direction. However, the centrifugal and centripetal arms of the loop usually remained close to one another, with a narrow angle between the delta vector and the long QRS vector. The data are summarized in Table XVII.

Type A Pre-excitation: 14 Cases. *Horizontal plane.* The rotation of the vectorial loop was clockwise in 5 cases, counterclockwise in 7 and

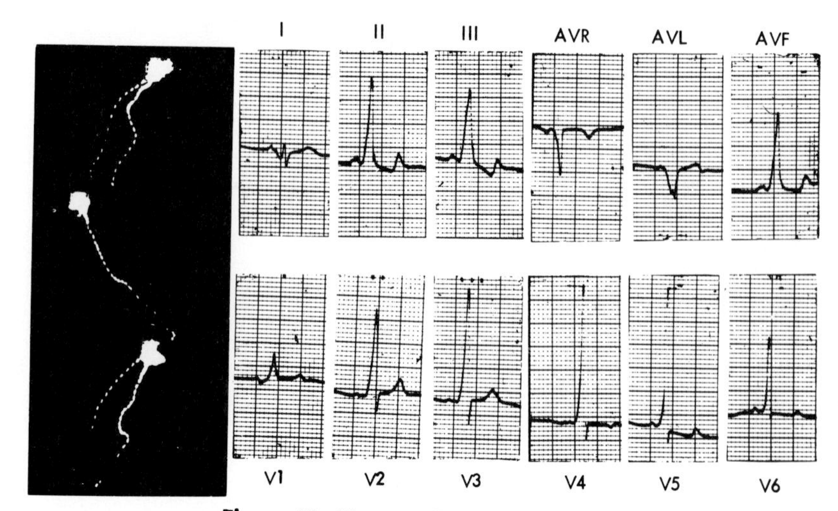

Figure 43. *Vectorcardiogram in pre-excitation type AI in a 27-year-old man.*

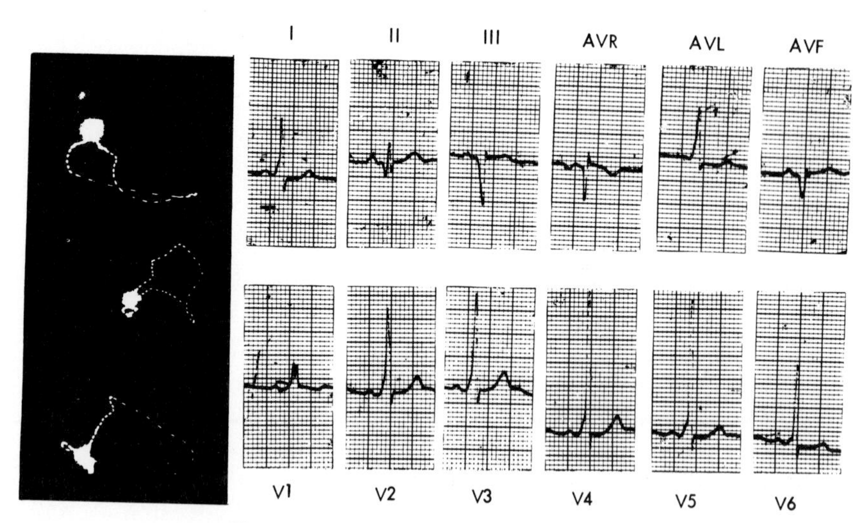

Figure 44. *Vectorcardiogram in pre-excitation type AS in a 40-year-old woman.*

figure eight–shaped in 2. There were 3 examples of slowing of the time marks in the final stage of the loop, and the ECG and VCG patterns were similar to RBBB.

Sagittal plane. In addition to its anterior direction, the delta vector was also oriented upward in 7 cases and downward in 7. The rotation of

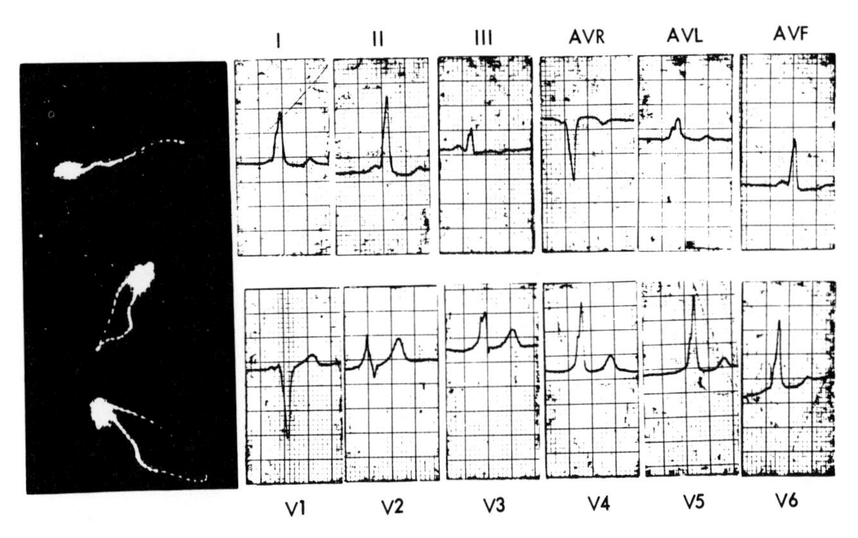

Figure 45. *Vectorcardiogram in pre-excitation type BI in an 18-year-old man.*

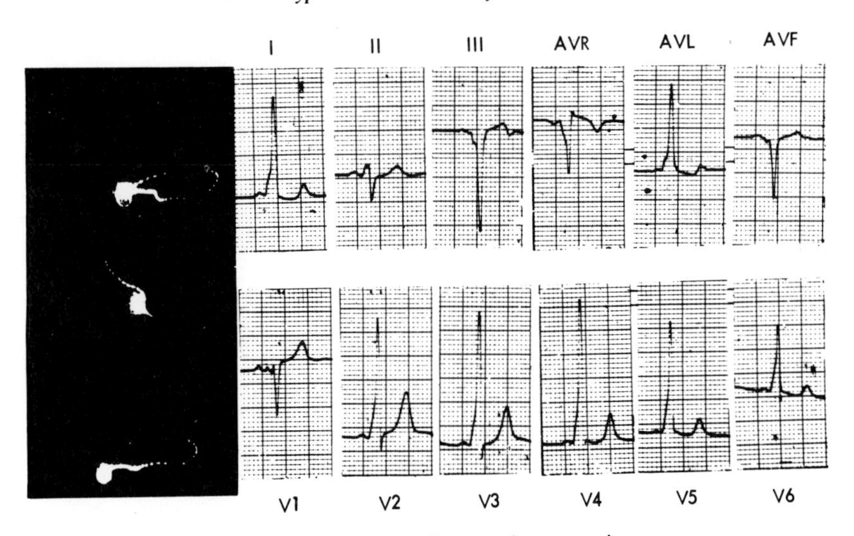

Figure 46. *Vectorcardiogram in pre-excitation type BS in a 23-year-old man.*

the QRS loop was clockwise in 7, counterclockwise in 5, and figure eight–shaped in 2 cases.

Frontal plane. Here again the delta vector faced upward in 7 cases. The QRS loop rotation in this plane was clockwise in 8, counterclockwise in 3 and figure eight–shaped in 3 cases.

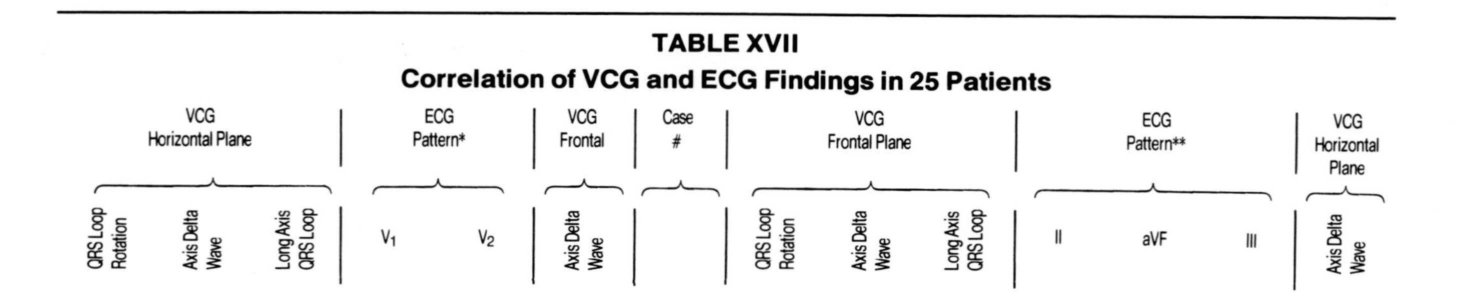

TABLE XVII

Correlation of VCG and ECG Findings in 25 Patients

VCG Horizontal Plane			ECG Pattern*		VCG Frontal	Case #	VCG Frontal Plane			ECG Pattern**			VCG Horizontal Plane
QRS Loop Rotation	Axis Delta Wave	Long Axis QRS Loop	V_1	V_2	Axis Delta Wave		QRS Loop Rotation	Axis Delta Wave	Long Axis QRS Loop	II	aVF	III	Axis Delta Wave

CW	+115°	+120°	R	R	+175°	84	8	+175°	+150°	R	R	R	+115°
CW	+107°	+120°	R	R	+105°	24	CW	+105°	+117°	R	Rs	R	+107°
CCW	+105°	+25°	Rr'	Rs	+100°	12	CCW	+100°	+10°	Rs	rs	rs	+105°
CCW	+100°	+20°	rs	Rs	+100°	52	CCW	+100°	+45°	r	r	Rs	+100°
CCW	+100°	+5°	Rs	Rs	−95°	39	CW	+80°	+70°	R	R	R	+75°
CCW	+90°	+50°	Rs	Rs	+30°	23	CW	+60°	+90°	Rs	Rs	R	+73°
8	+90°	+10°	Rsr'	Rs	−130°	6	CCW	+45°	+35°	R	R	R	−15°
CCW	+90°	+75°	R	R	−90°	26	CW	+35°	+35°	R	R	Rs	−10°
CW	+85°	+75°	Rs	Rs	−120°	21	CW	+30°	+50°	R	R	R	+90°
CCW	+75°	+55°	Rs	Rs	+80°	43	CCW	+27°	+20°	R	R	R	−30°
CW	+73°	+103°	R	R	+60°	61	CCW	+25°	+22°	R	R	rs	−35°
8	+57°	+33°	R	R	−55°	49	CCW	+20°	+26°	Rs	Rs	qrs	−20°
CCW	+50°	?	Rs	Rs	−55°	25	CCW	0°	+12°	Rs	rs	qrs	−5°
CW	+35°	+10°	rR'	R	−20°	27	CCW	−10°	−15°	rs	qs	qs	+5°
CCW	+30°	0°	rsR'	R	−60°	58	8	−20°	+10°	rs	rs	rs	−20°
CCW	+5°	−10°	rs	Rq	−10°	55	8	−20°	+15°	Rs	rs	rs	+35°
CW	−5°	0°	qrs	R	0°	20	CW	−35°	−10°	qr	qR	QS	−60°
CCW	−10°	+10°	qs	rs	+35°	62	CCW	−45°	−45°	qs	qs	qs	−12°
CCW	−12°	−12°	rs	R	−45°	22	CW	−55°	0°	qrs	qr	qr	+57°
CCW	−15°	−15°	rs	rs	+45°	28	CCW	−55°	−65°	rs	rs	qs	+50°
CCW	−20°	−25°	rs	rs	−20°	8	8	−60°	−40°	qr	QS	QS	+30°
8	−20°	+10°	QS	rs	+20°	9	CW	−90°	?	qr	Qr	Qr	+85°
8	−30°	−15°	QS	QS	+27°	10	8	−95°	−10°	qs	Qs	QS	+100°
8	−35°	−10°	rs	rs	+25°	7	CW	−120°	?	qrs	Qs	Qr	+85°
CW	−62°	−30°	rs	rs	−35°	37	CW	−130°	0°	qrs	QS	QS	+90°

The Tables are arranged according to the orientation of the delta wave in the Horizontal* and Frontal** planes.
Note: the multiplicity of direction of the delta wave in space; the acute delta wave–QRS wave angle in the majority of cases; the correlation between the orientation of the delta wave and QRS loop and the ECG patterns.

In the three instances in which the horizontal plane revealed a RBBB pattern, slowing of the last part of the vectorial loop was also seen in the frontal and sagittal planes (Figure 44).

The T loop was discordant or an angular deviation of greater than 60° from the QRS loop in all 3 planes in 6 cases, and in 2 of the 3 planes in 2 cases. The T loop was concordant in all 3 planes in 3 cases and concordant in 2 planes only in another 3 cases.

Type B Pre-excitation: 11 Cases. *Horizontal plane.* The QRS loop showed counterclockwise rotation in 7 cases, clockwise in one case and figure eight–shaped in 3 cases.

Sagittal plane. Apart from its left or backward direction, the delta vector faced upward in 5 cases and downward in 6 cases. QRS loop rotation in this plane was clockwise in 2, counterclockwise in 7 and figure eight–shaped in 2 cases.

Frontal plane. As in the sagittal plane the delta vector was directed upward in 5 cases, downward in 5 and was 0° in one. The rotation of the loop was clockwise in 3, counterclockwise in 7 and figure eight–shaped in 2 cases.

The T loop was discordant or at an angular deviation of greater than 60° from the QRS loop in all 3 planes in 6 cases, in 2 of the 3 planes in 4 cases; it was concordant in one case only.

Delta Vector—QRS Vector Angle Correlation. The vectorial loop continued in the direction of the delta wave, and the angle between the vector of this wave and the long vector of the entire QRS loop in all planes was acute (average 29° in the horizontal plane and 32° in the frontal plane). In 4 cases this angle was greater than 40°: 2 suffered from organic heart disease (mitral stenosis and myocardial infarction), one had a RBBB pattern, and there was no additional finding in the fourth.

ECG and VCG Correlation. *Precordial leads.* Fourteen of the patients were type A pre-excitation (Rosenbaum), in which there is an R or Rs pattern in V and V_2 and a large R in the other chest leads. The delta vector was directed anteriorly (in the horizontal plane) in all 14 cases (+115° to +30°), and as expected, the entire vectorial loop was oriented forward. In 5 cases the long axis direction of the entire QRS loop was from +5° to +25°, and in the remaining 9 cases it was directed anteriorly beyond +30°. Thus there was a strong correlation between Rosenbaum's[37] ECG type A and Bleifer's[10] VCG Group A.

Eleven of the cases were ECG type B (Rosenbaum), in which ECG patterns in V_1 were rS in 7 cases, and qRs or QS in 4 cases. In 4 of the former group of 7 cases, V_4R was also recorded and revealed a QS pattern. According to Rosenbaum's definition it is not necessary to find rS or QS in V_2 in order to include the case in type B, and indeed in 4 of our 11 cases the pattern in V_2 was R or Rs. In 10 of these 11 cases the delta vector was directed to the left or backward in the horizontal plane (+30° to −60°), and again, predictably, the rest of the vectorial loop followed the direction of the delta wave in the horizontal plane and faced leftward or backward (+10° to −30°). Again, we find a strong correlation between Rosenbaum's ECG and Bleifer's VCG groups.

Limb leads. In 12 of the 25 cases, R or Rs patterns were present in ECG leads II, III and aVF, and the delta vector in the frontal plane faced downward (0° to +175°), as did the rest of the vectorial loop. In 7 of these 12 cases the sagittal and frontal planes belonged to the VCG Group A (delta vector directed forward and downward) and 5 belonged to VCG Group B (delta vector leftward or backward and downward).

In the other 13 of the 25 cases, 8 showed qR or Qs in the ECG leads II, III and aVF (except for one patient who exhibited this only in III and aVF), and the delta vector faced upward (−30° to −130°) in the frontal plane, as did the rest of the vectorial loop. We can conclude that, in pre-excitation syndrome, when "q" is seen in leads II, III and aVF, the delta vector and the entire vectorial loop face upward more than −30°; and when "q" exists in only part of these leads, the delta vector and the vectorial loop are directed from 0° to −30°. Seven of these 13 cases belonged to Bleifer's VCG Group A (forward and upward) and 6 to VCG Group B (leftward or backward and upward).

Order of Depolarization in the Ventricles during Pre-excitation Conduction

In summarizing the most important findings of their vectorcardiographic investigations, Bleifer et al[10] pointed out that the delta portion of the QRS s E loop was usually oriented in the same direction as the rest of the loop, and determined its spatial orientation. This was confirmed by others in spatial vectorcardiography studies or vectorial interpretation of ECG tracings. Lowe et al,[29] for example, in stressing that the best correlation between the values of the delta and main QRS vectors was obtained in the horizontal plane, found in this plane a mean angle of 18° in 28 cases (range 0°-65°). Zao et al[38] also determined that the average angle between the delta and QRS vectors was about 30°, with a range of 0° to 90°. Tranchesi et al[39], without giving precise values, stated that "the small spatial angle obtained between the delta and the QRS vectors in the two types of WPW syndrome suggests a definite influence of the electrical forces that are developed during the premature activation upon the mean orientation of the total ventricular activation." A similar mean acute angle was found in our 25 cases between the long vector of the delta wave and the QRS loop, 29° in the horizontal and 32° in the frontal plane.

This marked closeness of the delta and QRS vectors may contribute valuable information regarding the order of depolarization of the ventricles during pre-excitation, and will be discussed further in Theories.

Classification of Pre-excitation Cases

In a recent paper, Lowe et al[29] concluded that much confusion still exists in the classifications from the ECG and VCG points of view because of the different approaches used in the two techniques. He was especially critical of the fact that while Rosenbaum's classification is based on the polarity of the major deflection of the QRS complex in the right precordial leads, others classify their cases as A or B according to the direction of the delta wave in these leads. He goes on to say that while Rosenbaum uses only the anteroposterior axis and believes that the frontal axis is in no way helpful in differentiating the cases, Grant et al[40] claim that the delta vector in the frontal plane is much more important than that in the horizontal plane for classifying cases. Lowe et al[29] and others[41] thus concluded that in pre-excitation a wide spatial spectrum of both delta and main QRS vectors can be observed, and they emphasized this point with a three-dimensional figure which illustrates schematically the direction of the delta vector in their 28 cases, and with a similar drawing by Gallagher et al.[41] Thus, VCG studies strengthen the appropriateness of our descriptive ECG classification, the clear-cut delta waves and the QRS loops seen in all three planes in the VCG tracings favor our division into 4 major groups (AI, AS, BI and BS—Figure 43, 44, 45, 46, and Diagram II).

Use of VCG in Determining the Presence of Bundle Branch Blocks or Other Pathological Conditions

Vectorcardiography is useful in diagnosis when pre-excitation and BBB occur simultaneously, or when the pre-excitation pattern simulates or mimics BBB (as described in Chapter IX). If a whole ventricle is depolarized before the other, as in BBB, there is a corresponding late imbalance of electrical forces.[29] Apparently, when RBBB occurs concomitant with pre-excitation, there is the initial slowing of the QRS loop characteristic of the pre-excitation (delta wave), and an additional slowing in the terminal portion of the QRS loop which is considered by many[23] to represent the RBBB. However, when a slowing of the time markers appears in the middle portion of the QRS loop (together with the initial slowing of pre-excitation), the coexistence of an LBBB should be considered.[24] Unfortunately, however, these diagnostic indications are not always considered pathognomonic for the coexistence of R or

LBBB and pre-excitation. Bleifer et al[10] commented that the altered initial and late forces of the QRS s E during pre-excitation appear so different from the normal that it is hard to conceive that the dissimilarity is due to the addition of a delta vector to the normal vector. These authors feel that the entire conduction through both ventricles was abnormal, and that the late QRS changes look not unlike partial or complete BBB of the contralateral ventricle. Indeed, it has long been known (and is described in Chapter IX) that type A pre-excitation can simulate RBBB and type B can simulate LBBB.[42]

The slowing of the middle or terminal time markers is thus not always an indication of an additional BBB. This conclusion was strengthened by Castellanos et al[15] who, while pacing the atria at increased rates in patients with pre-excitation, produced an increase in the delta vector component of the QRS which eventually gave rise to a broad bizarre QRS not unlike BBB. However, there is no doubt that the whole of ventricular conduction in their cases came from the single pre-excitation focus without any BBB component. Therefore, it cannot be definitely stated that a BBB coexists with pre-excitation, either from the ECG or from the VCG patterns, except when one of the two conduction disturbances is intermittent and an initial or later slowing of the QRS loop appears which was not previously present.[23,24]

We and other investigators have observed that the great majority of cases show a very narrow angle between the long axis of the delta vector and vector of the QRS loop (±30°) (Diagram II). As a result, it is felt that a great increase (or discordance) of this acute angle indicates, in addition to pre-excitation, a serious organic or functional disease or disorder. Miller and Victoria (quoted by Lowe et al[29]) studied a group of children, a high percentage of whom had serious congenital heart disease, and found that the delta and QRS vectors were widely divergent. Tranchesi et al,[39] Lowe et al,[29] Bleifer et al[10] and we observed similar divergencies in patients with associated heart diseases. Therefore, a delta QRS vector angle larger than 30° may serve as an indicator for the presence of additional cardiac disease in these patients.

Use of VCG in Determining the Location of an Accessory Pathway

Investigators once believed that vectorcardiography, with its three-dimensional presentation of the electrical forces in the heart, might shed light on the origin of the pre-excitation and the location of the pathological atrioventricular bypass. More modern studies have shown that neither a regular ECG tracing nor additional VCG figures can serve to locate an accessory bypass. In a recent study[43] VCG tracings in combina-

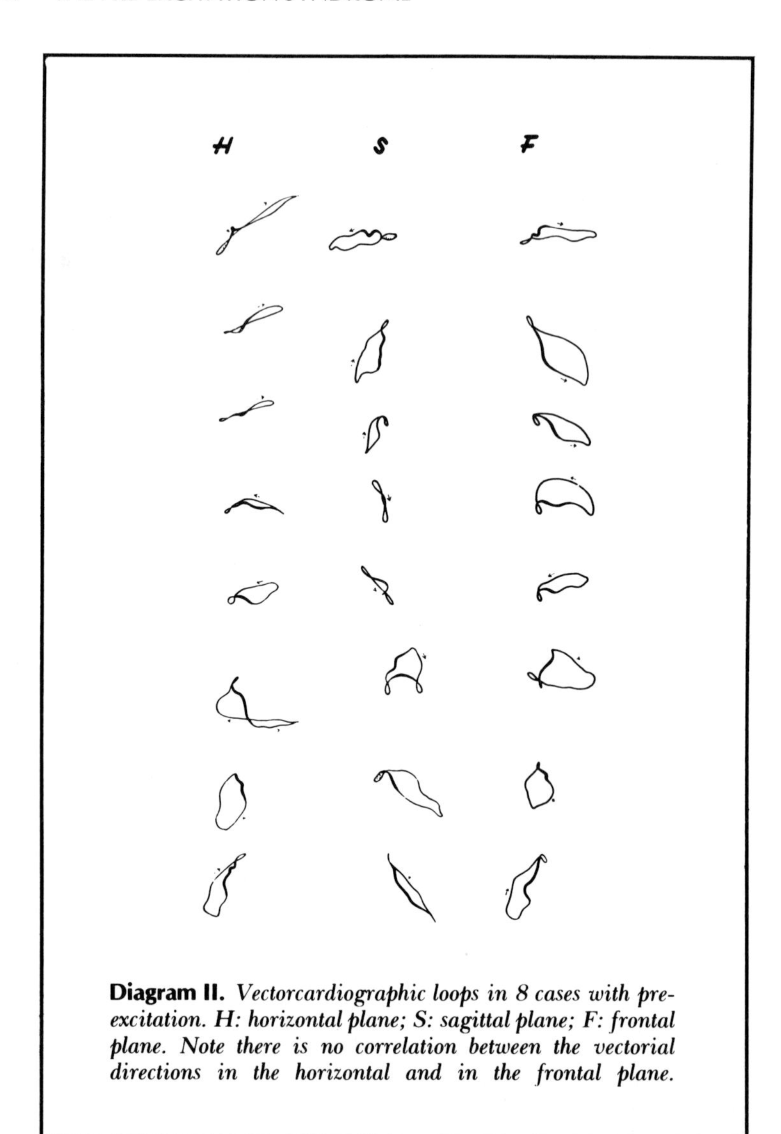

Diagram II. *Vectorcardiographic loops in 8 cases with pre-excitation. H: horizontal plane; S: sagittal plane; F: frontal plane. Note there is no correlation between the vectorial directions in the horizontal and in the frontal plane.*

tion with His bundle electrograms were used to differentiate septal from free wall connections. Maybe that combination of noninvasive and invasive methods will, in the future, help detect more precisely the location of accessory bundles. On the other hand, some forms of ventricular premature beats can reproduce the characteristic pattern of pre-excitation in VCG tracings in patients in whom this syndrome was never proved to exist (Figure 47).

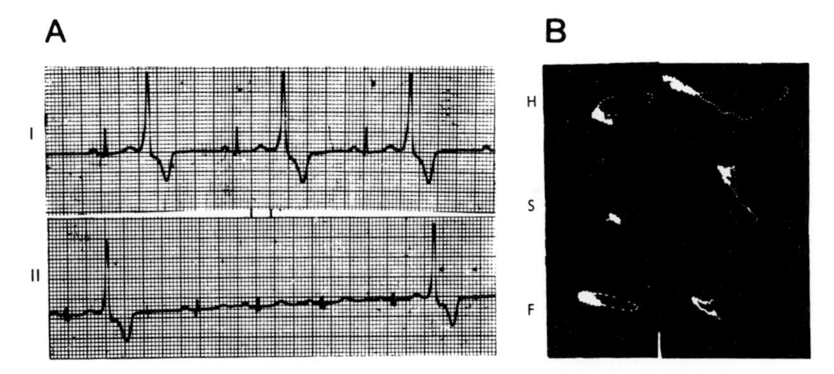

Figure 47. *Ventricular premature beats mimicking pre-excitation.* **a.** *ECG tracing, leads I, II. Note ventricular premature beats (VPB) with pre-excitation configuration.* **b.** *Vectorcardiogram of same patient. left: during normal AV conduction; right: during VPB's. Note similarity to pre-excitation vectorcardiogram loops (see Figure 43, 44, 45, 46).*

LONG-TERM AMBULATORY MONITORING

About half the patients suffering from pre-excitation report episodes of palpitations, dizziness and chest pain, with varying degrees of severity. These complaints are generally attributed to bouts of tachycardia, and preventive treatment is often based solely upon these subjective reports by patients. While this is an undesirable approach to the treatment of arrhythmia, routine ECG records only short time periods and usually the episode has passed before the patient even reaches his doctor. Advances in technology have produced a portable, ambulatory ECG monitoring system which provides an ideal solution to many problems of diagnosis and clarification of the pathophysiology of the syndrome.

Two studies using this new technique have been published to date,[44,45] one with 27 patients and one with 20 patients; in addition, 30 cases were monitored in the Tel Hashomer series. When comparing the three series one must bear in mind the different criteria and experimental designs employed. Hindman et al[46] monitored randomly chosen patients over a 24-hour period, while we and Isaeff et al[45] chose our cases from among those complaining of palpitations and for 10-12 hours only. These data showed the great unlikelihood of picking up any episode of arrhythmia during monitoring. Hindman's hope of finding a more precise method of treatment of the arrhythmias after their identification was not fulfilled. The results will be summarized according to Hindman's five objectives.

Incidence of Supraventricular Paroxysmal Tachycardia

Eight of Hindman et al's 27 patients reported episodes of palpitations during the 24 hours, but no episodes of paroxysmal tachycardia were recorded. One of our 30 cases had a short bout of supraventricular tachycardia, while 5 of Isaeff's 20 patients had such attacks. These findings are quite remarkable and serve to point out the part played by sheer chance in the detection of pre-excitation tachycardias.

Correlation of Symptoms Suggestive of Paroxysmal Supraventricular Tachycardia with Actual ECG Events

The long-term monitoring demonstrated that other events besides supraventricular tachycardias can be the cause of patients' complaints of palpitations or other unpleasant feelings. Hindman found that 8 of 10 episodes (each in a separate case) were correlated with ECG events that could explain the symptoms: 3 occurred during bursts of nonexertional sinus tachycardia (in one, at a maximum heart rate of 180 beats per minute); 2 during multiple ventricular premature contractions; one during repetitive atrial premature beats; one during a ventricular escape beat secondary to sinus arrest (she complained of irregular pulse and sudden dizziness); and another during both atrial and ventricular premature contractions. In the remaining 2 patients no unusual ECG event could be detected during the complaints. Among our 10 patients who complained of palpitations and/or other symptoms, 5 were during sinus tachycardia (with a maximal rate of 145), 3 during recurrent premature beats, one during escape beats, and in only one during a bout of true supraventricular tachycardia.

Isaeff did not mention whether there were other patients with symptoms during the monitoring in addition to the 5 patients who complained of palpitations and in whom a supraventricular tachycardia was found. (Paroxysmal atrial fibrillation was also discovered in one of these.)

The most surprising finding was the relatively high incidence of paroxysmal attacks of sinus tachycardia among pre-excitation patients. This, together with recent reports of sinus node dysfunctions in the syndrome, including sinus arrest, sinoatrial blocks, sinus bradycardia, *etc.* draws attention to the role played by the sinus node in the pathophysiology of the pre-excitation syndrome.

One must be careful in assessing the cause of ventricular tachycardia recorded during monitoring. Its appearance in one of our patients (Figure 48a and b) was probably due to his history of atherosclerotic heart disease and long-standing hypertension.

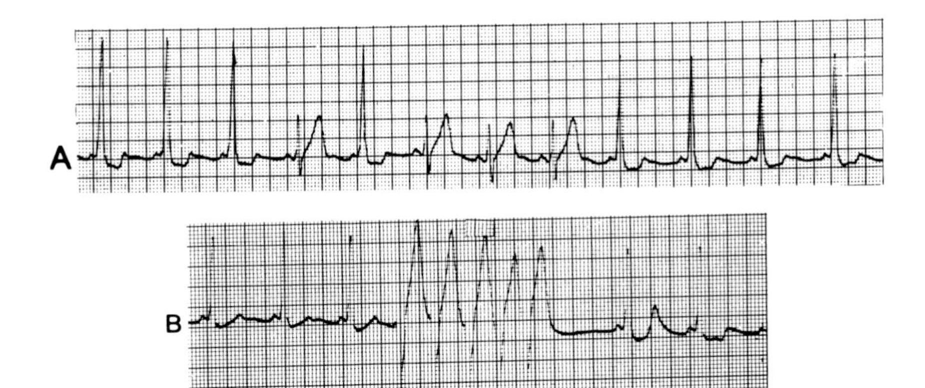

Figure 48. *Bout of probable ventricular tachycardia in a 60-year-old man, original syndrome.* **a.** *Intermittent pre-excitation.* **b.** *Bout of true ventricular tachycardia. Ambulatory ECG monitoring, dizziness and uneasy feeling during the bouts.*

Occurrence of Asymptomatic Episodes in Patients Without a History of Previous Symptoms

Some of Hindman's and of our cases exhibited recurrent premature beats during monitoring, but the patients expressed no complaints. Reactions seem to vary between patients; some are very alert, introspective and report every premature beat or sinus arrhythmia, while others are unaware of such minor events or do not pay attention to them. In Isaeff's series, all 5 patients noticed the attacks of supraventricular tachycardia.

Nature of Pre-excitation, Persistent or Intermittent, during Daily Normal Activity

Among our 215 pre-excitation patients, approximately 25% showed intermittency on routine ECG records (either in the same tracing or in two different tracings), and we assumed that the incidence of normalization (intermittency) is in fact much higher. This was borne out by Hindman's monitored series where no less than 18 of 27 patients (67%) demonstrated intermittent pre-excitation during the 24-hour period. The conduction normalized gradually with increasing heart rates in 9 of the patients, and suddenly in the other 9. Only 6 (20%) of our 30 monitored patients showed intermittency (Figure 48a), and this low figure may be explained by our monitoring period being only half as long as Hindman's. The percentage was similar to that found by us on routine ECG tracings.

Detection of Other Electrocardiographic Abnormalities

In 17 (63%) of Hindman's 27 patients, ectopic activity was common: 5 had ventricular premature beats, 6 atrial premature contractions and 6 both, meaning that in 12 of 17 cases the ectopic activity was supraventricular, while in 11 it was of ventricular origin. This finding supports our previous comments that not only is the incidence of premature beats much higher than in both normal and hospitalized populations, but most of the premature beats are of supraventricular origin in pre-excitation, and ventricular origin in the others. Among our patients we detected premature beats in only 6, but with one single lead system it was difficult to diagnose them as definitely supraventricular or ventricular in origin: atrial premature beats conducted through an anomalous electrical pathway can be very similar to ventricular premature beats. It is noteworthy that while premature beats are generally considered to be the trigger mechanism of the tachycardias in pre-excitation, not one of Hindman's cases developed a bout of tachycardia (including the 5 where antiarrhythmic medication was intentionally stopped before the test). This apparent contradiction, as well as the other arrhythmias discovered with the help of monitoring, will be discussed in Theories.

METHODS FOR DETECTION OF MECHANICAL PRECONTRACTION OF THE VENTRICULAR MUSCLE

It became clear that the most important feature of the pre-excitation syndrome—the early electrical depolarization of ventricular myocardium in unusual locations—depends on the site of the electrical bypass from the atria to the ventricles. This led to the hypothesis that the electrical phenomenon must produce a secondary early mechanical contraction of the myocardium in the same unusual location. While asynchronous electrical activation has been well documented, its influence upon mechanical events (and possibly hemodynamic changes) has not been clearly defined.[47-60] Many methods have been used to prove or define these premature contractions. The most common are phonocardiography,[61-72] apexcardiography,[64,73] electrokymography,[74] high speed cinematography,[75] roentgenkymography[76] and recently echocardiography.[77,78] The two invasive techniques of intracardiac ECG recording[79,80] are rarely used and are only mentioned here.

Ohnell[70] was the first to allude to the possible mechanical consequences of ventricular pre-excitation. A jugular phlebogram obtained during the transition from junctional rhythm without pre-excitation to sinus rhythm with pre-excitation showed some variations. Unfortunately, poor technical facilities and many artifacts lead one to question his results. Ohnell was also among the first to comment on the splitting of the second heart sound as a clue to local ventricular contractions.

Kossman and Goldberg[66] recorded phonocardiograms and carotid pulse tracings in a case of type B intermittent pre-excitation. During normal conduction the second sound was physiologically split with the pulmonic component following the aortic. During pre-excitation the second sound became single and the interval between the onset of the QRS complex and the upstroke of the carotid pulse tracing increased. The authors attributed this phenomenon to delayed left ventricular excitation; premature right ventricular excitation may also have contributed to the alteration in the second heart sound. March et al[68] used phonocardiography and hemodynamic tests to study 12 cases of pre-excitation, 3 with intermittency and 9 with fixed pre-excitation patterns. One of the 3, a case of type B pre-excitation, demonstrated paradoxical splittings of the second heart sound and an earlier rise in right ventricular pressure during pre-excitation (as compared to normal conduction). A second patient, with type A intermittent conduction, exhibited an early aortic valve closure as evidenced by a premature aortic component of the second heart sound and the early inscription of dicrotic notch on a carotid pulse tracing. The third case, type unspecified, did not show any phonocardiographic changes when conduction changed from normal to anomalous. Three patients without intermittency had a wide split second heart sound with delayed pulmonic components. Six others, also with diagnosed fixed pre-excitation, had normal phonocardiographic and carotid pulse tracings. Zuberbuhler and Bauersfeld[72] found a paradoxical splitting of the second heart sound in 3 of 4 patients with type B pre-excitation and attributed it to premature excitation of the right ventricle. The most extensive study on this subject was undertaken by Ishikawa et al,[64] who examined 9 type B and 11 type A cases with phonocardiography, apexcardiography and carotid pulse tracings. Some of the indicators tended to be shorter in these patients as compared to a group of normal controls, although the differences did not reach a level of statistical significance. Following normalization, these parameters were again measured and found to be the same as the controls.

Phonocardiographic studies in children performed by Rodriguez-Torres et al[71] and Memmer and Rautenberg[69] did not produce more definitive results than the studies described in adults. Ishikawa et al[64] summarized all these discrepancies in the phonocardiographic features, and concluded that the onset of some mechanical events was accelerated in pre-excitation, but to a much lesser degree than the acceleration of electrical events. The findings obtained in this way are simply an indirect index of cardiac action and do not always agree completely with actual mechanical events in the heart.

Prinzmetal et al[75] tried to find additional explanations for the puzzling and conflicting findings of phonocardiography in patients with the pre-excitation syndrome. While their method of producing artificial pre-excitation in animals may not be identical with the pathophysiology of human anomalous conductions, they felt that the mechanical consequences of both human and artificially produced fusion beats (pre-excitation) was very similar. Using high speed cinematography, they suggested a possible explanation of the conflicting findings in phonocardiography: "There is noted in the region surrounding the stimulating electrode a weak localized contraction of that part of the ventricle. This weak contraction does not open the semilunar valve and expel blood from the ventricle."[75] This simple observation, they stated, helped to explain the conflicting data already presented concerning the mechanical effects of pre-excitation. All of the techniques utilized in studying this problem were relatively insensitive to minor changes in the pattern of ventricular contraction. A "weak localized contracting" may not alter the jugular phlebogram or carotid upstroke tracings, since both techniques depend upon volume changes related to the displacement of blood. The timing and quality of the first[81] and second heart sounds[72] are intimately related to the motion of blood, and therefore may be unaffected by minor alterations in the sequence of ventricular excitation. The fact that some workers, using jugular phlebography, phonocardiograms, carotid artery pulse investigations, apexcardiography and other methods of investigation, have observed abnormalities in patients with pre-excitation, while others have not, suggests that the mass of ventricles undergoing pre-excitation varies among patients and "that a certain critical mass must be stimulated before abnormalities can be detected."[82]

The use of electrokymography[83] for timing aortic and pulmonary artery pulsations is subject to the same limitations.

Bandiera and Antognetti[76] circumvented the relative insensitivity of these techniques by using roentgenkymography to study the motion of the cardiac borders in pre-excitation. They were able to demonstrate areas of precocious contraction near the base of the left ventricle in type A pre-excitation and in the right ventricle with type B (in 8 of their 11 cases only). However, the persistence of these abnormalities at a time when ventricular pre-excitation disappeared (following intravenous procainamide administration) was not explained by the authors.

A recently introduced tool, echocardiography,[77,78,82,84-88] is capable of detecting alterations in ventricular wall motion rather than the effects of blood displacement. Several echocardiographic studies in patients with pre-excitation have been published.[77,78,82,84,85,88] All in all, including 51 of our 215 cases, 159 patients were investigated with this method: 55 belonged to Group A, and 104 to Group B, Rosenbaum et al's classification. The movements of the intraventricular septum were investigated in all the cases, while the movements of the posterior wall of the left ventricle were studied in only 3 series[77,78,82] and in our own 51 cases. Paradoxical movements of the interventricular septum were seen in 55 of the 104 type B cases, and in only one of the 55 type A cases. Anomalous movements of the posterior left ventricular wall were seen in 47 of the 55 type A cases, and in 11 of the 104 type B cases (partly as an isolated feature or in combination with interventricular abnormal movements). The rationale of this method of investigation is based on the assumption that where anomalous movements are observed, pre-excitation and precontraction must be present in the vicinity.

The number of investigations using echocardiography is too small to be able to draw any conclusions as to whether this method gives more precise information on precontraction areas in the ventricles in pre-excitation, and whether these asynchronizations in the pump of the ventricles have clinical effects on cardiac output and other hemodynamic parameters.

While most papers concentrate on precontraction and its probable location, a few studies not dealing with echocardiography do consider the hemodynamic consequences found during the tachycardia in the pre-excitation syndrome,[89,90] but they are probably no different from the ones encountered in regular tachycardia.

EXERCISE TESTS

Different forms of exercise are often used in medical examinations, including the Master 2-step exercise test, ergometer, treadmill, *etc.* In individuals with nonspecific chest pain or suspected anginal syndrome and normal ECG tracings, a characteristic depression of the ST-T segment during exercise indicates coronary insufficiency or ischemic response. Exercise tests in pre-excitation cases can reveal whether moderately heavy work triggers a paroxysmal tachycardia or possibly normalizes conduction. Pathological features which are usually masked by the pre-excitation QRS complexes, such as old myocardial infarction, ventricular hypertrophy, *etc*, may even be revealed.

Considering the practical clinical and basic scientific value of exercise tests, it is surprising how few investigations have been undertaken in this field. A review of the literature revealed only 9 reports.[5,91-98] Since the use of different forms of exercise and different criteria for interpreting the same methods made it difficult to compare the studies, we performed exercise tests on a large number of the Tel Hashomer patients.

Work Capacity and Physical Fitness

While many pre-excitation patients complain of fatigue and inability to complete their daily activities, it is questionable whether they really have a low work capacity or whether the complaints can be explained by iatrogenic influences or emotional fixation on their cardiac disorder. Ahlborg et al[91] found a normal work capacity in all 16 young male soldiers with pre-excitation who underwent the test; they attributed these results to the young age and homogeneity of their sample, compared with the hospital populations studied by others. Lars Sandberg,[96] on the other hand, found that only 8 of his 19 cases were within the normal range of work capacity, and the other 11 were below normal.

In our group of 178 patients with pre-excitation who were subjected to exercise tests, work capacity was evaluated in 157; the remainder were excluded due to the presence of organic cardiac diseases or because they were highly trained athletes. The mean capacity was found to be slightly lower than normal (81.9%, normal being 100%), with a score of 90% or more in the younger group and 60-75% in the older group. We concluded that there is a slightly lower than normal work capacity in almost all cases with pre-excitation (the same for men and women), the incapacity increasing with advancing age. This overall lower work capacity might be explained by the slight asynchronization of the normal sequence of ventricular contraction, with perhaps some hemodynamic consequences; in the older age group, additional atherosclerotic involvement of the cardiovascular system may account for the even lower work capacity.

ECG Signs of Coronary Insufficiency

Some pre-excitation patients complain of chest pains which are characteristic of anginal syndrome. When using exercise tests to evaluate the cause of this pain, the criteria for diagnosing coronary insufficiency must be the same for patients with and without pre-excitation (*ie,* a horizontal depression of the ST-T segment of more than 1 mm during or after exercise). Sandberg[96] administered an exercise test using the ergometer to a group of 5 pre-excitation patients over the age of 50, 4 of whom had had either a myocardial infarction or suffered precordial pains of an angina pectoris nature, and to a second group of 14 cases under the age of 29 with no complaints of chest pain and no suspected coronary heart disease. In all 19 cases the ECG tracings during exercise showed pronounced plateau-shaped depressions or inversions of T waves, characteristic of coronary disease, equally pronounced in both groups.

Sandberg's conclusion, subsequently confirmed by others, was that it is very difficult to establish the existence of coronary heart disease on the basis of the ST-T changes during an exercise test, in the presence of the pre-excitation syndrome. It is generally felt that these changes are a direct reflection of the basic conduction disturbances seen in this syndrome. In our own series, 94% of the 178 patients who underwent the exercise test showed ST-T changes characteristic of coronary insufficiency, without any preference for age or sex (Figure 49). In some cases

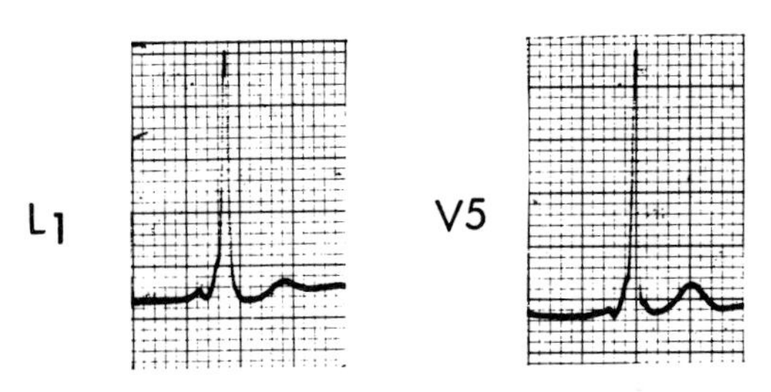

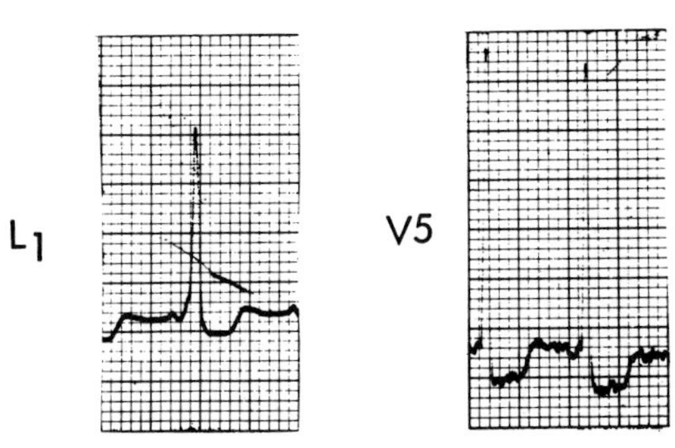

Figure 49. *False positive exercise test in a 17-year-old boy with pre-excitation. No complaints, no cardiac disease detected. Note the marked S-T depression in leads I and V_5 after exercise.*

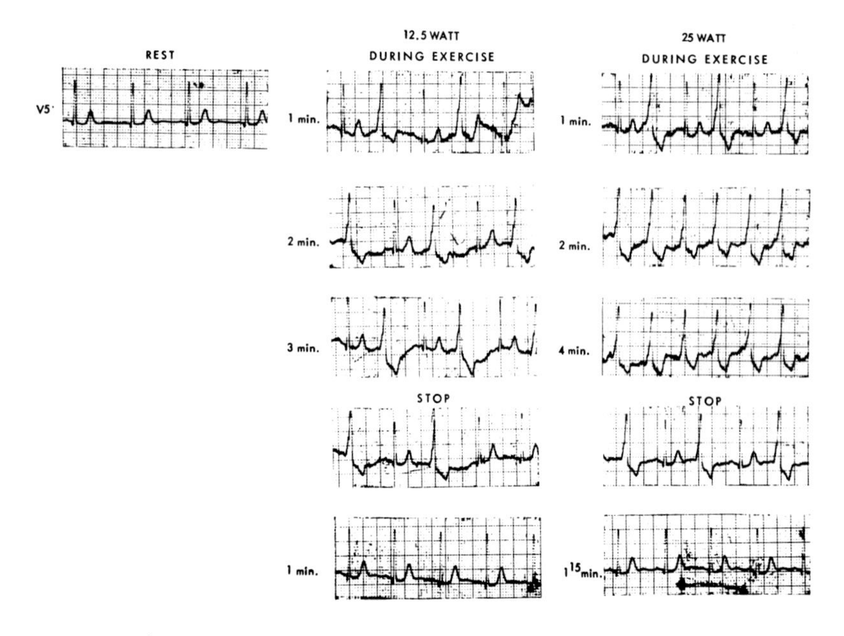

Figure 50. *Intermittent pre-excitation during exercise test in a 50-year-old man with pre-excitation. Normal ECG tracing at rest (lead V_5), appearance of pre-excitation during exercise, and normal AV conduction after rest (bottom).*

with intermittent pre-excitation, "pathological" changes were seen only in the beats with anomalous conduction, while positive T waves and no depression were observed in normally conducted beats (Figure 50).

Relationship between Effort and Attacks of Paroxysmal Tachycardia

Physicians sometimes instruct their pre-excitation patients to avoid physical exertion in order not to bring on an attack of paroxysmal tachycardia. The necessity for such a limitation can be assessed by exercise tests. In most papers no such tachycardia was mentioned, and we therefore deduced that it did not occur. This deduction was strengthened by the surprising finding that not one of our 178 patients (performing tasks of increasing difficulty and duration) developed an attack of paroxysmal tachycardia, neither during the test nor during the recuperation period. Some premature beats were observed in 7 patients (4%), but this is the same percentage as in the general population and lower than that observed by us in the entire pre-excitation series.

It can be stated that effort is not a trigger mechanism for paroxysmal tachycardia in the pre-excitation syndrome, and reduction of physical activity is not necessary unless indicated by other diseases or pathological conditions.

Physical Effort and Normalization of the Conduction

It became clear early in the investigations of pre-excitation that the strange ECG patterns can obscure other pathological ECG features essential for a correct diagnosis of the patient's disease. It was also known that tachycardia can induce intermittency, beginning with White's first patient in whom the ECG pattern normalized after he ran a long distance. In addition, most tachycardias seen in patients with the syndrome exhibited normal QRS complexes during the paroxysmal atrial tachycardia. Thus, it was hypothesized that normal conduction might be produced by inducing a marked tachycardia with an exercise test. The objective findings were very different from the expectations, however. Wolff and White,[98] using the Master 2-step test, succeeded in producing normalization in only one of 10 cases, while Master himself[99] failed completely. Averill et al[5] and Gazes,[95] also using the Master 2-step test, both observed normalization in only 2 cases, the former in 60 cases and the latter in 23 cases. A higher percentage of normalization was observed using ergometry (Sandberg in 8 of 28 cases, Czapliki in 3 of 22 cases), and it was assumed that the more advanced technique for exercising patients was the cause. However, our investigations using the ergometer proved the inaccuracy of this thesis, as only 17 out of 178 cases showed normalization (less than 10%).

Not only did the pre-excitation pattern disappear in a relatively small fraction of tested patients during exercise and tachycardia, in some cases the pre-excitation pattern appeared during exercise under similar conditions (Averill et al[5] in one case, and in 8 of our series (Figure 50)). It is noteworthy that the pre-excitation pattern appeared during exercise mainly in elderly patients in whom some degree of atherosclerotic heart disease could not be excluded. We would also emphasize that it is not always easy to differentiate between true and pseudonormalization; Ahlborg et al,[91] for instance, stated that in 8 of 16 cases they saw "partial or total normalization of their ECG" without being able to clearly separate the two groups.

Additional ECG Findings

The "partial normalization" or "pseudonormalization" observed in many of our patients during exercise can be summarized as follows: 1) a reduction in size of the delta wave which sometimes becomes biphasic;

2) a constant P-delta interval, not changed during exercise as compared with the pre-exercise tracing; 3) a narrowing of the QRS duration in many cases with a reduction of the delta-R and delta-J intervals; and 4) in some cases the appearance of a broad S wave after the R, suggesting some ventricular conduction disturbances (RBBB-like).

All these findings, appearing during sinus tachycardia, may look like the "concertina effect" described by Ohnell. However, his criteria of the "pulling out" and "pushing in" phenomenon, expressed mainly by changes in the P-delta interval, were not observed in our material. We prefer therefore to call this a "pseudo-concertina" effect.

Finally, we call attention to the appearance during exercise of a higher peaked P wave ("P pulmonale"). Since this finding was not present in all cases, we may exclude a secondary reaction to the sinus tachycardia (a "compression effect" of the P waves), and attribute it rather to a change in the usual atrial depolarization vector.

Body Surface Isopotential Mapping

Recently a new noninvasive method to determine the localization of the accessory pathway was suggested: body surface isopotential mapping. At present this approach is still in its embryologic stage, and only few reports have been published.[100-104] Eighty-five to 230 electrodes[100,101] are used to construct the isopotential maps. As with so many other methods, the lack of a unified technical approach and classification as well as the lack of correlation, in most cases, with surgical and/or pathological findings, do not permit us to draw conclusions as to the advantages of this method.

REFERENCES

1. **Donzelot E:** *Traité des Cardiopathies Congénitales,* p 1025, Paris: Masson, 1954.
2. **Addarii F, Lancellotti A, Labriola E:** Microelectrocardiographic (M-ECG) and microvectorcardiographic (M-VCG) examinations. I. M-ECG observations. (Ital) *Mal Cardiovasc* 10:759, 1969.
3. **Alboni P, Malacarne C, Tomasi AM:** Electro- and vector-cardiographic considerations on the Wolff-Parkinson-White pre-excitation. *G Ital Cardiol* 4:710, 1974.
4. **Alboni P, Rusconi L, Nava A:** Pathogenetic hypothesis on the left anterior block in cases of WPW. (Eng Abstr) *G Ital Cardiol* 3:846, 1973.
5. **Averill KH, Fosmoe RJ, Lamb LE:** Electrocardiographic findings in 67,375 asymptomatic subjects. IV. Wolff-Parkinson-White syndrome. *Am J Cardiol* 6:108, 1960.
6. **Ballarino M, Bruno L, Lehrer-Grego E:** Syndrome of ventricular pre-excitation (WPW) and branch block. Electrocardiographic and vectorcardiographic study. (Ital) *Cuore e Circul* 53:263, 1969.
7. **Belletti DA, Gould L, Lyon AF:** Vectorcardiographic features of Wolff-Parkinson-White syndrome with and without associated myocardial infarction. *Angiology* 21:5, 1970.
8. **Bilger R, So CS:** Über das Elektrokardiogramm und Vektorkardiogramm des Wolff-Parkinson-White syndrom. *Arch Kreislaufforsch* 37:111, 1962.

9. **Bilger R, So CS:** Über das Vektorkardiogramm und Elektrokardiogramm des Wolff-Parkinson-White syndrom. *Cardiologia* 42:189, 1963.
10. **Bleifer S, Kahn M, Grishman A, et al:** Wolff-Parkinson-White syndrome. A vectorcardiographic, electrocardiographic and clinical study. *Am J Cardiol* 4:321, 1959.
11. **Burch GE, DePasquale NP:** Electrocardiographic and vectorcardiographic detection of heart disease in the presence of the pre-excitation syndrome (Wolff-Parkinson-White syndrome). *Ann Intern Med* 54:387, 1961.
12. **Fernandez-Caamano F, Heller J, Lenègre J:** Association of Wolff-Parkinson-White type B with right branch block; electrovectorcardiographic study of a case by Fernandez. *Arch Mal Coeur* 60:571, 1967.
13. **Cabrera E, Feldman J, Olinto FC:** Estudio electro y vectorcardiográfico en un caso de Wolff-Parkinson-White con bloqueo de rama derecha. *Arch Inst Cardiol Mex* 29:404, 1959.
14. **Castellanos A Jr, Agha AS, Portillo B, et al:** Usefulness of vectorcardiography combined with His bundle recording and cardiac pacing in evaluation of the pre-excitation (Wolff-Parkinson-White) syndrome. *Am J Cardiol* 30:623, 1972.
15. **Castellanos A Jr, Mayer JW, Lemberg L:** The electrocardiogram and vectorcardiogram in Wolff-Parkinson-White syndrome associated with bundle branch block. *Am J Cardiol* 10:657, 1962.
16. **Cinotti G, Filorizzo GS, Cabani S, et al:** Vectorcardiographic aspects of the Wolff-Parkinson-White syndrome associated with branch block. (Ital) *Boll Soc Ital Cardiol* 14:36, 1969.
17. **Denes P, Goldfinger P, Rosen KM:** Left bundle branch block and intermittent type A preexcitation. *Chest* 68:356, 1975.
18. **Desiatov AI:** Vectorcardiographic characteristics and several problems in the pathogenesis of Wolff-Parkinson-White syndrome. (Rus) *Ter Arkh* 42:109, 1970.
19. **Disertori M:** Electro-vectorcardiographic study of a case of Wolff-Parkinson-White syndrome with a right branch block. (Ital) *Cardiol Prat* 22:429, 1971.
20. **Duchosal PW, Grosgurin JR:** *Atlas d'Electrocardiographie et de Vectorcardiographie. Etude du Ventriculogramme dans les Conditions Normales et Pathologiques.* Basel, New York: S Karger, 1959.
21. **Dyk T, Kozlowski W:** On the problems of administration of digitalis in Wolff-Parkinson-White syndrome. (Pol) *Pol Tyg Lek* 23:1280, 1968.
22. **Fischer DM, Gornacchia D, Jacopi F:** Electro and vectorcardiographic analysis of a case of LBBB associated with unstable type B WPW syndrome. (Eng Abstr) *G Ital Cardiol* 3:833, 1973.
23. **Gamboa R, Penaloza D, Sime F, et al:** The role of the right and left ventricles in the ventricular pre-excitation (WPW) syndrome. An experimental study in man. *Am J Cardiol* 10:650, 1962.
24. **Gandjour A, Stoermer J, Arnold A:** Vektorkardiographische Befunde beim WPW-Syndrom mit und ohne Schenkblock. *Z Kreislaufforsch* 59:577, 1970.
25. **Homola D, Nevrtal M:** The spatiocardiographic picture of the WPW syndrome. *Scr Med Fac Med Brunensis* 38:13, 1965.
26. **Koike T, Okumura M, Okajima T, et al:** Proceedings: Frank's vectorcardiography of Wolff-Parkinson-White syndrome without complications. *Jpn Circ J* 39:882, 1975.
27. **Kuppardt H, Israel G, Kuppardt B, et al:** Case report on WPW syndrome. (Ger) *Z Gesamte Inn Med* 27:748, 1972.
28. **Longhini C, Pinelli G, Martelli A, et al:** Polycardiographic study of ventricular pre-excitation. *Minerva Cardioangiol* 21:89, 1973.
29. **Lowe KG, Esmile-Smith D, Ward C, et al:** Classification of ventricular pre-excitation. Vectorcardiographic study. *Br Heart J* 37:9, 1975.
30. **Parameswaran R, Ohe T, Nakhjavan FK, et al:** Spontaneous variations in atrioventricular conduction in pre-excitation. *Br Heart J* 38:427, 1976.
31. **Polu JM, Gilgenkrantz JM:** Vectorcardiography in the Wolff-Parkinson-White syndrome. *Ann Cardiol Angeiol* (Paris) 21:335, 1972.
32. **Sakamoto Y, Hiroki T, Kunio S, et al:** Spatial velocity electrocardiograms in WPW

syndrome and complete left bundle branch block. *Jpn J Clin Med* 31:3094, 1973.

33. **Tranchesi J, Reisin LH, Pileggi F, et al:** The vectorcardiogram in the Wolff-Parkinson-White syndrome. *Arq Bras Cardiol* 17:89, 1964.
34. **Ueda H, Harumi K, Shimomura K, et al:** A vectorcardiographic study of WPW syndrome. *Jpn Heart J* 7:255, 1966.
35. **Warembourg H, Ducloux G, Flament G:** Wolff-Parkinson-White syndrome. 3. Vectorcardiography. (Fr) *Lille Med* 14:992, 1969.
36. **Warembourg H, Flament G, Ducloux G, et al:** Vectorcardiogram of the Wolff-Parkinson-White syndrome. Changes induced by ajmaline. Association with bundle branch block. *Ann Card Angeiol* (Paris) 25:315, 1975.
37. **Rosenbaum FF, Hecht HH, Wilson FN, et al:** The potential variations of the thorax and esophagus in anomalous atrioventricular excitation (Wolff-Parkinson-White syndrome). *Am Heart J* 29:281, 1945.
38. **Zao ZZ, Herrman GR, Hejtmancik MR:** A vector study of the delta wave in "nondelayed" conduction. *Am Heart J* 56:920, 1958.
39. **Tranchesi J, Guimaraes AC, Teixeira V, et al:** Vectorial interpretation of the ventricular complex in Wolff-Parkinson-White syndrome. *Am J Cardiol* 4:334, 1959.
40. **Grant RP, Tomlinson FB, Van Buren JK:** Ventricular activation in pre-excitation syndrome (Wolff-Parkinson-White). *Circulation* 18:355, 1958.
41. **Gallagher JJ, Svenson RH, Sealy WC, et al:** The Wolff-Parkinson-White syndrome and the preexcitation dysrhythmias. *Med Clin North Am* 60:101, 1976.
42. **Hejtmancik MR, Herrman GR:** The electrocardiographic syndrome of short P-R interval and broad QRS complex. A clinical study of 80 cases. *Am Heart J* 54:708, 1957.
43. **Knippel M, Pioselli D:** Pre-excitation syndrome studied by His bundle electrogram. (Eng Abstr) *G Ital Cardiol* 3:685, 1973.
44. **Hindman M, Last J, Rosen K:** The Wolff-Parkinson-White syndrome: clinical and electrocardiographic observations using continuous portable monitoring. *Circulation* 48 (suppl IV): 177, 1973.
45. **Isaeff DM, Harper Gaston J, Harrison DC:** Wolff-Parkinson-White syndrome. Long-term monitoring for arrhythmias. *JAMA* 222:449, 1972.
46. **Hindman MC, Last JH:** Wolff-Parkinson-White syndrome observed by portable monitoring. *Ann Intern Med* 79:654, 1973.
47. **Batalov Z:** Phasic analysis of the mechanical systole of the left ventricle in the syndrome of Wolff-Parkinson-White (preliminary report). *Folia Med* (Plovdiv) 8:264, 1966.
48. **Batalov Z:** Comparative investigations of the phase analysis between bundle-branch heart block and Wolff-Parkinson-White syndrome. *Folia Med* (Plovdiv) 11:343, 1969.
49. **Battro A, Braun-Menendez E, Orias O:** Asincronismo de la contracción ventricular en el bloqueo de rama: su demonstración mediante el registro optico de los fenomenos mecanicos de la actividad cardiaca. *Rev Argent Cardiol* 3:325, 1936.
50. **Beller BM:** Influence of abnormalities of ventricular excitation on right to left shun flows in atrial septal defect. *Am J Cardiol* 20:583, 1967.
51. **Cagán S, Caganová A:** Chronometry of cardiac systole in patients with the WPW syndrome. (Slo) *Cas Lek Cesk* 109:1109, 1970.
52. **Cagán S, Caganová A:** Chronometry of cardiac systole in patients with inconstant or compensated WPW syndrome. *Cor Vasa* 13:208, 1971.
53. **Cagan S, Caganová A:** Phases of the heart systole in persons with the WPW syndrome. (Eng Abstr) (Cz) *Bratisl Lek Listy* 56:456, 1971.
54. **Grassi T, Conte S:** Considerations on the hemodynamics in a case of alternating Wolff-Parkinson-White syndrome. Analysis of leads I and II of the arteriograms and apical cardiograms. (Ital) *Minerva Cardioangiol* 11:631, 1963.
55. **Jebavy P, Hurych J, Bergman K, et al:** Hemodynamics and deformation of the hemodilution curve in paroxysmal tachycardia in a patient with WPW syndrome. (Cz) *Cas Lek Cesk* 109:965, 1970.
56. **Pomerantsev VP, Korolenko AB, Kubyshkin VF:** An analysis of the cardiac cycle

phases in the Wolff-Parkinson-White electrocardiographic syndrome (according to polycardiographic and dynamocardiographic findings). (Rus) *Ter Arkh* 37:111, 1965.
57. **Rogel S, Berkoff H, Kaplinsky E:** Hemodynamic consequences of experimental ventricular pre-excitation. *Am Heart J* 67:516, 1964.
58. **Rozenblit J, Krauze T:** Parameters of heart dynamics in Wolff-Parkinson-White syndrome: a polycardiographic analysis of 10 cases. (Pol) *Pol Tyg Lek* 27:1890, 1972.
59. **Spangenberg JJ, Vedoya R, Gonzalez-Videla J:** Un caso de QRS ancho y mellado con PR acortado (sindrome de Wolff-Parkinson-White). Ausencia de asincronismo ventricular. *Rev Argent Cardiol* 4:244, 1938.
60. **Zakov N, Schleicher I:** Zur Haemodynamik des Herzens bei verkuerzter Ueberleitungszeit. *Z Kreislaufforsch* 35:413, 1943.
61. **Bergland JM, Rucker WR, Reeves JT, et al:** Pre-excitation as a cause of appearance and increased intensity of systolic murmurs. *Circulation* 33:131, 1966.
62. **Bortolan-Pirona G, Colό G, de Luca G, et al:** Phonocardiographic peculiarities in a case of infantile Wolff-Parkinson-White syndrome. *Minerva Pediatr* 27:113, 1975.
63. **Cohen SI, Lau SH, Haft JI, et al:** The intensity of the first heart sound in the Wolff-Parkinson-White syndrome. *Mt Sinai J Med* 37:17, 1970.
64. **Ishikawa H, Kagoshimai T, Hoshika Y, et al:** Wolff-Parkinson-White syndrome: mechanocardiographic study on the mechanical consequences of ventricular pre-excitation. *Am Heart J* 90:35, 1975.
65. **Ishikawa H, Kigawa Y, Masuda T, et al:** Phonocardiographic studies on Wolff-Parkinson-White syndrome. *J Nara Med Assoc* 23:491, 1972.
66. **Kossmann CE, Goldberg HH:** Sequence of ventricular stimulation and contraction in a case of anomalous atrioventricular excitation. *Am Heart J* 33:308, 1947.
67. **Libretti A, Schwartz PJ, Grazi S:** A cardiac murmur depending on the Wolff-Parkinson-White syndrome. *Am Heart J* 83:532, 1972.
68. **March HW, Selzer A, Hultgren HN:** The mechanical consequences of anomalous atrioventricular excitation (WPW syndrome). *Circulation* 23:582, 1961.
69. **Menner K, Rautenberg HW:** Apropos of asynchronous ventricular contraction in Wolff-Parkinson-White syndrome. (Ger) *Z Kreislaufforsch* 54:369, 1965.
70. **Ohnell RF:** Pre-excitation, a cardiac abnormality. *Acta Med Scand* (suppl 152), 1944.
71. **Rodriguez-Torres R, Yao AC, Lynfield J:** Significance of split heart sounds in children with Wolff-Parkinson-White syndrome. *Bull NY Acad Med* 44:511, 1968.
72. **Zuberbuhler JR, Bauersfeld SR:** Paradoxical splitting of the second heart sound in the Wolff-Parkinson-White syndrome. *Am Heart J* 70:595, 1965.
73. **Bodrogi G, Bereczky A, Kovacs G:** Mechanical consequences of the WPW syndrome. *Acta Cardiol* (Bruxelles) 21:145, 1966.
74. **Segers M, Hendrickx J:** Exploration electrokymographique d'un cas de WPW intermittent. *Acta Cardiol* 8:643, 1953.
75. **Prinzmetal M, Kennamer R, Corday E, et al:** *Accelerated Conduction. The Wolff-Parkinson-White Syndrome and Related Conditions.* New York: Grune & Stratton, 1952.
76. **Bandiera G, Antognetti PF:** Ventricular precontracting area in the Wolff-Parkinson-White syndrome. Demonstration in man. *Circulation* 17:225, 1958.
77. **DeMaria AN, Vera Z, Neumann A, et al:** Alterations in ventricular contraction pattern in the Wolff-Parkinson-White syndrome. Detection by echocardiography. *Circulation* 53:249, 1976.
78. **Hishida H, Sotobata I, Koike Y, et al:** Echocardiographic patterns of ventricular contraction in the Wolff-Parkinson-White syndrome. *Circulation* 54:567, 1976.
79. **Aravanis C, Lekos D, Vorides E, et al:** Wolff-Parkinson-White syndrome. Right ventricular precontracting area proved by cardiac catheterization. *Am J Cardiol* 13:77, 1964.
80. **Köhler JA, Sternitzke N:** On hemodynamics and the intercavitary electrocardiogram in the WPW syndrome. (Ger) *Z Kreislaufforsch* 56:26, 1967.
81. **Antlitz AM, Byers WS:** The Wolff-Parkinson-White syndrome and the first heart sound. Report of a case. *Dis Chest* 48:654, 1965.
82. **Gimbel KS:** Left ventricular posterior wall motion in patients with the Wolff-Parkin-

son-White syndrome. *Am Heart J* 93:160, 1977.
83. **Schumann G, Jansen HH, Anschütz F:** On the pathogenesis of the WPW syndrome. (Ger) *Virchows Arch (Path Anat)* 349:48, 1970.
84. **Chandra MS, Kerber RE, Brown DD, et al:** Echocardiography in Wolff-Parkinson-White syndrome. *Circulation* 53:943, 1976.
85. **Francis GS, Theroux P, O'Rourke RA, et al:** An echocardiographic study of interventricular septal motion in the Wolff-Parkinson-White syndrome. *Circulation* 54:174, 1976.
86. **McDonald IG, Feigenbaum H, Cheng S:** Analysis of left ventricular wall motion by reflected ultrasound. *Circulation* 46:14, 1972.
87. **Ticzon AR, Damato AN, Caracta AR, et al:** Echographic evaluation of the intraventricular septal motion during preexcitation and normal conduction in the WPW syndrome. (Abstr) *Clin Res* 23:210, 1975.
88. **Ticzon AR, Damato AN, Caracta AR, et al:** Intraventricular septal motion during pre-excitation and normal conduction in Wolff-Parkinson-White syndrome. *Am J Cardiol* 37:840, 1976.
89. **Ferrer MI, Harvey RM, Weiner MH, et al:** Hemodynamic studies in two cases of Wolff-Parkinson-White syndrome with paroxysmal AV nodal tachycardia. *Am J Med* 6:725, 1949.
90. **Jebavý P, Hurych J, Bergmann K, et al:** Haemodynamic changes during paroxysmal tachycardia in a patient with Wolff-Parkinson-White syndrome. *Br Heart J* 33:157, 1971.
91. **Ahlborg B, Atterhög JH, Ekelund LG, et al:** Pre-excitation in young men. Incidence, some anthropometric data and physical work capacity. *Acta Med Scand* 196:275, 1974. 196:275, 1974.
92. **Bordia A, Lodha SM:** Wolff-Parkinson-White syndrome effect of change of posture and of exercise in eight patients. *J Assoc Physicians India* 19:787, 1971.
93. **Drory Y, Sherf L, Fleishman F, et al:** Work capacity in Wolff-Parkinson-White syndrome. (Hebr) (Eng Abstr) *Harefuah* 84:433, 1973.
94. **Feil H, Brofman BL:** Effect of exercise on electrocardiogram of bundle branch block. *Am Heart J* 45:665, 1953.
95. **Gazes PC:** False-positive exercise test in the presence of the Wolff-Parkinson-White syndrome. *Am Heart J* 7:13, 1969.
96. **Sandberg L:** Studies on electrocardiographic changes during exercise tests. *Acta Med Scand* (suppl 365): 88, 1961.
97. **Willems D, Klepzig H:** Exertion ECG in bundle branch block and WPW syndrome with respect to the nitrate test. (Ger) *Z Kreislaufforsch* 59:315, 1970.
98. **Wolff L, White PD:** Syndrome of short P-R interval with abnormal QRS complexes and paroxysmal tachycardia. *Arch Intern Med* 82:446, 1948.
99. **Master AF, Jaffe HL, Dack S:** Atypical bundle branch block with short P-R interval in Grave's disease effect of thyroidectomy. *J Mt Sinai Hosp* 4:100, 1937.
100. **de Ambroggi L, Taccardi B, Macchi E:** Body-surface maps of heart potentials. Tentative localization of pre-excited areas in forty-two Wolff-Parkinson-White patients. *Circulation* 54:251, 1976.
101. **Yamada K, Toyama J, Wada M, et al:** Body surface isopotential mapping in Wolff-Parkinson-White syndrome: noninvasive method to determine the localization of the accessory atrioventricular pathway. *Am Heart J* 90:721, 1975.
102. **Cobb FR, Blumenschein SD, Sealy WC, et al:** Successful surgical interruption of the bundle of Kent in a patient with Wolff-Parkinson-White syndrome. *Circulation* 38:1018, 1968.
103. **Rossi L, Knippel M, Taccardi B:** Histological findings, His bundle recordings and body-surface potential mappings in a case of Wolff-Parkinson-White syndrome: an anatomoclinical comparison. *Cardiology* 60:265, 1975.
104. **Tatematsu H, Wada M, Okajima M, et al:** On-line conversional mode processing system for body surface mapping as designed for clinical application. An example: WPW syndrome. *Adv Cardiol* 10:20, 1974.

VI

Invasive Methods of Investigation

Since its "discovery", the pre-excitation syndrome has been a challenge to cardiologists—a Rosetta stone, whose deciphering might illuminate any number of electrophysiological events recorded in the human heart. Numerous theories were postulated for its interpretation, some of which were either strengthened or discarded with technological advances in cardiology. Now, after a decade of sophisticated investigations, questions are still emerging[1,2] and contradictory results are multiplying.[3,4] One outstanding feature of the progress in cardiology has been the possibility of surgery in cases of pre-excitation with intractable tachycardias, undertaken on the basis of invasive electronic methods of investigation. Surgery has failed to corroborate these findings in a number of cases, however, and surgical treatment has become a failure.

The invasive methods can be divided into two major groups: 1) intracardiac recordings and stimulations; and 2) epicardial mapping.

INTRACARDIAC RECORDINGS AND STIMULATIONS

Intracardiac recordings and stimulations include His bundle electrocardiography, and atrial and ventricular pacing. These methods are generally used in combination and the findings complement each other. A great number of papers on this subject have been published in the last decade.[1,3-76] The findings will be discussed according to the principal aims of the techniques.

Determination of the Type of Anomalous AV Connection

Since Holzman's and Scherf's classic paper was published in 1932,[77] it has been generally accepted that the only cause of pre-excitation is a conducting muscular bridge (Kent bundle) directly connecting the atrium with the ventricle, therefore shortcutting the passage of the electrical stimulus through the AV node and the rest of the normal conduction system. Attempts to explain the pre-excitation syndrome as a physiological dysfunction rather than a structural connection (Prinzmetal,[78] Sodi-Pallares[79]) were only short-lived, with most cardiologists soon returning to their previous belief. Outstanding pathologists used the most thorough methods of investigation and were unable to find a muscular AV bridge[80], but even this fact did not discourage the believers. In 1963 the "AV nodal bypass fibers of James"[81] were described and suggested as a possible normal AV nodal bypass route for the electrical stimulus descending from the atria to the ventricles. With the introduction of intracardiac registration and pacing methods, it finally became completely clear that there can be more than one single explanation for pre-excitation and "short P-R normal QRS syndromes". This was an important factor in many decisions regarding surgery.[82]

Intracavitary electrocardiograms from the specialized conduction system, recorded simultaneously with surface leads during both sinus rhythm and atrial pacing at increasing rates, reveal 4 types of ventricular pre-excitation[13] and "short P-R normal QRS syndromes." In order to classify a case, the P-R interval (as measured conventionally) is divided into three subgroups: 1) *the P-A interval* (normal values 20-40 msec), which gives a rough measure of conduction time from the sinus node to the low right atrium in the vicinity of the AV node; 2) *the A-H interval,* which in normal healthy people is 55-120 msec and gives an estimate of AV nodal conduction time; and 3) *the H-V interval* (35-55 msec) which is the value of the His ventricular conduction time.[10] The normal response to atrial pacing at increasing rates is a progressive prolongation of the stimulus–ventricular interval, at the expense of the A-H interval without any changes in the H-V interval[13] (Diagram III). The 4 types of ventricular pre-excitation (and one of "short P-R normal QRS") are described below.

1. A *direct AV bypass outside the conduction system* (Kent bundle in Rossi's nomenclature, RAB, LAB and SAB in our Table XVI; see Diagram IV). In this group a muscular accessory bundle completely outside the regular conduction system is assumed to exist (Kent bundle). Thus, the electrical stimulus descending from the sinus node divides into two depolarization fronts. One passes through the accessory bundle, short-circuits the normal slowing down of the conduction in the AV node and arrives at an unusual site in the ventricles' myocardium much earlier than usual, producing the characteristic short P-R interval and the delta wave. With intracardiac registration it manifests in a short A-V (low atrium-delta wave) interval but a normal A-H interval (low atrium-His spike). This phenomenon is due to the fact that the second depolarization front of the electrical stimulus passes through the AV node, where it undergoes the normal slowing of conduction. Therefore, a

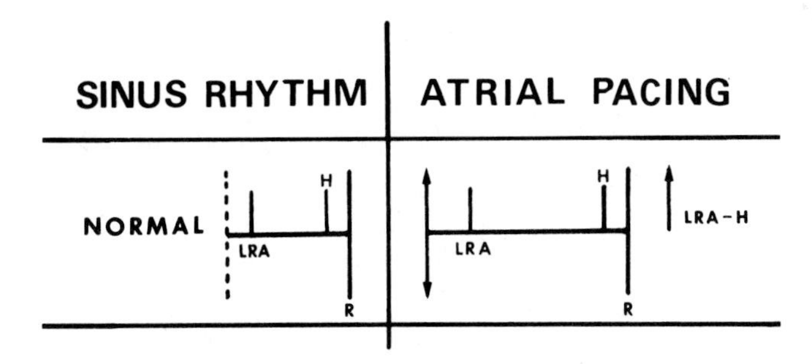

Diagram III. *Effects of atrial stimulation in patients without pre-excitation. Broken vertical lines indicate the beginning of atrial depolarization. Bidirectional arrows represent the stimulus artifact delivered to the right atrium. LRA: low right atrium; H: His bundle deflection; R: beginning of ventricular depolarization. (Modified from Castellanos et al.[10])*

third characteristic feature appears—namely, that the H deflection and the delta wave are inscribed almost simultaneously. During atrial pacing at critical rates the A-V interval (delta) does not change, because accessory bundle transmission time remains the same. On the other hand, the normal AV nodal delay appears and, consequently, the A-H interval increases so that the H deflection appears progressively after the beginning of ventricular depolarization. The inscription of a forward H deflection within the QRS complex is the hallmark of accessory bundle conduction. While this occurs, more and more ventricular muscle is activated through the accessory bundle,[13] resulting in a proportional prolongation (and "distortion") of the ventricular complexes until there is exclusive accessory bundle conduction.[13]

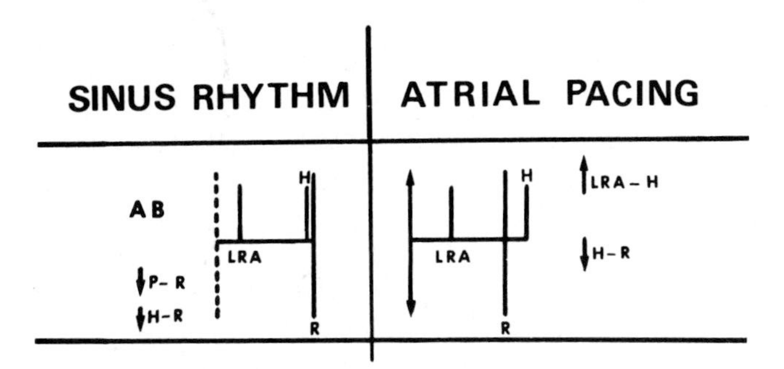

Diagram IV. *Effects of atrial stimulation in patients with an accessory bundle directly connecting the atrium with the ventricle. AB: accessory bundle (see Diagram III). (Modified from Castellanos et al.[10])*

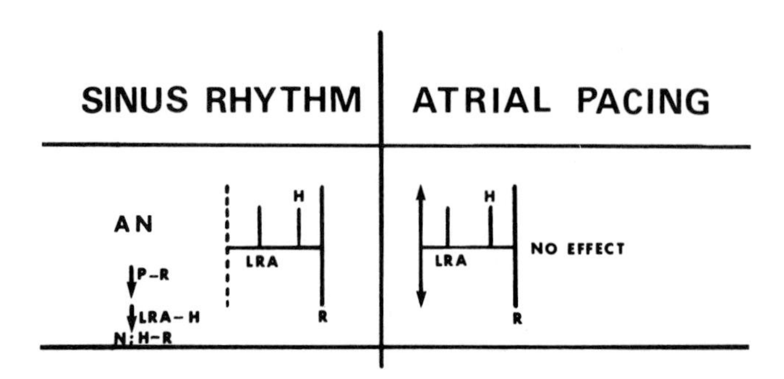

Diagram V. *Effects of atrial pacing in patients with actively conducting AV nodal bypass fibers (James' fibers). AN: conducting James' fibers (see Diagram III). Modified from Castellanos et al.[10])*

2. *A mediated bypass through the conduction system—inlet connections* (James' fibers in Rossi's classification, AN in our own) (Diagram V). Since no delta wave is produced, this second group does not belong to the "pre-excitation syndrome", but it is important for the understanding of all the short P-R normal QRS syndromes and for didactic purposes. The bypass fibers described by James[81,83] as a part of the posterior internodal tract (PIT) are assumed to carry the electrical stimulus earlier than usual and before the anterior (AIT) and middle internodal tracts (MIT); normally, the depolarization order is the opposite: AIT–MIT–PIT. What functional (or structural) factor produces this reversal in the order of atrial conduction sequence is, at present, still a matter of speculation. In cases of this type, during sinus rhythm a short P-R interval and normal QRS complex are inscribed on the surface ECG. The electrical stimulus passes first through the atrial–low AV nodal connection, thereby bypassing all or most of the AV node before proceeding downstream through the rest of the normal conduction system. With intracardiac registrations, the consequence of this AV nodal bypass phenomenon is a short A-H time together with a normal H-V interval. During atrial pacing the A-H interval does not increase as usual (or increases only slightly) and remains at more or less fixed values.[13] It is important to emphasize that not all patients with short P-R intervals and narrow QRS complexes have AV bypasses of the James' fibers type.[13] On the other hand, when a primary sinus node–low AV nodal conduction over the James' bypass fibers is established in a case where intraventricular conduction disturbances coexist, a puzzling surface ECG may result since the short P-R interval of this type is not followed by the normal narrow QRS, but rather by wide QRS complexes (sometimes even mimicking findings of pre-excitation).[10,13]

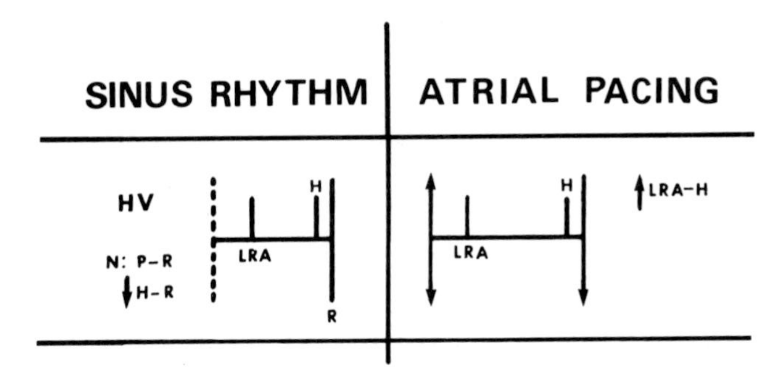

Diagram VI. *Atrial stimulation in patients with Mahaim paraspecific fibers. HV: connection between His bundle and ventricular septum (see Diagram III). (Modified from Castellanos et al.[10])*

3. *A mediated bypass through the conduction system—outlet connections* (Mahaim fibers in Rossi's classification, NV or HV in our own) (Diagram VI). Myocardial connections are assumed to exist between the lower AV node or the His bundle and the ventricular muscle. These are the paraspecific fibers described first by Mahaim.[84] Their existence and conduction through them may be suspected when a normal P-R interval in a regular ECG tracing is followed by a characteristic pre-excitation QRS complex with a clear delta wave. His bundle electrograms reveal a normal A-H time (since the conduction in these cases is exclusively through the AV node), but a shortened H-V interval. (The electrical stimulus reaches the ventricular myocardium ahead of time because the conduction is through the anomalous His bundle-ventricular connections.) During atrial pacing the A-H interval gradually becomes prolonged, as expected in cases where the electrical stimulus is passing through the AV node. The H-V remains the same during pacing, obviously, but the entire A-V interval becomes prolonged. Neither the morphology nor the QRS duration in these cases are affected by electrical stimulation of the atria.

4. A *combination of 2 and 3* (Diagram VII). This type of pre-excitation is produced when electrical conduction along the AV nodal bypass fibers (James' fibers) takes place together with the continuation of the stimulus conduction through an anomalous His bundle-ventricular connection. In these cases the surface ECG is indistinguishable from that of an accessory bundle (AB) completely outside of the normal conduction system (Kent bundle)[85]: the P-R interval is short and the QRS characteristic of pre-excitation. During intracardiac registrations, however, conduction time from the low atrium to the His bundle (A-H interval) is short because the AV node is bypassed. The decrease in the duration of

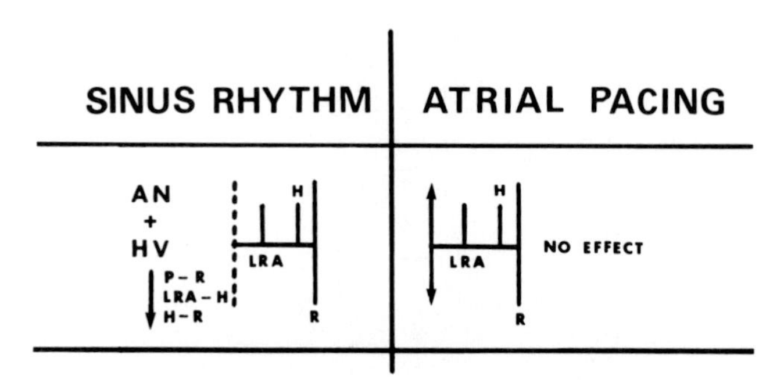

Diagram VII. *Atrial pacing in patients with both conducting AV nodal bypass fibers and paraspecific Mahaim connections. AN: conducting James' fibers; HV: conducting paraspecific fibers (see Diagram III). (Modified from Castellanos et al.[10])*

the H-V interval and the initial delta wave are a consequence of the His bundle-ventricular communication.[13] Atrial pacing does not prolong the A-H interval since the impulse is not delayed in the AV node. This type of pre-excitation can be further differentiated from the accessory bundle type by two additional findings: 1) the QRS complex is maximally distorted during sinus beats and does not change in morphology or duration over a wide range of atrial paced heart rates; and 2) the H spike is always located before the beginning of the ventricular depolarization and never after it.

5. *Multiple accessory bundles or combinations of group 1 with 2 and 3.* Most of the above described electrophysiological investigations are used to select patients with life threatening tachycardias for cardiac surgery. The rationale for surgery is to stop the tachycardia by sectioning an accessory bundle which lies completely outside the normal conduction system, thereby interrupting the "circus movement" of the tachycardias. It is clear that only cases falling into group 1 are candidates for this kind of surgery, while those in groups 2, 3 and 4 are not suitable. In some cases multiple accessory pathways may be present,[85,86] or a combination of an accessory pathway with an additional conducting atrium–low nodal or His bundle–(AN or AH) ventricular connection may coexist.[8] In one series,[85] for instance, among 135 patients with pre-excitation, electrophysiological evidence for multiple anomalous conducting AV pathways was demonstrated in 20. Five had two distinct accessory AV connections, 12 had a single accessory connection associated with "enhanced AV conduction" (corresponding to our groups 2 and 4 or James' bypass fiber conduction), and one had an additional His bundle-ventricular connec-

tion (our group 3, Mahaim fibers conduction). Thirteen of these 20 patients underwent surgery. Intraoperative mapping studies confirmed the preoperative predictions in 11. In 2 patients, however, even these sophisticated investigations were not sufficient and the presence of a secondary accessory AV connection was documented only after the ablation of the first.[85] Findings of multiple conducting AV pathways were also described by others.[8] Practically every electrophysiological study which demonstrates an accessory bundle (our group 1), combined with the unusual finding of a short A-H and/or V-H interval, should arouse suspicion of multiple pathways.

Location of the Accessory Pathway

Knowledge of the type of AV bypass pathway is far from sufficient to allow surgeons to operate in cases of intractable tachycardias which show a well documented accessory bundle completely outside the normal conduction system. The exact location of the accessory bundle in the heart must be defined. Electrocardiographic anatomic correlations have generally demonstrated accessory muscular connections between the right atrium and ventricle in the hearts of patients with type B ECG[81] and left-sided connections in patients with type A ECG. Despite these correlations, Gallagher et al[3] stated that in their series of operated patients the ECG was useful as a first approximation of the site of pre-excitation, but did not allow discrimination of free wall accessory pathways (18 patients) from septal accessory pathways (11 patients). Additional studies are thus required to locate the accessory pathway prior to surgical intervention. Sometimes a conventional ECG can even be misleading. In a case of Wellens',[1] for instance, the ECG was classified as pre-excitation type B; but introducing atrial premature beats with increasing prematurity led to an increasingly larger ventricular area being excited over the accessory pathway, and it became clear that type A pre-excitation and not type B was present. In order to determine the exact site of the anomalous conduction, the standard ECG should be combined with His electrograms during atrial pacing. Electrograms should be recorded from the coronary sinus, and determination made of the site of initial atrial retrograde excitation during supraventricular tachycardias. Some details of these techniques are as follows:

When atrial pacing is performed close to an assumed AV bypass, the pre-excitation area will be larger than during pacing far away from the bypass.[1]

Left and right ventricular pre-excitation can be distinguished by recording activity from the base of the left ventricle during pre-excitation, via a catheter electrode in the distal coronary sinus. Such records have consistently revealed left ventricular activity closely coinciding with the onset of the delta wave in patients with type A and left ventricular pre-excitation. In patients with type B, activity from the base of the left ventricle was recorded well after the beginning of the delta wave (usually during the terminal phases of the QRS complex).[3]

Still another method for determining the exact location of an accessory bundle is to study the pattern of retrograde atrial activation during supraventricular tachycardia.[3,88] Simultaneous electrograms from the low lateral right atrium, the low septal right atrium (via the His bundle catheter) and the lateral left atrium (via a coronary sinus catheter) are recorded during paroxysmal atrial tachycardia. Since atrial reactivation in this situation proceeds from the point of insertion of the accessory pathway into the atrium (after having depolarized the ventricles), identification of the earliest point of retrograde atrial activation furnishes the presumed location of the pathway.

In normal subjects, the earliest retrograde atrial activation occurs on the low septal right atrium (His bundle catheter location), simultaneous with, or followed closely by, activation of the os of the coronary sinus. This information can be gained by continuous intracardiac registration during withdrawal of a coronary sinus catheter from its distal position back to the orifice. In patients with an accessory bundle between the atria and ventricles, retrograde excitation of the atria over this anomalous pathway changes the normal sequence of atrial retrograde activation and results in an eccentric atrial depolarization during paroxysmal atrial tachycardia (PAT).[3]

Similar information can be obtained when introducing single premature ventricular beats or continuous ventricular pacing (but not producing paroxysmal atrial tachycardia, PAT). While normally, retrograde activation of the atrium is initiated in the atrial septum and spreads laterally to both atria, in cases of laterally situated accessory bundles the first atrial point of activation after the ventricular stimulation will be eccentric, as it was with PAT.[4]

If, however, the accessory pathway is located in the septal area and not laterally, the situation is much more complicated. In these cases special efforts must be made to exclude the possibility of a conventional AV nodal reentry, since they both have very similar starting points of retrograde atrial activation.

Finally, pre-excitation is never manifested (no delta wave) when an assumed accessory pathway is capable of only retrograde conduction. Demonstration that an accessory pathway is the cause of any paroxysmal supraventricular tachycardia in this situation requires intensive electrophysiological investigations (concealed pre-excitation).

Participation of the Assumed Anomalous Pathway in the Tachycardias

The most common tachycardia encountered in the pre-excitation syndrome is a paroxysmal supraventricular tachycardia. Gallagher et al[3] found that this can be induced by appropriately timed premature atrial or ventricular beats. The rhythm produced is called paroxysmal atrial tachycardia (PAT) when the QRS complex is preceded by a His bundle deflection at an interval of at least 30 msec, and each QRS complex is associated with a single atrial depolarization. In most patients the initiation of tachycardia is typically related to a "block" of a premature beat in the accessory bundle: this happens when the effective refractory period of the accessory bundle is long compared to that of the AV node, so that the initiating premature beat may propagate through the AV node. After activation of the ventricles, re-excitation of the atrium occurs over the accessory bundle. Continuation of this pattern results in classical PAT. This "circus movement" of the electrical stimulus was considered for a long time to be the only mechanism involved in the production of PAT in pre-excitation.

Recently, however, it was demonstrated that the accessory bundle does not necessarily have to be incorporated in the tachycardia circuit; the arrhythmia can be confined to the atrium, the ventricles[2,89] and especially to the AV node.[1,90] In this latter kind of tachycardia, longitudinal dissociation occurs within the AV node with development of two functional pathways, each with its own conduction properties. It was postulated that the premature stimulus inducing the tachycardia was blocked in one AV nodal pathway, which recovered during slow conduction in the other pathway and was available for retrograde conduction. Episodes of reentrant tachycardia could then occur using one AV nodal pathway as an antegrade route and the other as a retrograde route.[53] The chief practical problem for the investigator, clinician or surgeon is to distinguish tachycardias in which the accesssory pathway is included in the circuit, from those with a reciprocating mechanism within the AV node. This is of major importance both for the basic understanding of the syndrome and its rhythm disturbances and for decisions regarding treatment, such as surgery in cases with life threatening tachycardias.

The two forms of PAT can be differentiated as follows: if the tachycardia can be induced by a premature beat which conducts with no special antegrade conduction delay, then a bypass tract is most probably involved in the reentry circuit; on the other hand, if PAT occurs at critical AH prolongation intervals or AV nodal Wenckebach induced by pacing, then AV nodal reentry is the operating mechanism.[3] Another observation which favors the participation in the PAT of an accessory pathway is the demonstration of an eccentric retrograde activation during the tachycardia. When, however, the retrograde atrial activation is near the atrial septum, it is difficult to distinguish retrograde accessory bundle conduction in the septum from retrograde conduction over the His-AV node system. In such cases, when the retrograde His spike appears after atrial depolarization or ventricular premature depolarization during PAT can be induced, which retrogradely re-excites the atria without disturbing the preceding His deflection, an accessory pathway participation is suggested. Slowing of PAT associated with the appearance of a bundle branch block suggests also that a bypass tract, located in the ventricle showing evidence of block, is participating in the PAT.

In the case of atrial flutter fibrillation, the accessory pathway is more readily implicated, since rapidly conducted beats manifest pre-excitation if conducted over an accessory bundle.[3]

Determination of Effective Refractory Period of Anomalous AV Connection, Compared to Normal Conduction System

The effective refractory period (ERP) as defined by Gallagher et al[4] is the measurement of the time required for excitable tissues which have been depolarized to again become excitable. It is generally determined by introducing progressively earlier premature beats during a basic rhythm, until a coupling interval is found at which the premature beat fails to propagate through the structure being tested. Some investigators of the pre-excitation syndrome undertook to establish the ERP of the assumed accessory pathway as compared to that of the AV node. In contrast to the AV node, where the ERP lengthens with increasing driving rates, it usually shortens in the accessory bundle. This fact is of great importance because it shows that the AV node takes on a protective role toward the ventricles during driving, while the accessory bundle may lack this protective capability.[1,91] Based on this reasoning, Durrer[92] presented a methodology in which determination of the antegrade refractory period of an accessory bundle also allowed prediction of the maximum rate at which the accessory pathway will conduct impulses to the ventricle during atrial fibrillation.[4] Others[15,16] used this method for the same purpose, with discrepancies in the results. For instance, Gallagher et al[3] found that the correlation of the shortest antegrade ERP of the accessory pathway with the cycle length of the ventricular response during spontaneous atrial fibrillation was not as strong as that reported by Wellens and Durrer.[93] "Usually, long ERP's have been associated with moderate rates of ventricular response, while short ERP's have been associated with rapid rates. However, some patients with short ERP's developed only moderately fast ventricular rates during atrial fibrillation with most beats conducting normally."[3] Gallagher et al felt that the explanation may lie in the phenomenon of concealment in the accessory pathway.

It is possible to experimentally induce rapid atrial fibrillation by rapid atrial stimulation in patients with a short ERP of the accessory pathway. However this procedure is not without danger; direct current cardioversion was reported necessary in three instances due to a ventricular fibrillation.[3] An additional method for measuring ERP was introduced by Wellens and Durrer[69] in 1974: by using the single test stimulus technique during ventricular pacing, the patterns of ventriculoatrial conduction were investigated. They described four VA conduction patterns: 1) via the accessory pathways only; 2) via both the accessory pathway and the His–AV nodal axis; 3) via the H-AV pathway only; and 4) absence of any VA conduction. This kind of ventricular pacing yielded additional information on the electrophysiological behavior of the two different pathways (the normal and the anomalous) in cases with pre-excitation, which was used for clinical assessment. This technique also allowed detection of cases with an assumed accessory bundle but without signs of pre-excitation.[47,70]

Assessment of the Effect of Drugs

Programmed electrical stimulation also provides the possibility of directly studying the effect of drugs on the electrophysiological properties of the two atrioventricular pathways.[1] This aspect should be investigated only after establishing some basic factors, such as the existence of congenital or acquired heart diseases (which by themselves may provoke or intensify tachycardias), the type of tachycardia (especially "circus movement" tachycardia *versus* AV nodal reentrant tachycardia), and the ERP's of the normal and anomalous pathways. Studies have shown that drugs like procainamide, quinidine and ajmaline, and probably also lidocaine tend to increase the ERP of an accessory bundle but have no effect on that of the AV node.[1,2,94–97] In contrast, digitalis, beta-blockers and verapamil have a positive effect on the ERP of the AV node, but not on the accessory bundle.[1,2,94-97] It has also been shown that digitalis can be a very dangerous drug in patients with a short refractory period in the accessory bundle because it may further shorten the refractory period and produce a very rapid ventricular response, which in turn may induce ventricular fibrillation (Chapter VIII, Treatment).

Identification of High Risk Patients

Wellens[1,93] recently described a new approach for the identification of high risk patients: those having a proven accessory pathway with a very short refractory period in which atrial fibrillation or flutter can become a life threatening tachycardia.[3,4,98] The method consists of determining the length of the ERP of the accessory pathway (by a single test stimulus), and comparing it with the ventricular rate during spontaneous (or experimentally induced) atrial fibrillation. Wellens et al[1,93] reported a high correlation between the length of the ERP of the accessory pathway and the shortest R-R interval and mean ventricular rate during atrial fibrillation. As previously mentioned, other investigators,[3] although confirming Wellens et al's findings, were not able to demonstrate such a good correlation. In principle, it is now accepted that high risk cases are considered those which exhibit attacks of atrial fibrillation and/or flutter.

Postoperative Evaluation

Two surgical techniques for patients suffering from severe disabling and life threatening tachycardia are used at present: sectioning of an accessory bundle or sectioning of the His bundle. In both, postoperative investigation and reassessment of the findings are essential.[1] When a successful interruption of the accessory pathway results in a complete cure, evidenced by disappearance of the pre-excitation patterns in the ECG tracings, the patient should be reevaluated postoperatively since sometimes the accessory pathway which is functioning in sinus rhythm might not be the one active during tachycardias (AV nodal reentry tachycardias). In other patients in whom the AV node–His bundle was proven to be part of the "circus movement" tachycardias, section of the His bundle may also result in a relief from tachycardia.[99-101] After this kind of surgical intervention, postoperative investigation demonstrates: 1) an inability to initiate any more "circus movement" tachycardia by a premature beat; and 2) no changes in the pre-excitation QRS configuration following atrial pacing. This latter finding suggests that ventricular excitation continues by way of the accessory bundle only.

EPICARDIAL MAPPING

Numerous studies on epicardial mapping have been published.[3,45,60,67,71,75,85,99-126] In 1967 Durrer and Roos[110] published the case of a 21-year-old woman who underwent surgery to correct an atrial septal defect and type B pre-excitation. Their treatment was based on years of working with a technique for determining the activation sequence over the epicardial surface in experimental animals. They noted that the earliest point of ventricular activation was recorded at the right lateral portion of the AV groove, and that this electrical activity closely followed the P wave and occurred within the first 10 msec of the delta wave.[67,110] Later in the same year, Burchell et al[106] performed epicardial mapping in a patient with type B pre-excitation; pressure over the AV groove, where the earliest activation occurred, interrupted the attacks of tachycardia, and an injection of procainamide in the same area abolished the delta wave. They then made the first attempt to divide an assumed accessory bundle in the light of epicardial mapping findings. Unfortunately there was no clinical improvement, but the way was clear for cardiac surgery in cases with pre-excitation and intractable tachycardias, and the important role of epicardial mapping prior to surgical intervention was established. A year later Cobb et al[107] reported the first successful interruption of an accessory bundle. Since then more than 80 patients have undergone surgical treatment and, in almost all of them, epicardial mapping was used to establish the point of ventricular pre-excitation (see Chapter VIII, Treatment, Surgery). The reference points from which to time the ECG recorded by the exploring electrode have varied among investigators. Cobb et al[107] used surface P waves, Lister et al[37] and Wyndham et al[75] used an atrial electrogram, while others, Neutze et al[118] and Lindsay et al,[116] took an arbitrary point in the surface QRS complex.

The method most commonly used today is that described by Gallagher et al[3] from Duke University, where over half of all the published surgical interventions for tachycardias in pre-excitation have been performed. The epicardial mapping is performed after cannulation for cardiopulmonary bypass. Maximal pre-excitation is achieved by pacing the atrium at a site close to the presumed bypass tract. The precise site is determined by preoperative intracardiac studies and by slowly passing a stimulating probe around the atrioventricular rings until maximum pre-excitation is noted. A plaque electrode is then sutured to the ventricle as close as possible to the site, and serves as a reference for timing and ventricular pacing. Activation is explored by measurements of approximately 55 epicardial sites and the data plotted on drawings of the heart, yielding the point of earliest epicardial activation. According to Gallagher et al,[3] the correlation between the timing of this event and that of the delta wave in the conventional ECG has proved valuable in localizing the accessory pathway: in patients with free wall AV connections, the earliest epicardial activation occurred before the onset of the delta wave, while in cases with septal connections the area of earliest epicardial activation occurred 5-15 msec after the onset of the delta wave.

Upon completion of the antegrade map, Gallagher et al[3] used ventricular pacing through the ventricular electrode and mapped the adjacent ventricular and atrial points around the atrioventricular ring in order to determine the earliest point of retrograde atrial excitation. In all cases with free wall connections, antegrade and retrograde studies corresponded closely.

With all these thoroughly planned approaches and sophisticated techniques, we are still unable to solve all the problems of the exact location of the AV bypass. The Duke University group[85] recently reported that out of 13 patients who underwent surgery, the intraoperative mapping was confirmed in only 11 of them. In the other 2 cases, additional accessory bypasses, previously unsuspected, were revealed after the division of the presumed AV connection.

REFERENCES

1. **Wellens HJ:** Contribution of cardiac pacing to our understanding of the Wolff-Parkinson-White syndrome. *Br Heart J* 37:231, 1975.
2. **Wellens HJ, Durrer D:** Effect of procaine amide, quinidine, and ajmaline in the Wolff-Parkinson-White syndrome. *Circulation* 50:114, 1974.
3. **Gallagher JJ, Gilbert M, Svenson RH, et al:** Wolff-Parkinson-White syndrome. The problem, evaluation and surgical correction. *Circulation* 51:767, 1975.
4. **Gallagher JJ, Svenson RH, Sealy WC, et al:** The Wolff-Parkinson-White syndrome and the pre-excitation dysrhythmias. *Med Clin North Am* 60:101, 1976.
5. **Akhtar M, Damato AN, Lau SH, et al:** Clinical uses of His bundle electrocardiography. Part III. *Am Heart J* 91:805, 1976.
6. **Bisset JK, de Soyza N, Kane JJ, et al:** Altered refractory periods in patients with short P-R intervals and normal QRS complex. *Am J Cardiol* 35:487, 1975.
7. **Castellanos A:** Letter: Accelerated AV-conduction and complete A-V block. *Am Heart J* 89:545, 1975.
8. **Castellanos A, Agha AS, Befeler B, et al:** Double accessory pathways in Wolff-Parkinson-White syndrome. *Circulation* 51:1020, 1975.
9. **Castellanos A Jr, Agha AS, Portillo B, et al:** Usefulness of vectorcardiography combined with His bundle recording and cardiac pacing in evaluation of the pre-excitation (Wolff-Parkinson-White) syndrome. *Am J Cardiol* 30:623, 1972.
10. **Castellanos A, Aranda J, Agha AS, et al:** "Mechanisms of A-V conduction in ventricular pre-excitation," in *Advances in Electrocardiography* (Schlant RC, Hurst JW, Eds), vol 2, p 173, New York: Grune & Stratton, 1976.
11. **Castellanos A Jr, Castillo CA, Agha AS, et al:** Functional properties of accessory AV pathways during premature atrial stimulation. *Br Heart J* 35:578, 1973.
12. **Castellanos A Jr, Castillo CA, Cunha D, et al:** Value of His bundle recordings in the evaluation of the pre-excitation (Wolff-Parkinson-White) syndrome. *Proc K Ned Akad Wet (Biol Med)* 75:402, 1972.
13. **Castellanos A, Castillo CA, Martinez A, et al:** "Mechanisms of AV conduction in ventricular pre-excitation," in *Advances in Electrocardiography* (Schlant RC, Hurst JW, Eds), p 249, New York: Grune & Stratton, 1972.
14. **Castellanos A Jr, Chapunoff E, Castillo C, et al:** His bundle electrograms in two cases of Wolff-Parkinson-White (pre-excitation) syndrome. *Circulation* 41:399, 1970.
15. **Castellanos A, Levy S, Befeler B, et al:** Mechanisms determining the ventricular rate in Wolff-Parkinson-White arrhythmias. *Adv Cardiol* 14:221, 1975.
16. **Castellanos A Jr, Myerburg RJ, Craparo K, et al:** Factors regulating ventricular rates during atrial flutter and fibrillation in pre-excitation (Wolff-Parkinson-White) syndrome. *Br Heart J* 35:811, 1973.

17. **Castillo CA, Castellanos A Jr:** His bundle recordings in patients with reciprocating tachycardias and Wolff-Parkinson-White syndrome. *Circulation* 42:271, 1970.
18. **Coumel P, Attuel P, Motté G, et al:** Paroxysmal nodal tachycardias. Determination of the inferiormost junction point of the reentry circuit. Dissociation of the so-called 'reciprocal internodal rhythms.' *Arch Mal Coeur* 68:1255, 1975.
19. **Coumel P, Attuel P, Slama R, et al:** 'Incessant' tachycardias in Wolff-Parkinson-White syndrome. II. Role of atypical cycle length dependency and nodal-His escape beats in initiating reciprocating tachycardias. *Br Heart J* 38:895, 1976.
20. **Coumel P, Waynberger M:** The tachycardia due to reciprocal rhythm in the course of Wolff-Parkinson-White syndrome. *Coeur Med Interne* 11:97, 1972.
21. **Coumel P, Waynberger M, Garnier JC, et al:** Ventricular pre-excitation syndrome associating short P-R and delta wave, without QRS widening (3 cases of W-P-W with narrow complexes). *Arch Mal Coeur* 64:1234, 1971.
22. **Coumel P, Waynberger M, Slama R, et al:** Value of the recording of His potentials in Wolff-Parkinson-White syndrome. 6 cases. (Fr) *Acta Cardiol* (Bruxelles) 26:188, 1971.
23. **Dini P, Santini M, Milazzotto F, et al:** The use of intracavitary electrograms and of electric stimulation in the ventricular pre-excitation syndromes. I. Wolff-Parkinson-White syndrome with paroxysmal attacks of arrhythmia. (Eng Abstr) *G Ital Cardiol* 3:703, 1973.
24. **Durrer D, Wellens HJ:** The Wolff-Parkinson-White syndrome anno 1973. *Eur J Cardiol* 1:347, 1974.
25. **Fontaine G:** Ventricular pre-excitation syndromes (Wolff-Parkinson-White syndrome). New concepts. *Nouv Presse Med* 1:2735, 1972.
26. **Fontaine G, Guiraudon G, Vachon J, et al:** Kent's bundle section in a case of A-B type Wolff-Parkinson-White syndrome. I. Pre-operative electrophysiological investigations. *Arch Mal Coeur* 65:905, 1972.
27. **Gleichmann U, Seipel L, Loogen F:** Bundle of His electrography in the pre-excitation syndrome. *Verh Dtsch Ges Kreislaufforsch* 39:322, 1973.
28. **Gomes JAC, Haft JI:** Wolff-Parkinson-White syndrome type B with His depolarization occurring after the QRS. Further evidence that WPW-QRS is a fusion beat. *Chest* 67:445, 1975.
29. **Grolleau R, Puech P, Cabasson J, et al:** Particularitiés de la conduction auriculo-ventriculaire dans un syndrome de Wolff-Parkinson-White. *Arch Mal Coeur* 67:13, 1974.
30. **Haft JI, Gomes JAC:** The Wolff-Parkinson-White syndrome: the value of His bundle electrogram. *Cathet Cardiovasc Diagn* 2:113, 1976.
31. **Ichise S, Takag M:** Proceedings: tachycardia in Wolff-Parkinson-White syndrome (electrocardiographic observation on the bundle of His). *Jpn Circ J* 39:861, 1975.
32. **Ishikawa H, Kigawa Y, Morisato F, et al:** Electrocardiographic studies on the Wolff-Parkinson-White syndrome. (Jap) *J Nara Med Assoc* 23:455, 1972.
33. **Josephson ME, Seides SF, Damato AN:** Wolff-Parkinson-White syndrome with 1:2 atrioventricular conduction. *Am J Cardiol* 37:1094, 1976.
34. **Knippel M, Pioselli D:** Pre-excitation syndrome studied by His bundle electrogram. (Eng Abstr) *G Ital Cardiol* 3:685, 1973.
35. **Krikler D, Curry P, Attuel P, et al:** 'Incessant' tachycardias in Wolff-Parkinson-White syndrome. I. Initiation without antecedent extrasystoles or PR lengthening, with reference to reciprocation after shortening of cycle length. *Br Heart J* 38:885, 1976.
36. **Lau SH, Stein E, Kosowsky BD, et al:** Atrial pacing and atrioventricular conduction in anomalous atrioventricular excitation (Wolff-Parkinson-White syndrome). *Am J Cardiol* 19:354, 1967.
37. **Lister JW, Worthington FX, Gentsch TO, et al:** Pre-excitation and tachycardias in Wolff-Parkinson-White syndrome, type B. A case report, *Circulation* 45:1081, 1972.
38. **Mark AL, Basta LL:** Paroxysmal tachycardia with atrioventricular dissociation in a patient with a variant of pre-excitation syndrome. *J Electrocardiol* 7:355, 1974.
39. **Masini G, Dianda R, Gherardi G:** The recording of His bundle potentials in the ventricular pre-excitation. (Eng Abstr) *G Ital Cardiol* 3:695, 1973.

40. **Masini G, Gherardi G, Dianda R, et al:** Disorders of rhythm and mechanisms of re-entry in ventricular pre-excitation. *G Ital Cardiol* 4:450, 1974.
41. **Massumi RA:** His bundle recordings in bilateral bundle-branch block combined with Wolff-Parkinson-White syndrome. Antegrade type II (Mobitz) block and 1:1 retrograde conduction through the anomalous bundle. *Circulation* 42:287, 1970.
42. **Massumi RA, Mason DT, Vera Z, et al:** The Wolff-Parkinson-White syndrome. *Postgrad Med* 53:49, 1973.
43. **Massumi RA, Vera Z:** Patterns and mechanisms of QRS normalization in patients with Wolff-Parkinson-White syndrome. *Am J Cardiol* 28:541, 1971.
44. **Massumi RA, Vera Z, Mason DT:** The Wolff-Parkinson-White syndrome, a new look at an old problem. *Mod Concepts Cardiovasc Dis* 42:41, 1973.
45. **Moore EN, Spear JF, Boineau JP:** Recent electrophysiologic studies on the Wolff-Parkinson-White syndrome. *NEJM* 289:956, 1973.
46. **Murayama K, Nakata Y, Okada R, et al:** His bundle electrogram in a patient, Ebstein's anomaly complicated with WPW syndrome (type B)—mechanism of WPW (type B). *Respir Circ* (Tokyo) 24:63, 1976.
47. **Narula OS:** Retrograde pre-excitation. Comparison of antegrade and retrograde conduction intervals in man. *Circulation* 50:1129, 1974.
48. **Neuss H, Nowak F, Schlepper M:** Changes of conduction properties of anomalous pathways in cases with WPW syndrome. Overdrive suppression of conductivity. *Z Kardiol* 62:489, 1973.
49. **Neuss H, Schlepper M, Spies HF:** Double ventricular response to an atrial extrasystole in a patient with WPW syndrome type B. A possible mechanism triggering tachycardias. *Eur J Cardiol* 2:175, 1974.
50. **Neuss H, Spies HF, Grosser KD:** Left atrial lead and stimulation in WPW syndrome type A. *Dtsch Med Wochenschr* 100:17, 1975.
51. **Pritchett EL, Tonkin AM, Dugan FA, et al:** Ventriculo-atrial conduction time during reciprocating tachycardia with intermittent bundle-branch block in Wolff-Parkinson-White syndrome. *Br Heart J* 38:1058, 1976.
52. **Roelandt J, Schamroth L, Draulans J, et al:** Functional characteristics of the Wolff-Parkinson-White bypass. A study of six patients with His bundle electrocardiograms. *Am Heart J* 85:260, 1973.
53. **Rosen KM:** A-V nodal reentrance. An unexpected mechanism of paroxysmal tachycardia in a patient with preexcitation. *Circulation* 47:1267, 1973.
54. **Salerno JA, Tavazzi L, Massacci E, et al:** Electric stimulation and recording of intracavitary potentials used in the identification and localization of abnormalities in the Wolff-Parkinson-White syndrome. *Boll Soc Ital Cardiol* 18:1154, 1973.
55. **Santini M, Dini P, Milazzotto F, et al:** The intercavitary electrograms and the electric stimulation in the pre-excitation syndromes. II. WPW syndrome associated with conduction defects. (Eng Abstr) *G Ital Cardiol* 3:838, 1973.
56. **Sobrino JA, Rico J, Mate I, et al:** Electrocardiogram of the bundle of His in the Wolff-Parkinson-White and Lown-Ganong-Levine syndromes. *Rev Esp Cardiol* 28:487, 1975.
57. **Spurrell RA, Krikler DM, Sowton E:** Problems concerning assessment of anatomical site of accessory pathway in Wolff-Parkinson-White syndrome. *Br Heart J* 37:127, 1975.
58. **Sugimoto K, Wada J:** Study of Wolff-Parkinson-White syndrome using the recording technique of His bundle electrograms. *Jpn Circ J* 37:187, 1973.
59. **Sung RJ, Castellanos A, Gelband H, et al:** Mechanism of reciprocating tachycardia initiated during sinus rhythm in concealed Wolff-Parkinson-White syndrome: report of a case. *Circulation* 54:338, 1976.
60. **Svenson RH, Gallagher JJ, Sealy WC, et al:** An electrophysiologic approach to the surgical treatment of the Wolff-Parkinson-White syndrome: two cases utilizing catheter recording and epicardial mapping techniques. *Circulation* 49:799, 1974.
61. **Svenson RH, Miller HC, Gallagher JJ, et al:** Electrophysiological evaluation of the Wolff-Parkinson-White syndrome. Problems in assessing antegrade and retrograde

conduction over the accessory pathway. *Circulation* 52:552, 1975.

62. **Theisen K, Grohmann HW, Jahrmärker H:** Changing A-V conduction induced by atrial pacing in man. *Klin Wochenschr* 49:366, 1971.
63. **Tonkin AM, Miller HC, Svenson RH, et al:** Refractory periods of the accessory pathway in the Wolff-Parkinson-White syndrome. *Circulation* 52:563, 1975.
64. **Touboul P, Clement C, Porte J, et al:** Etude comparée des effets de la stimulation auriculaire gauche et droite dans le syndrome de Wolff-Parkinson-White. *Arch Mal Coeur* 66:1027, 1973.
65. **Touboul P, Clement C, Roques JC, et al:** Wolff-Parkinson-White syndrome. Evidence of an accessory auriculoventricular tract obtained by recording of the potential of the bundle of His. *Arch Mal Coeur* 64:638, 1971.
66. **Touboul P, Tessier Y, Magrina J, et al:** His bundle recording and electrical stimulation of atria in patients with Wolff-Parkinson-White syndrome type A. *Br Heart J* 34:623, 1972.
67. **Wallace AG, Boineau JB, Davidson RM, et al:** Symposium on electrophysiologic correlates of clinical arrhythmias. 3. Wolff-Parkinson-White syndrome: a new look. *Am J Cardiol* 28:509, 1971.
68. **Watabe H, Sugimoto K, Hakuno K, et al:** Proceedings: clinical application of His bundle electrogram recording—Wolff-Parkinson-White syndrome. *Jpn Circ J* 39:882, 1975.
69. **Wellens HJ, Durrer D:** Patterns of ventriculo-atrial conduction in the Wolff-Parkinson-White syndrome. *Circulation* 49:22, 1974.
70. **Wellens HJ, Durrer D:** The role of an accessory atrioventricular pathway in reciprocal tachycardia. Observations in patients with and without the Wolff-Parkinson-White syndrome. *Circulation* 52:58, 1975.
71. **Wellens HJ, Janse MJ, Van Dam RT, et al:** Epicardial mapping and surgical treatment in Wolff-Parkinson-White syndrome type A. *Am Heart J* 88:69, 1974.
72. **Wellens HJ, Schuilenberg RM, Durrer D:** Electrical stimulation of the heart in patients with Wolff-Parkinson-White syndrome, type A. *Circulation* 43:99, 1971.
73. **Wellens HJ, Schuilenberg RM, Durrer D:** Electrical stimulation of heart in study of patients with the Wolff-Parkinson-White syndrome type A. *Br Heart J* 33:147, 1971.
74. **Wirtzfeld A, Kiefhaber S, Baedeker W:** Studies on the initiation and termination of paroxysmal tachycardias in the pre-excitation syndrome. *Z Kardiol* 63:339, 1974.
75. **Wyndham CR, Amat-y-Leon FL, Denes P, et al:** Posterior left ventricular pre-excitation. Report of a case. *Arch Intern Med* 134:243, 1974.
76. **Zipes DP, Rothbaum DA, DeJoseph RL:** Pre-excitation syndrome. *Cardiovasc Clin* 6:209, 1974.
77. **Holzmann M, Scherf D:** Über Elektrokardiogramme mit verkürtzter Vorhof-Kammer-Distanz und positiven P-Zacken. *Z Klin Med* 121:404, 1932.
78. **Prinzmetal M, Kennamer R, Corday E, et al:** *Accelerated Conduction. The Wolff-Parkinson-White Syndrome and Related Conditions,* New York: Grune & Stratton, 1952.
79. **Sodi-Pallares D, Galder RM:** *New Bases of Electrocardiography,* St. Louis: Mosby, 1956.
80. **Homola D, Srnova V:** The diagnosis of combined heart conduction disorders with ajmaline. (Ger) *Z Kreislaufforsch* 58:89, 1969.
81. **James TN:** The connecting pathways between the sinus node and A-V node and between the right and the left atrium in the human heart. *Am Heart J* 66:498, 1963.
82. **Miller HC, Svenson RH, Gallagher JJ, et al:** Proceedings: preoperative assessment of patients with Wolff-Parkinson-White syndrome. *Br Heart J* 37:558, 1975.
83. **James TN:** Morphology of the human atrioventricular node, with remarks pertinent to its electrophysiology. *Am Heart J* 62:756, 1961.
84. **Mahaim I:** Kent's fibers and the A-V paraspecific conduction through the upper connections of the bundle of His-Tawara. *Am Heart J* 33:651, 1947.
85. **Gallagher JJ, Sealy WC, Kasell J, et al:** Multiple accessory pathways in patients with the Wolff-Parkinson-White syndrome. *Circulation* 54:571, 1976.
86. **Denes P, Amat-y-Leon F, Wyndham C, et al:** Electrophysiologic demonstration of bilateral anomalous pathways in a patient with Wolff-Parkinson-White syndrome

(type B preexcitation). *Am J Cardiol* 37:93, 1976.

87. **Befeler B, Castellanos A Jr, Castillo CA, et al:** Arrival of excitation at the right ventricular apical endocardium in Wolff-Parkinson-White syndrome type B. *Circulation* 48:655, 1973.
88. **Sellers TD Jr, Gallagher JJ, Cope GD, et al:** Retrograde atrial preexcitation following premature ventricular beats during reciprocating tachycardia in the Wolff-Parkinson-White syndrome. *Eur J Cardiol* 4:282, 1976.
89. **Neuss H, Schlepper M:** Unusual re-entry mechanisms in patients with Wolff-Parkinson-White syndrome. *Br Heart J* 36:880, 1974.
90. **Mandel WJ, Laks MM, Obayashi K:** Atrioventricular nodal reentry in the Wolff-Parkinson-White syndrome. *Chest* 68:321, 1975.
91. **Takagi M, Ichinose S, Tsuruha Y, et al:** Effective refractory period of the accessory pathway of WPW syndrome and its clinical significance. (Eng Abstr) *Respir Circ* (Tokyo) 23:157, 1975.
92. **Durrer D:** The attacks of paroxysmal tachycardia in the Wolff-Parkinson-White syndrome. *G Ital Cardiol* 2:150, 1972.
93. **Wellens HJ, Durrer D:** Wolff-Parkinson-White syndrome and atrial fibrillation. Relation between refractory period of accessory pathway and ventricular rate during atrial fibrillation. *Am J Cardiol* 34:777, 1974.
94. **Mandel WJ, Laks MM, Obayashi K, et al:** The Wolff-Parkinson-White syndrome: pharmacologic effects of procaine amide. *Am Heart J* 90:744, 1975.
95. **Spurrell RA, Krikler DM, Sowton E:** Effects of verapamil on electrophysiological properties of anomalous atrioventricular connexion in Wolff-Parkinson-White syndrome. *Br Heart J* 36:256, 1974.
96. **Wellens HJ:** Effect of drugs in the Wolff-Parkinson-White syndrome. *Adv Cardiol* 14:233, 1975.
97. **Wellens HJ, Durrer D:** Effect of digitalis on atrioventricular conduction and circus-movement tachycardias in patients with Wolff-Parkinson-White syndrome. *Circulation* 47:1229, 1973.
98. **Dreifus LS, Haiat R, Watanabe Y, et al:** Ventricular fibrillation. A possible mechanism of sudden death in patients and Wolff-Parkinson-White syndrome. *Circulation* 43:520, 1971.
99. **Dreifus LS, Nichols H, Morse D, et al:** Control of recurrent tachycardia of Wolff-Parkinson-White syndrome by surgical ligature of the A-V bundle. *Circulation* 38:1030, 1968.
100. **Dunaway MC, King SB, Hatcher CR, et al:** Disabling supraventricular tachycardia of Wolff-Parkinson-White syndrome type A, controlled by surgical A-V block and a demand pacemaker after epicardial mapping studies. *Circulation* 45:522, 1972.
101. **Edmonds JH Jr, Ellison RG, Crews TL:** Surgically induced atrioventricular block as treatment for recurrent atrial tachycardia in Wolff-Parkinson-White syndrome. *Circulation* 39 (suppl 105): 11, 1969.
102. **Boineau JP, Moore EN, Sealy WC, et al:** Epicardial mapping in Wolff-Parkinson-White syndrome. *Arch Intern Med* 135:422, 1975.
103. **Bourdillon PJ, Bentall HH, Freyer R, et al:** Surface and epicardial mapping in the Wolff-Parkinson-White syndrome. *Proc R Soc Med* 66:391, 1973.
104. Surgery in Wolff-Parkinson-White syndrome. *Br Med J* 4:547, 1974.
105. **Burchell HB:** Surgical approach to the treatment of ventricular pre-excitation. *Adv Intern Med* 16:43, 1970.
106. **Burchell HB, Frye RL, Anderson MW, et al:** Atrioventricular and ventriculoatrial excitation in Wolff-Parkinson-White syndrome (type B). Temporary ablation at surgery. *Circulation* 36:663, 1967.
107. **Cobb FR, Blumenschein SC, Sealy WC, et al:** Successful surgical interruption of the bundle of Kent in a patient with Wolff-Parkinson-White syndrome. *Circulation* 38:1018, 1968.
108. **Cole JS, Wills RE, Winterscheid LC, et al:** The Wolff-Parkinson-White syndrome: problems in evaluation and surgical therapy. *Circulation* 42:111, 1970.

109. **Coumel P, Waynberger M, Fabiato A, et al:** Wolff-Parkinson-White syndrome: problems in evaluation of multiple accessory pathways and surgical therapy. *Circulation* 45:1216, 1972.
110. **Durrer D, Roos JP:** Epicardial excitation of the ventricles in a patient with Wolff-Parkinson-White syndrome (type B). *Circulation* 35:15, 1967.
111. **Fontaine G, Frank F, Bonnet M, et al:** Méthode d'étude expérimentale et clinique des syndromes de Wolff-Parkinson-White et d'ischémie myocardique par cartographie de la polarisation ventriculaire épicardique. *Coeur Med Interne* 12:105, 1973.
112. **Fontaine G, Guiraudon G, Bonnet M, et al:** Kent's bundle section in a case of A-B Wolff-Parkinson-White syndrome. II. Epicardial cartographies. *Arch Mal Coeur* 65:925, 1972.
113. **Holzmann M:** New diagnostic and therapeutic developments in Wolff-Parkinson-White syndrome. (Ger) *Schweiz Med Wochenschr* 101:494, 1971.
114. **Knippel M, Pioselli D, Rovelli F, et al:** Tachicardie ribelli nella sindrome de preeccitazione. Trattamento chirurgico di cinque casi. *G Ital Cardiol* 4:657, 1974.
115. **Latour H, Puech P, Grolleau R, et al:** Surgical treatment of severe atrtacks of paroxysmal tachycardia in Wolff-Parkinson-White syndrome and its limitations. *Arch Mal Coeur* 63:977, 1970.
116. **Lindsay AE, Nelson RM, Abildskov JA, et al:** Attempted surgical division of the pre-excitation pathway in the Wolff-Parkinson-White syndrome. *Am J Cardiol* 28:581, 1971.
117. **Meijne NG, Mellink HM, Van Dam RT, et al:** Surgical treatment of ventricular preexcitation. *J Cardiovasc Surg* 14:232, 1973.
118. **Neutze JM, Kerr AR, Whitlock RL:** Epicardial mapping in a variant type A Wolff-Parkinson-White syndrome. *Circulation* 48:662, 1973.
119. **Rossi L, Knippel M, Raccardi B:** Histological findings, His bundle recordings and body-surface potential mappings in a case of Wolff-Parkinson-White syndrome; an anatomoclinical comparison. *Cardiology* 60:265, 1975.
120. **Sealy WC:** Surgical treatment of Wolff-Parkinson-White syndrome. *Bull Soc Int Chir* 29:252, 1970.
121. **Sealy WC, Boineau JP, Wallace AG:** The identification and division of the bundle of Kent for premature ventricular excitation and supraventricular tachycardia. *Surgery* 68:1009, 1970.
122. **Sealy WC, Hattler BG, Blumenschein SD, et al:** Surgical treatment of Wolff-Parkinson-White syndrome. *Ann Thor Surg* 8:1, 1969.
123. **Sealy WC, Wallace AG:** Surgical treatment of Wolff-Parkinson-White syndrome. *J Thor Cardiov Surg* 68:757, 1974.
124. **Sealy WC, Wallace AJ, Ramming KP, et al:** An improved operation for the definitive treatment of the Wolff-Parkinson-White syndrome. *Ann Thor Surg* 17:107, 1974.
125. **Tonkin AM, Dugan FA, Svenson RH, et al:** Coexistence of functional Kent and Mahaim-type tracts in the pre-excitation syndrome. Demonstration by catheter techniques and epicardial mapping. *Circulation* 52:193, 1975.
126. **Wallace AG, Sealy WC, Gallagher JJ, et al:** Surgical correction of anomalous left ventricular pre-excitation: Wolff-Parkinson-White (type A). *Circulation* 49:206, 1974.

VII

Prognosis

The pre-excitation syndrome is generally considered a benign disease, involving only the inconvenience of repeated attacks of palpitations[1-3] in approximately half the patients. However, several cases of sudden death have been reported over the years,[4-9] making the prognostic outlook less optimistic.[10]

A general assessment of the prognosis in pre-excitation can be gleaned from three sources, each of which should reflect some knowledge about the natural history of the disease: 1) longitudinal studies from the literature; 2) premiums assessed by life insurance companies to patients with the disorder; and 3) the status of such cases viewed by the military (especially the air force and civil aviation authorities). The first source is most disappointing and confusing: of more than one thousand published papers on pre-excitation, only a few include long-term follow-up data,[6,11-21] and these are hardly informative because of different methods and approaches.

There is also a general lack of information reflected in the different attitudes of life insurance companies towards these patients. Most of the American companies (Smith[20] interrogated 12) consider that the mortality rate is increased by 25-30% for people below the age of 35 with pre-excitation and no tachycardia, and by 100% for people over the age of 35. In the presence of paroxysmal tachycardias it is considered increased by 60-300%, depending on the number, duration and nature of the attacks.[20] In contrast to this strict policy, in Sweden neither frequent nor prolonged attacks of tachycardia cause a premium increase for a client with pre-excitation. In Denmark, the premium for individuals with pre-excitation, with or without tachycardia, is the same as that for people who are 5 years older, provided there are no signs of heart disease.[14,18,20]

Similar contradictions are found in the civilian aviation and air force regulations regarding such subjects in different countries.[22-24] In Israel, cases with paroxysmal tachycardia are not accepted into the armed forces, and those without this condition are accepted in the army but not in the air force. In France, cases of pre-excitation, both with and without a history of palpitations, are not accepted in either the military or civil aviation. In the United States and Canada, pre-excitation patients without rhythm disturbances are accepted as navigators but not as pilots. Smith[20] proposed a more liberal policy based on his long experience in the United States with flying personnel; he believes that "asymptomatic aviators and aviation candidates with WPW anomaly could be allowed to continue in an unrestricted flying status."

MORBIDITY

Without Tachycardias

A distinction must be made between cases of pre-excitation alone and those with signs of associated heart disease. Differences between adults and children must also be noted.

Isolated Pre-excitation

People with a characteristic ECG of pre-excitation and no attacks of tachycardia or associated heart disease behave as normals,[25-29] without complaints or symptoms that can be directly related to the syndrome. It is our policy to explain to these patients the nature of the anomaly, and to stress that it is in no way a heart disease and requires no change in life style. We sometimes explain that it is like having one brown and one green eye from birth. This policy was adopted because the ECG tracings are sometimes misinterpreted by other physicians as a major heart affliction. After ergometric testing we recommend sport activities in order to prove to them that nothing is wrong with their hearts. We also explain to an asymptomatic patient that in half the cases with pre-excitation the electrical anomaly may produce palpitations, and we teach them some Valsalva's maneuvers. This is important because some patients who were originally free of paroxysmal tachycardia experience attacks of palpitation later on and become terrified (sometimes irreversibly) if they have not been forewarned.

The morbidity in pre-excitation is the same in children (up to age 16) as in adults.[30-46] Parents should be informed about the nature of the disorder and the eventuality of tachycardia. In no case should any natural activity of the child be restricted, including gymnastics and sports. The school physician should be informed.

With Associated Heart Disease

About 30% of people with pre-excitation have associated heart disease (Tables VI and XVIII). In cases where there are no paroxysmal tachycardias, the health state is a function only of the status of the associated heart disease. Pre-excitation without tachycardia has never

TABLE XVIII
Studies with Long-term Follow-up

Study	# Cases	Lost to Follow-up	Years to Follow-up	Associated Heart Diseases	Tachycardias	Death: General	Death: Paroxysmal Tachycardia	Death: Sudden	Intermittency
Flensted-Jensen[14]	42	—	25	9	29	19	2?	2	54%
Berkman and Lomel[13]	128	—	5-28	6	17	3	—	—	47%
Nasser et al[16]	29	3	0-10	15	12	4	—	—	24%
Orinius[18]	75	6	20	22	54	30	1?	—	
Otto[17]	37	?	0-10	10	1?	—	—	—	24%
Smith[20]	50	—	—	—	18	3	—	—	—
Giardina et al[6]	62	—	0-20	20	35	8	1 (Fallot)	—	17%
Swiderski et al[21]	48	—	0-11	20	28	4	1+3?	—	31%
Friedman et al[15]	31	—	0-15	9	24	—	?	—	39%
Tel Hashomer	215	29	0-25	50	109	20	1?	1	26%
Total	717	38 5.29%	0-28	161 22.45%	327 45.60%	91 12.69%	9 1.25%	3 0.41%	

been reported to influence the course of other heart diseases, and vice versa. No reciprocal adverse effects of this clinicopathologic combination were reported in our 215 patients or in published series. Many patients underwent surgery, but neither the anesthetic nor the operation itself was affected by the pre-excitation.[47]

The major difference between children and adults lies in the nature of the associated heart disease, which is congenital (or occasionally a rheumatic valvular disease) in children, and usually atherosclerotic, hypertensive or rheumatic in adults. The clinical status, especially in newborns, is usually more grave, and cardiac emergencies are seen more often in this age group than in adults, depending on the nature of the congenital deformation. But as previously stated, neither condition—the pre-excitation or the heart disease—was found to adversely affect the course of the other.[48]

With Tachycardias

The single factor which transforms the pre-excitation syndrome from an electrocardiographic curiosity into a disease is the presence of recurrent paroxysmal tachycardias,[49-52] which occur in about half the

TABLE XIX
Age at First Attack of Tachycardia in the Tel Hashomer Series

Age in years	# Patients
0-10	8
11-20	32
21-30	20
31-40	12
41-50	6
51-60	6
61 +	2
Not clear	12
	98
Lost to follow-up	11
Total	109

cases. An accurate picture of the prognosis of such cases is difficult to obtain due to the paucity of longitudinal studies in the literature, the unpredictable nature of tachycardia in pre-excitation, and the necessary reliance on the subjective reports of patients or their parents for information.

Table XVIII summarizes 9 series from the literature and our own group in which there was long-term follow-up. Three[6,15,21] deal with children only (including newborns and infants), and one mainly with elderly people;[17] the rest contains material from general populations.[13,14,16,18,20]

Age at Onset of Paroxysmal Tachycardias

It is generally accepted that the pre-excitation syndrome can be discovered at any age, and that the palpitations often associated with the disorder can make their first appearance at any time (Table XIX).

Pre-excitation behaves differently in newborns than in older children,[53-69] in whom the syndrome behaves pretty much as in adults. There is a high incidence of tachycardia in children under the age of 10 months, ranging from 77-84% depending on the study,[70-72] while in older children and adults it is 40-60% (see Chapter III, The Electrocardiogram in the Pre-excitation Syndrome, Tachycardias).

Giardina[6] reported 62 cases of pre-excitation in children, 35 of whom had paroxysms of tachycardia: 29 were 2 months old or less when the first episode occurred, and two were less than 24 hours old. Fifteen already had a history of one or multiple episodes of tachycardia, inadequately controlled before referral; the others were in their first attack. Twenty-eight had no associated heart disease. Although the mechanisms of the tachycardias in these infants could not be accurately assessed, a reciprocating type could be identified in some.[6,73] Swidersky et al[21] reported 48 children with pre-excitation, 28 of whom had tachycardias; 16 had their first attack in infancy. Thirteen of the 16 had no associated heart disease, a picture similar to that described by Giardina[6] and characteristic of pre-excitation in children. In older children and adults the numbers of tachycardias in patients with and without associated heart disease are more or less equal.

The number of infants where pre-excitation is discovered because of attacks of tachycardias is much higher than in all other age groups. Lundberg, for instance, found that as many as 50% of all the infants seen by him who were suffering from paroxysmal tachycardias also had pre-excitation (as compared to 5-10% of the older children and adults). This high incidence of infants with pre-excitation discovered during attacks of tachycardia, and the high incidence of tachycardias in infants with pre-excitation, set this age group of newborns and infants apart from all other pre-excitation patients. There is a general feeling among investigators that these features together with other clinical findings which will be discussed later in this chapter (such as the kind of tachycardia which is almost always a PAT, the excellent response to treatment with digitalis, and the short duration of the periods of tachycardia), are so different from pre-excitation in adults[74] that they indicate a different pathological mechanism of both the pre-excitation and the tachycardias.

Regarding the age at which the first attack of tachycardia occurs in elderly patients, in Flensted-Jensen's[14] series only one person had his first attack when past the age of 45. We found 8 in whom the onset occurred when past the age of 50 (Table XIX). This late onset of rhythm disturbances raises the question of whether they are related to the pre-excitation or to a basic degenerative heart disease such as atherosclerosis, which may be present even in a completely asymptomatic patient and in persons where clinical investigations in this direction have been negative. This question cannot be answered definitively at present, but one can assume a direct connection to pre-excitation if the documented tachycardia is of the pseudoventricular type where the QRS complexes are broad and deformed. The same apparently can be said for repeated PAT in the absence of other kinds of tachycardias. But when the attacks are simple paroxysmal atrial fibrillations, they are just as likely related to the basic heart disease as to pre-excitation. From the practical standpoint this question is irrelevant, since the therapeutic approach to both conditions is similar (see Chapter VIII, Treatment).

TABLE XX
Number of Years of Follow-up for Paroxysmal Tachycardias in the Tel Hashomer Series

Years of Follow-up	# Patients
0-10	34
11-20	28
21-30	11
31-40	2
41-50	1
51-60	—
61 +	—
Not clear	22
	98
Lost to follow-up	11
Total	109

Duration of Suffering from Tachycardias

There is also a striking difference in the length of time during which infants and adults suffer from paroxysms of tachycardia. In Giardina's series,[6] 85% of infants with tachycardias had no recurrences of paroxysms after 6-12 months. Others found recurrence of PAT beyond 18 months of age to be very rare.[21] Thus, the prognosis concerning continuation of tachycardia into childhood seems to be excellent, and the period of palpitations short.[70,71,73,75-82]

In children and adults, paroxysms of palpitations usually continue for years and even decades. In one longitudinal study[14] covering more than 20 years, 19 patients died, and of the remaining 28, 18 suffered from tachycardias throughout the entire period. In the Tel Hashomer series, 109 patients who complained of palpitations were followed for from less than one to more than 20 years: 42 of them suffered from paroxysms for more than 11 years (Table XX). These attacks did not seem to influence their general health status, and most learned to live with them. The palpitations were rarely disabling and interfered with the patients' normal activities only in a very few cases, and then usually only for a short period of time. This was also the impression of other investigators.[13,14] Moreover, most of them did not need or take any medication (see Chapter VIII, Treatment). Repeated physical and roentgenographic examinations, and ECG tracings during normal conduction in patients with intermittent pre-excitation, revealed no signs of deteriora-

TABLE XXI
Frequency of Paroxysms: Patients Suffering 1-5 Paroxysms of Tachycardia in the Tel Hashomer Series

Frequency	# Patients
During one year	32
During one month	22
During one week	12
A mixture of the above	22
Stopped	9
Unclear	1
	98
Lost to follow-up	11
Total	109

tion which could indicate that the repeated attacks of tachycardia had worsened the health status of the patients. Attempts to determine whether years of repeated attacks of tachycardia may have had an unfavorable influence on the natural history of cases with associated heart diseases (congenital rheumatic valvular or atherosclerotic) also produced no evidence of change which could be considered secondary to the tachycardias.

Frequency and Duration of Attacks of Tachycardia

One of the most salient features of the pre-excitation syndrome is the unpredictability of the appearance of tachycardias and the variation

TABLE XXII
Duration of Attacks of Paroxysmal Tachycardia in the Tel Hashomer Series

Duration	# Patients
Hours	10
Minutes	35
Seconds	7
Unclear	22
A mixture of the above	24
	98
Lost to follow-up	11
Total	109

TABLE XXIII
Patients' Assessment of Subjective Feeling in the Tel Hashomer Series

Self-assessment	# Patients
Improvement	23
Deterioration	6
No change	49
Stopped	9
Unclear	11
	98
Lost to follow-up	11
Total	109

in their duration. Sometimes the attacks are very short and can disappear for months or years, only to return and continue for hours or days without any apparent change in the activities of the patient.[83]

The 98 patients in the Tel Hashomer group who suffered paroxysmal tachycardias and who were available for follow-up were questioned as to the frequency and duration of these attacks. The data, summarized in Tables XXI and XXII, show that the largest group reported between one and 5 attacks in a year (32 patients); only 12 had the same number in one week.

The paroxysms of tachycardia continued for only several seconds or minutes in almost half, or 42, of these 98 patients. In 24 patients the pattern was mixed—sometimes seconds, sometimes hours, and sometimes the patient could not recall the duration. In a minority of cases (10 patients) the palpitations continued for hours and even days, and they usually required medication in the hospital.

These 98 patients were asked at the last examination to evaluate how they felt in relation to their attacks of tachycardia (Table XXIII). Forty-nine reported no noticeable change; 23 reported feeling better because the paroxysms became shorter or less frequent; 9 stated that the palpitations had stopped completely; and 6 were convinced that they felt worse.

Summarizing the information in Tables XXI, XXII and XXIII, it can be stated that in most patients the attacks of tachycardia were relatively short, appeared not more than a few times a month or even a year, and that these patients felt better with the passing of years. Most of them reported, however, that their condition was rather changeable and they

TABLE XXIV
Type of Tachycardia in the Tel Hashomer Series

	# Patients
PAT (narrow QRS)	13
Atrial flutter (narrow QRS)	2
PAT (broad QRS)	3
Atrial fibrillation (narrow QRS)	4
Atrial fibrillation (broad QRS)	11
Junctional tachycardia	1
Sinus tachycardia	6
Anamnesis of palpitation only	58
	98
Lost to follow-up	11
Total	109

had difficulty making a clear summary of the behavior of the tachycardias.

Types of Tachycardias

No ECG was registered during attacks of palpitations in 58 of the 98 Tel Hashomer patients (Table XXIV), mostly because of their short duration. When ECG documentation was obtained in 40 patients, 16 were found to suffer atrial tachycardia (PAT) with narrow (13) or broad (3) QRS complexes. Atrial fibrillation was also seen in 15 patients, with narrow ventricular complexes in 4 and broad ventricular complexes in 11. Atrial flutter was documented in 2 patients, and sinus tachycardia in 6.

Symptoms and Signs during Tachycardia

The patients in the Tel Hashomer series whose paroxysms of tachycardia generally last only seconds or minutes reported that they did not pay very much attention to them, knowing from experience that they were not dangerous and would be over very soon. Most complained of an "uneasy feeling" in the chest and looked for a place to lie down. Usually the attack stopped before they were able to find a suitable spot. Some of them used Valsalvas maneuvers. The usual activities of the day were not interrupted by these short periods of palpitations. In some cases the description applied to more than one or a few premature beats than to an actual bout of tachycardia, and was documented in some instances as such.[84]

When the attack of paroxysmal tachycardia in adults continued for more than 15 minutes, a series of complaints and symptoms appeared. Smith,[20] in a series of 50 cases, found that most patients described the feeling as "palpitations" (20); other descriptions included "faintness without loss of consciousness" (8), apprehension (6) and "chest pain" (4). In a few cases he described syncope, nausea and dyspnea. Nasser et al[16] described almost the same order of complaints: palpitations, faintness, apprehension, dyspnea, syncope and one case of impaired vision. In our patients, palpitations were the most frequently described symptom, followed by anxiety, chest pain and dyspnea, and syncope in only one case. It should be emphasized, however, that many patients have very vague and mild complaints while the ECG tracings show tachycardias up to 250-300 beats per minute. Some authors[85] suggested that in cases with atrial fibrillation or flutter with broad QRS complexes, when a differential diagnosis between true and pseudoventricular tachycardia is critical, the general status of the patient, which is much better in the case of pre-excitation, may give an important diagnostic clue. In some cases both pseudo and true ventricular tachycardia may coexist in the same patient; in such cases (Figure 48b), the signs and symptoms during the latter are much more impressive—perspiration, hypotension and pallor.

In one case of pure mitral stenosis (Figure 8), no deterioration occurred during a follow-up of more than 20 years; even during the paroxysmal attacks of tachycardia the patient did not complain of more symptoms than did others suffering from isolated pre-excitation and tachycardia. In some cases of anginal syndrome, however, characteristic precordial pain may appear during a prolonged paroxysm of tachycardia.

Wolff and White[83] stated that effort, fatigue, stress, trauma, anger, excessive drinking and general disease were precipitating factors in tachycardia. We could not confirm this from our own experience, since the start of a true attack of tachycardia was completely unpredictable and no causal connection could be found between behavior or feeling of the patient and the beginning of a paroxysm. Sometimes nonparoxysmal tachycardias are the result of anger, anxiety or other emotions experienced by the patients; under such circumstances, however, a patient with pre-excitation behaves much the same as a patient without pre-excitation. Excessive physical exertion can also produce a secondary physiological tachycardia in patients both with and without pre-excitation.

Attacks of paroxysmal tachycardia in children can be much more dramatic, particularly in infants and newborns. An attack of PAT in these age groups is typically associated with vomiting, irritability, and an ash-gray color. If the attack continues, signs of cardiac failure appear with rapid, difficult respiration and a cough. The temperature usually rises but may be also subnormal. In addition to tachycardia, examination at this time reveals an enlarged heart, rales in the lungs and hepatomegaly. Later peripheral edema and ascites occur, and the baby may die in congestive heart failure or in peripheral vascular collapse.[71,86]

In older children the symptoms are usually chest pain, dizziness, sweating and palpitations.[70] In children with associated heart disease, a prolonged attack can sometimes have a fatal outcome.[6]

"Burned Out" Cases

Because of the intermittent nature and great unpredictability of this electrical disorder of the heart, one must be very cautious in labeling a case "burned out," even when the characteristic ECG pattern has disappeared completely for several years.[87] Nevertheless this does occur, particularly in infants where the disappearance has been explained by a pathological mechanism different from those suggested for adults (see Chapter X, Theories).[6,73,82] In infancy the tachycardias tend to disappear together with the anomalous QRS tracings.

"Burned out" cases have also been observed in adults. Unlike children, however, there were recurrent attacks of tachycardia in some of these patients, although not of the broad QRS complexes type (pseudoventricular tachycardia).

MORTALITY

The introduction of surgery in the treatment of patients with pre-excitation changed this syndrome from a simple academic discussion of its pathological mechanism and electrophysiologic background into an acute clinical problem requiring critical decisions. Surgery in pre-excitation can be summarized as an attempt to help patients with intractable and crippling arrhythmias and/or to prevent the sudden and unexpected death directly related to these tachycardias. Since these unexpected deaths have become one of the "raisons d'être" for surgery, they must be carefully analyzed in order to clarify whether the mortality rate in the disease itself is smaller or greater than that in surgery. The results may necessitate a second look at what the optimal treatment should be in a given case—conservative or surgical.

Definitive statistics on the mortality rate in pre-excitation cannot be derived from the literature, because of lack of data and insufficient follow-up. Nevertheless, an impression may be gained from a few published follow-up studies.

Flensted-Jensen[14] calculated the expected mortality in several patients from the year of diagnosis of pre-excitation. He found a significantly larger number of the patients had died than he expected: 17 as compared to 10 ($P < 0.05$). When the data were broken down into age groups, he found the actual mortality to be the same as the expected mortality for the over 50 group; no patients died of tachycardia related to pre-excitation. The difference was in the younger age group, where 6 died between the ages of 28 and 44 years. Of these 6, two died of totally unrelated causes and neither of them had had tachycardia or other cardiac symptoms. Of the remaining 4 patients, 3 had additional serious cardiopathy (cardiomyopathy, familial cardiomyopathy and SBE), and death in each case could have resulted from the basic disease. Death, in our opinion, could be connected directly to tachycardias in only 2 of the 4 cases. The results of Flensted-Jensen's study raise the question of whether there is an increased mortality in pre-excitation. Two other studies on the subject tend to confirm our conclusion about Flensted-Jensen's work. Orinius[18] followed 50 patients with pre-excitation and paroxysmal tachycardia for 21 years and did not find a higher mortality rate when calculating their life expectancy rate. The Aetna Life and Casualty Company[8] followed 49 patients with pre-excitation covering 314 patient years. A precise statistical evaluation is pending, but at the moment the mortality rate appears to be within the predicted normal range.

It can be concluded from these three studies that there is no definite proof of an increased mortality rate in pre-excitation.

Deaths Associated with Tachycardia

There is no argument that in some patients prolonged periods of tachycardia were the only or the main contributing factor to death. The question is only how often this occurs. A search of the literature for cases of death directly resulting from tachycardia yielded the following: 23 cases reported by Okel,[8] including Wilson's famous case in 1915; 10 cases by Dreifus et al;[255] 8 others found by us in the literature;[6,7,88-90] and two patients from our series of 215, bringing the total to 43. Fourteen died in shock or congestive heart failure due to the long periods of tachycardia, and 29 died a sudden death.

While this tabulation gives a figure for the number of cases who died as a direct result of pre-excitation, it is of little help in illuminating the incidence of such deaths in the whole population of pre-excitation cases. First, the exact number of cases is unknown (it is estimated somewhere between 1,500 and 2,000); second, many of the reported cases suffered from additional heart diseases which by themselves may be complicated by tachycardias, such as mitral stenosis or cardiomyopathies. Thus, it may be more accurate to analyze long-term follow-up studies in which data on tachycardias, basic cardiac disease and death are indicated (Table XVIII).

All in all, 717 patients were followed from one to 25 years, most for more than 10 years, of which 327 suffered from palpitations and tachycardias (45.60%); 91 are known to have died. Twelve were suspected to have died as a direct result of the pre-excitation syndrome: 9 during attacks of tachycardia and 3 by sudden death. Death during tachycardia was certain in only 2 of the 9 suspected cases: an infant with Fallot's tetralogy who died during repeated attacks of PAT,[6] and one patient of Swidersky.[21] In the other 7 cases it was more an assumption than a proven fact.

Sudden Death

Sudden death is generally considered one of the forms of death directly connected with the pre-excitation syndrome. In some cases ventricular fibrillation[88] was documented, and probably accounted for the sudden death.[5,91,92] The ventricular fibrillation usually appears in the ECG tracing as a continuation of paroxysm of atrial fibrillation with broad QRS complexes (pseudoventricular tachycardia) at a very fast rate.

As in the situation of death due to prolonged tachycardia, it is also questionable whether sudden death is directly connected with pre-excitation. Among the 43 cases of death in patients with pre-excitation collected from the literature described above, 29 died a sudden death. In 17 cases accompanying factors were present—mainly additional heart diseases and some medications—which by themselves can cause sudden death. At least 8 of them[7,14,93,94] and one of our cases suffered from a cardiomyopathy, a not fully understood form of myocardial disease known to produce sudden death in cases in which pre-excitation was never documented.[95] The same can be said about Ebstein's anomaly of the tricuspid valve, which was present in at least one of the 17 cases. In 3 of the 17 cases other cardiopathies were found, each of which can be ultimately responsible for sudden death: myocarditis,[18] mitral stenosis,[5] and lipoma of the interatrial septum plus an abnormal moderator band in the right ventricle.[8] One of the cases[8] received intravenous digitalis during an attack of pseudoventricular tachycardia. This same kind of treatment was reported to produce ventricular fibrillation under similar circumstances,[5,91] and fortunately the patients were revived with electrical defibrillation. Furthermore, digitalis has been shown experimentally to enhance the electrical conduction over an accessory bundle and to be capable of transforming a fast pseudoventricular tachycardia into a ventricular fibrillation.[96,97] Thus, some cases of sudden death may have been produced iatrogenically by the use of digitalis during attacks of paroxysmal tachycardia in pre-excitation. This treatment is regarded as contraindicated by most authors at the present time (see Chapter VIII, Treatment).

For calculating the mortality incidence we can use only the cases collected in Table XVIII, but not the others reported above. If we consider all 12 of the cases in Table XVIII as representing death secondary to features of the pre-excitation syndrome, we find a mortality incidence of 1.67% in these 10 series. But since it is obvious that not all the patients died solely as a result of their electrical disturbance, the incidence of mortality in pre-excitation seems to be closer to that calculated by Lepeshkin,[2] namely 1%.

REFERENCES

1. **Lepeschkin E:** "The P-Q-R-S-T-U- Complex," in *Modern Electrocardiography,* vol 1, Baltimore: Williams & Wilkins, 1951.
2. **Lepeschkin E:** "The Wolff-Parkinson-White syndrome and other forms of pre-excitation in cardiology," in Luisada AL: *Clinical Cardiology,* New York: McGraw-Hill, 1962.
3. **Rubin IL:** Recurrent paroxysmal tachycardia in patients with WPW syndrome. *JAMA* 229:83, 1974.
4. **Coskey RL, Danzig R:** Cardiac arrest due to extreme tachycardia with Wolff-Parkinson-White syndrome. *West J Med* 120:319, 1974.
5. **Dreifus LS, Haiat R, Watanabe Y, et al:** Ventricular fibrillation. A possible mechanism of sudden death in patients and Wolff-Parkinson-White syndrome. *Circulation* 43:520, 1971.
6. **Giardina AC, Ehlers KH, Engle MA:** Wolff-Parkinson-White syndrome in infants and children. A long-term follow-up study. *Br. Heart J* 34:839, 1972.
7. **Martin-Noel P, Denis B, Grunwald D, et al:** 2 lethal cases of Wolff-Parkinson-White syndrome. *Arch Mal Coeur* 63:1647, 1970.
8. **Okel BB:** The Wolff-Parkinson-White syndrome, report of a case with fatal arrhythmia and autopsy findings of myocarditis, inter-atrial lipomatous hypertrophy, and prominent right moderator band. *Am Heart J* 75:673, 1968.
9. **Silverman JJ, Werner M:** Fatal paroxysmal tachycardia in a newborn infant with Wolff-Parkinson-White syndrome. *J Pediatr* 37:765, 1950.
10. **Puech P:** Severe forms of the Wolff-Parkinson-White syndrome. (Fr) *Ann Cardiol Angeiol* (Paris) 17:25, 1968.
11. **Bach F:** Paroxysmal tachycardia of forty-eight years' duration and right bundle branch block. *Proc R Soc Med* 22:412, 1929.
12. **Bădărău G, Cosovanu A:** Observations on the prognosis of the ventricular pre-excitation syndrome (WPW). (Rum) *Rev Med Chir Soc Med Nat Iasi* 16:623, 1972.
13. **Berkman NL, Lamb LE:** The Wolff-Parkinson-White electrocardiogram. A follow-up-study of five to twenty-eight years. *NEJM* 278:492, 1968.
14. **Flensted-Jensen E:** Wolff-Parkinson-White syndrome. A long-term follow-up of 47 cases. *Acta Med Scand* 186:65, 1969.
15. **Friedman S, Wells RE, Amiri G:** The transient nature of Wolff-Parkinson-White anomaly in childhood. *J Pediatr* 74:296, 1969.
16. **Nasser WK, Mishkin ME, Tavel ME, et al:** Occurrence of organic heart disease in association with the Wolff-Parkinson-White syndrome. Analysis of 29 cases. *J Indiana State Med Assoc* 64:111, 1971.

17. **Otto H:** The WPW syndrome in older people. (Ger) *Dtsch Med J* 18:723, 1967.
18. **Orinius E:** Pre-excitation. Studies on criteria, prognosis and heredity. *Acta Med Scand* (suppl 465), 1966.
19. **Poveda Sierra J, Pajaron A, Medrano GA:** Unfavorable course of the ventricular pre-excitation syndrome. Apropos of 3 cases. (Sp) *Arch Inst Cardiol Mex* 43:837, 1973.
20. **Smith RF:** The Wolff-Parkinson-White syndrome as an aviation risk. *Circulation* 29:672, 1964.
21. **Swiderski J, Lees MH, Nadas AS:** The Wolff-Parkinson-White syndrome in infancy and childhood. *Br Heart J* 24:561, 1962.
22. **Desiatov AI:** Medical expert examination of flight personnel with Wolff-Parkinson-White syndrome. (Rus) *Voen Med Zh* 8:68, 1964.
23. **Miolin A:** WPW syndrome, nodal rhythm and right branch block as a military medical problem. (Cro) *Vojnosanit Pregl* 26:608, 1969.
24. **Moretti A:** On a case of ventricular pre-excitation (Wolff-Parkinson-White syndrome) in the medicolegal evaluation for selection of flight personnel. (Ital) *Minerva Med* 61:3929, 1970.
25. **Hipp R, Marcinkowski Z, Skiba D:** The Wolff-Parkinson-White syndrome in the course of labour. (Pol) *Wiad Lek* 24:451, 1971.
26. **Kumazawa T, Yasuda K, Amau T, et al:** WPW syndrome and general anesthesia. (Jap) *Jpn J Anesthesiol* 19:68, 1970.
27. **Mellerowicz H:** Zur Frage der sportartzlichen Beratung beim WPW Syndrom. Antesystolie bei einem Spetzensportler. *Dtsch Med Wochenschr* 79:184, 1954.
28. **Oyama T, Kimura K, Matsuki A, et al:** Experience with halotane anesthesia in a patient with WPW syndrome. (Jap) *Jpn J Anesthesiol* 16:582, 1967.
29. **S'Jongers JJ, Dirix A, Jolie P, et al:** Wolff-Parkinson-White syndrome and sports. *Brux Med* 53:381, 1973.
30. **Averill KH, Lamb LE:** Electrocardiographic findings in 67,375 asymptomatic subjects. I. Incidence of abnormalities. *Am J Cardiol* 6:76, 1960.
31. **Bielecki W:** Wolff-Parkinson-White syndrome in children (in the light of 4 personal cases). (Pol) *Wiad Lek* 23:843, 1970.
32. **Bielén E, Wasowicz Z:** 3 cases of Wolff-Parkinson-White syndrome in children. (Pol) *Wiad Lek* 22:397, 1969.
33. **Fedele F, Pedrinazzi RC:** Considerations on the Wolff-Parkinson-White syndrome in childhood. Description of two cases. (Ital) *Minerva Med* 54:2741, 1963.
34. **Ferents VP, Ivanova NA, Petrova LN, et al:** Wolff-Parkinson-White syndrome in children. (Rus) (Eng Abstr) *Pediatriia* 52:53, 1973.
35. **Garello L, Ribaldone D:** Effects of oculocompression on the electrocardiographic patterns in the Wolff-Parkinson-White syndrome in children. (Ital) *Arch E Maragliano Patol Clin* 25:55, 1969.
36. **Gorelov NS:** Characteristics of Wolff-Parkinson-White syndrome in children. (Rus) *Pediatriia* 52:54, 1973.
37. **Knorr D:** Das Wolff-Parkinson-White Syndrom in Kindesalter. *Arch Kinderheilk* 154:28, 1957.
38. **Loh TF:** The Wolff-Parkinson-White syndrome (a report of two cases in children and review of recent advances). *J Singapore Paediatr Soc* 12:66, 1970.
39. **Losekoot G, Lubbers JL, Wellens HJ:** Supraventricular tachycardia in infancy and childhood. *Bull Assoc Cardiol Pediatr Eur* 8:50, 1972.
40. **Magiton RN:** Wolff-Parkinson-White syndrome in children. (Rus) *Vop Okhr Materin Dets* 12:77, 1967.
41. **Nitsch K:** Das WPW-Syndrom im Kindesalter. *Pädiatr Prax* 3:195, 1964.
42. **Paul O, Harrison CJ:** Wolff-Parkinson-White syndrome in an infant. JAMA 149:363, 1952.
43. **Risbourg B, Bens JL, Tribouilloy M, et al:** Wolff-Parkinson-White syndrome in a young infant. *Pediatrie* 27:303, 1972.
44. **Romano C, Gemme G, Pongiglione R, et al:** Wolff-Parkinson-White syndrome in childhood. (Ital) *Minerva Pediatr* 23:1637, 1971.

45. **Sauerbrei HU:** WPW Syndrom in Kindesalter. *Kinderärtzl Prax* 22:8, 1954.
46. **Welento C, Sadkiewicz A:** A case of Wolff-Parkinson-White syndrome in a 5-year-old child. *Pediatr Pol* 41:859, 1966.
47. **Delineau MA, Mériel P:** Repercussions of general anesthesia on the Wolff-Parkinson-White syndrome. Apropos of a clinical case. (Fr) *Cah Anesthesiol* 19:7, 1971.
48. **Chernova IV:** Disturbances of heart rhythm in congenital heart defects in children. (Rus) *Vopr Okhr Materin Det* 9:17, 1964.
49. **Badarau G, Cosovanu A:** Serious rhythm disorders in the Wolff-Parkinson-White syndrome. *Med Intern* (Bucur) 25:25, 1973.
50. **Durrer D:** The attacks of paroxysmal tachycardia in the Wolff-Parkinson-White syndrome. *G Ital Cardiol* 2:150, 1972.
51. **Marti García JL, Candel Delgado JM, Guijarro Morales A, et al:** Auricular fibrillation and Wolff-Parkinson-White (WPW). (Sp) *Rev Clin Esp* 136:555, 1975.
52. **Martin RH, Cofer TN Jr:** Atrial fibrillation in Wolff-Parkinson-White syndrome. *Missouri Med* 64:311, 1966.
53. **Arvola A:** Paroxysmal tachycardia and pre-excitation in a newborn infant. *Ann Med Interne Fenniae* 36:376, 1947.
54. **Bellu C, Russe E, Dirina N, et al:** Paroxysmal tachycardia with Wolff-Parkinson-White syndrome in infants. (Rum) *Pediatria* (Bucur) 19:449, 1970.
55. **Bernstein ED:** Paroxysmal tachycardia in newborns and infants. A report of six cases including intrauterine tachycardia, WPW syndrome and chronic ectopic atrial tachycardia. *Ann Paediatr* (Basle) 205:161, 1965.
56. **Durante M, De Cicco N:** Paroxysmal tachycardia in a newborn infant with Wolff-Parkinson-White syndrome. *Pediatria* 80:289, 1972.
57. **Fydryk J, Kraszewska Z:** Paroxysmal tachycardia in an infant with Wolff-Parkinson-White syndrome. (Pol) *Pediat Pol* 41:209, 1966.
58. **Guerricchio G:** Wolff-Parkinson-White syndrome and paroxysmal tachycardia in infants. (Apropos of a case). *Minerva Cardioangiol* 20:27, 1972.
59. **Joseph R, Ribierre M, Najean Y:** Le syndrome de Wolff-Parkinson-White dans la première enfance. Ses rapports avec la tachycardia paroxystique du nourrisson. *Sem Hop Paris* 34:552, 1958.
60. **Lanzavecchia C, Frattoni P, Barbieri P:** On paroxysmal tachycardia with Wolff-Parkinson-White syndrome in infants. Characteristics of a case. (Ital) *Minerva Pediatr* 15:1370, 1963.
61. **Mannheimer E:** Paroxysmal tachycardia in infants. *Acta Paediatr Scand* 33:383, 1945.
62. **Pilotti G, Ricci C:** Wolff-Parkinson-White syndrome and paroxysmal tachycardia in infancy. Case report and review of literature. *Minerva Pediatr* 26:426, 1974.
63. **Richmond JB, Moore HR, Callen IR:** The Wolff-Parkinson-White syndrome in an infant. *Pediatrics* 1:635, 1948.
64. **Schieve JF:** Paroxysmal tachycardia in an infant with Wolff-Parkinson-White syndrome. *Am J Dis Child* 77:474, 1949.
65. **Schott A:** Wolff-Parkinson-White syndrome (short P-R intervals associated with disturbances of intraventricular conduction) with attacks of paroxysmal tachycardia in an infant aged eight months suffering from probable congenital heart disease. *Proc R Soc Med* 40:472, 1947.
66. **Wolff GS, Han Y, Curran J:** Wolff-Parkinson-White syndrome in the neonate. *Am J Cardiol* 41:559, 1978.
67. **Vacheron P:** Sur la fréquence du Wolff-Parkinson-White chez les nourrissons, à propos d'un cas clinique personnel. *Arch Mal Coeur* 47:345, 1954.
68. **Walsh SZ:** Wolff-Parkinson-White syndrome in a healthy two-hour-old infant without paroxysmal tachycardia. *JAMA* 186:14, 1963.
69. **Zawora S:** Wolff-Parkinson-White syndrome and paroxysmal tachycardia in a 16-month-old girl. (Pol) *Wiad Lek* 20:1081, 1967.
70. **Dreifus LS, Arriaga J, Watanabe J, et al:** Recurrent Wolff-Parkinson-White tachycardia in an infant. Successful treatment by a radio-frequency pacemaker. *Am J Cardiol* 28:586, 1971.

71. **Engle MA:** Wolff-Parkinson-White syndrome in infants and children. *Am J Dis Child* 84:692, 1952.
72. **Seganti A, Varcasia E:** Quattro casi pediatrics della sindrome di Wolff-Parkinson e White. *Pediatria Internaz* 4:67, 1954.
73. **Moene RJ, Roose JP:** Transient Wolff-Parkinson-White syndrome and neonatal reciprocating tachycardia. *Circulation* 48:443, 1973.
74. **Jansa MA:** Wolff-Parkinson-White syndrome. Report of two cases in infants. *J Pediatr* 50:207, 1957.
75. **Babin JP, Martin C:** Ectopic supraventricular paroxysmal tachycardia in the newborn. Report of three cases. *Sem Hop Paris* 48:785, 1972.
76. **Gleckler WJ, Lay JVM:** Wolff-Parkinson-White syndrome and paroxysmal tachycardia in infancy; report of case. *JAMA* 150:683, 1952.
77. **Kaye HH, Reid DS, Tynan M:** Studies in a newborn infant with supraventricular tachycardia and Wolff-Parkinson-White syndrome. *Br Heart J* 37:332, 1975.
78. **Kreidberg MB, Dushan TA:** Paroxysmal auricular tachycardia associated with Wolff-Parkinson-White syndrome in a newborn infant. *J Pediatr* 43:92, 1953.
79. **Martin C, Babin JP, Navarro C, et al:** Paroxysmal tachycardia of the newborn due to Wolff-Parkinson-White syndrome. (Fr) *Arch Fr Pediatr* 27:555, 1970.
80. **Schiebler GL, Adams P Jr, Anderson RC:** Wolff-Parkinson-White (pre-excitation) syndrome in infancy and childhood. *Univ Minn Med Bull* 30:94, 1958.
81. **Sondergaard G:** The Wolff-Parkinson-White syndrome in infants. *Acta Med Scand* 145:386, 1953.
82. **Wanderman KL, Faber I:** Spontaneous disappearance of Wolff-Parkinson-White syndrome in an infant. (Hebr) (Eng Abstr) *Harefuah* 84:438, 1973.
83. **Wolff L, White PD:** Syndrome of short P-R interval with abnormal QRS complexes and paroxysmal tachycardia. *Arch Intern Med* 82:446, 1948.
84. **Hindman MC, Last JH:** Wolff-Parkinson-White syndrome observed by portable monitoring. *Ann Intern Med* 79:654, 1973.
85. **Yahini YH, Zahavi I, Neufeld HN:** Paroxysmal atrial fibrillation in Wolff-Parkinson-White syndrome simulating ventricular tachycardia. *Am J Cardiol* 14:248, 1964.
86. **Nadas AS, Daeschner CW, Roth A, et al:** Paroxysmal tachycardia in infants and children. Study of 41 cases. *Pediatrics* 9:167, 1952.
87. **Sapinski A:** Periodical disappearance of the Wolff-Parkinson-White syndrome. (Pol) *Wiad Lek* 19:567, 1966.
88. **Castillo-Fenoy A, Goupil A, Offenstadt G, et al:** Syndrome de Wolff-Parkinson-White et mort subite. *Ann Med Interne* 124:871, 1973.
89. **Fasth A:** Wolff-Parkinson-White syndrome. A fatal case in a girl with no other heart disease. *Acta Paediatr Scand* 64:138, 1975.
90. **Touche M, Touche S, Jouvet M, et al:** Prognostic elements in the Wolff-Parkinson-White syndrome. *Presse Med* 76:567, 1968.
91. **Gallagher JJ, Gilbert M, Svenson RH, et al:** Wolff-Parkinson-White syndrome. The problem, evaluation, and surgical correction. *Circulation* 51:767, 1975.
92. **Lim CH, Toh CC, Chia BL:** Ventricular fibrillation in type B Wolff-Parkinson-White syndrome. *Aust NZ J Med* 4:515, 1974.
93. **Hejtmancik MR, Herrman GR:** The electrocardiographic syndrome of short P-R interval and broad QRS complex. A clinical study of 80 cases. *Am Heart J* 54:708, 1957.
94. **Westlake RE, Cohen W, Willes WH:** Wolff-Parkinson-White syndrome and familial cardiomegaly. *Am Heart J* 64:314, 1962.
95. **Kariv I, Sherf L, Solomon M:** Familial cardiomyopathy with special consideration of electrocardiographic and vectorcardiographic findings. *Am J Cardiol* 13:6, 1964.
96. **Gomes de Carvalho A:** WPW syndrome and tachycardia. *NEJM* 282:874, 1970.
97. **Vakil RJ:** Transitory WPW aberration after intravenous strophanthin. *Br Heart J* 17:267, 1955.

VIII

Treatment

The great majority of paroxysmal tachycardia episodes, being very brief and terminating spontaneously,[1,2] requires no special treatment. Occasionally, however, the paroxysms are longer and may produce considerable hemodynamic deterioration and symptoms. In such cases treatment is indicated.[3-9]

MEDICAL MANAGEMENT

The effects of the various drugs employed in the treatment of tachycardia in pre-excitation are not always predictable by their pharmacological action. This is because the underlying physiology and pathoanatomy of the pre-excitation syndrome are not completely understood.[10] There may be variation in the conduction and refractory periods in single or a combination of abnormal pathways, or there may be involvement of different kinds of myocardial cells in the structure of the abnormal pathways. Simple working myocardial cells,[11,12] P cells like the ones seen in the sinus or AV node,[13] and Purkinje-like cells,[14] all of which were described in accessory bundles, act differently under the influence of a given drug.

Except for a few medical centers where intracardiac investigations were undertaken in order to determine the optimum drug for a given patient,[15] medical treatment in pre-excitation was and remains mostly empirical and in many cases a matter of trial and error. Nevertheless, the knowledge compiled from the many intracardiac experiments[15-30] provides valuable information on the pathophysiology of the tachycardias and the ways drugs influence them.

During Paroxysmal Attacks of Tachycardia

The main types of tachycardia seen in the pre-excitation syndrome are paroxysmal supraventricular tachycardia and atrial fibrillation. The former type, especially paroxysmal atrial tachycardia with narrow QRS complexes, at rates of 120-230 beats per minute, constitutes 70-80% of all the tachycardias.[2,31,32] It usually occurs when the normal and accessory AV conduction become functionally dissociated. Such a situation arises when a critical timed premature atrial, junctional or ventricular beat blocks either the accessory pathway or the AV node. After propagation to the atrium or ventricle over one pathway, the impulse may find the previously blocked pathway no longer refractory, and may be conducted back over it to the chamber of origin. When there is a critical balance between conduction velocities and refractoriness, repetitive reentry sustains the tachycardia.[2] Generally, the circle is via the AV node, His bundle and bundle branches, while the retrograde conduction is over the abnormal bypass, resulting in normal QRS complexes. Recently, some cases of paroxysmal atrial tachycardias turned out to be reentrant tachycardias inside the AV node rather than via the accessory pathway when investigated with special intracardiac methods.[33,34] The cause and management of this type of tachycardia are still unknown but, since it cannot be distinguished from the other types on a simple routine ECG tracing, the treatment remains the same (see Chapter XII).

The second type of supraventricular tachycardia, seen in approximately 20-30% of the documented cases, is atrial flutter fibrillation.[2,31,32] The mechanism of its initiation is less clear than that of paroxysmal atrial tachycardia. Apparently a number of mechanisms can trigger it, one being the reentry of impulses into the atrium during its vulnerable period.[2]

We shall deal first with the medical treatment of atrial fibrillation, since these attacks are usually longer in duration, faster in rate, more resistant to drugs, and may even cause death by inducing ventricular fibrillation.[35,36]

The rationale of treatment of atrial fibrillation is to prolong the refractory period of the anomalous bypass, which is usually very short, permitting a great number of electrical stimuli (up to 300 per minute) to be transferred to the ventricles. Procainamide and quinidine are particularly effective in prolonging the refractory period of the bypass tract.[29,37–41] Other agents such as ajmaline[42–48] and amiodarone[1,17,22,26] also have major effects on the refractoriness of the accessory pathway; however, their use is limited in some countries, including the United States, and only a small number of medical reports are available on their influence on pseudoventricular tachycardias.

Parenteral procainamide was found to be the drug of choice against attacks of atrial fibrillation with fast cardiac rates and broad QRS complexes in our patients and others.[31] We administer the drug intravenously, 50 mg every 2-3 minutes to a total of 1 gm or 10/mg/kg. (If this does not stop the tachycardia, we give quinidine, 0.2 gm every hour [alone or in combination with a beta-blocker], and usually this combination ends the attack within 2-3 hours.) Our patients did not exhibit any significant drop in blood pressure as a result of parenteral procainamide.

Until recently, many physicians[1,10] preferred to administer digitalis intravenously in cases of pseudoventricular tachycardia, considering the paroxysm a simple rapid atrial fibrillation for which digoxin is known to be the most effective drug. Recent evidence, however, suggests that this treatment may acutely decrease the refractory period of the accessory pathway by 10-50 msec, thereby increasing the already fast cardiac rate of the tachycardia.[28] This very fast atrial fibrillation has been known to induce ventricular fibrillation, with fatal outcome in some patients[49] and successful resuscitation in others.[31,36] It is generally agreed at the present time that digitalis preparations are strictly contraindicated in such conditions.[2]

Lidocaine[50-52] injected intravenously as a bolus followed by a continuous infusion can sometimes suppress the tachycardia which is resistant to other drugs.[1] Although lidocaine ordinarily has little effect on supraventricular arrhythmias, it may be effective in pre-excitation by acting on the accessory conduction tract or by interrupting the "circus movement" at the ventricular level.[1] Treatment of the more common form of tachycardia, paroxysmal atrial tachycardia, is essentially the same as for atrial fibrillation, with the exception of an initial attempt to stop the tachycardia by a vagal maneuver in the former. The "circus movement" is usually in a direction opposite to that seen in pseudoventricular tachycardia—via AV node and His and Purkinje axis. The prolongation of the refractory period in the bypass tract by drugs such as procainamide, quinidine, ajmaline or lidocaine results in interruption of the "circus movement" by elimination of one of the components of the chain: the accessory tract.

We have seen good results with intravenously injected verapamil.[53] In one case of atrial flutter with 1:1 conduction and a cardiac rate of 300 (Figure 33b), 5 mg verapamil promptly stopped the attack. This agent belongs, with digoxin and beta blocker, to the group of drugs which does not act at all, or only a little, on the anomalous bypass tract, but has a beneficial influence on supraventricular tachycardias (and especially paroxysmal atrial tachycardia) by drastically prolonging the refractory period of the AV node.

Digitalis products are still used in suppressing paroxysmal atrial tachycardias because of their influence on the AV node.[54-56] The results are generally good. In our hospital, however, intravenous digoxin is considered contraindicated in all forms of tachycardia connected with the pre-excitation syndrome because of the unpredictable course some forms of the tachycardia may take. We have seen not only different forms of paroxysmal tachycardia at different times in one patient, but also one case of alternating pseudoventricular (atrial fibrillation) and paroxysmal atrial tachycardia in the same tracing (Figure 36d). In light of the danger of using digitalis in the atrial fibrillation form of pre-excitation tachycardia,[57-58] we prefer to eliminate parenteral digitalis in all cases.

This statement applies only to adult patients and older children, and not to newborns and infants. As outlined in Chapter VII, Prognosis, the clinical course, behavior, form and response to medication of the tachycardias in this age group are very different from the adult type, and the danger of irreversible heart failure and death in these infants is much greater.[59] This point must be strongly emphasized, since the tachycardia in infants is always the reentrant type (PAT). (To the best of our knowledge no pseudoventricular tachycardia has been reported in an infant.) Thus, the drug of choice of pediatricians and cardiologists is digitalis.[59-63] Giardina[64] found that all but one of the infants seen by her during an attack of paroxysmal tachycardia responded to digitalization. Digitalis in the form of parenteral digoxin or digitoxin was administered over a 12-20 hour period as the initial therapy. Of the patients treated with digitoxin, 22% converted to normal sinus rhythm with less than 0.03 mg/kg (two of these patients were under one year of age and 3 were older than 2 years), 61% needed 0.04 mg/kg (these patients were all under one year), and 11% required as much as 0.09 mg/kg in 24 hours for conversion (2 patients, 8 months and 2 months). Of the patients treated with digoxin, 36% needed 0.025 mg/kg or less administered parenterally for conversion to normal sinus rhythm (all under one year of age), 36% needed 0.06-0.08 mg/kg intramuscularly for conversion (6 patients under one year and one over 2 years). Thus, these infants responded to digitoxin or digoxin in the usual therapeutic range.[64]

A number of other pharmacological agents have been used recently for treatment of paroxysmal tachycardias, but few clinical trials and reports are available. These drugs are: ajmaline;[17,22,26,29,65] tiprenolol, a new beta-blocking agent;[23] propafenon;[21] aprindine guanethidine;[18,66,67] antazoline;[68,69] and prednisolone.[70]

Preventive Treatment to Avoid Tachycardia

Most paroxysms of tachycardia are short (seconds to minutes), end spontaneously or by a simple Valsalva's maneuver, appear quite seldom (1-5 attacks per year), do not produce troublesome symptoms or signs, and do not even in the long run lead to deterioration of the general health status (see Chapter VII, Prognosis, and Tables XXI-XXIII). Therefore, preventive treatment is often considered unnecessary.[1,2] Of our 98 patients included in a follow-up study, preventive treatment was used in 48 (47%) and the remainder were taught only some simple vagotonic procedures like holding their breath. Preventive treatment is considered necessary when patients suffer repeated attacks of tachycardia which last a long time and interfere with their normal activities.

According to Gallagher et al,[31] paroxysmal tachycardia in pre-excitation occurs when: 1) there is a premature beat which may be atrial, junctional or ventricular in origin; 2) it comes at a time when conduction over the AV node and bypass can be dissociated; and 3) after this beat is conducted over one pathway from its chamber of origin, the other pathway must be excitable to permit initiation of "circus movement." These authors concluded that pharmacological therapy should be aimed at modifying one or more of these links in the circuit by: 1) reducing the number of premature beats; 2) narrowing the "window" or interval in which premature beats can dissociate the two pathways; and 3) prolonging refractoriness so that the returning impulse is blocked either in the accessory pathway or in the AV node. The ideal medical therapy and that which is most widely used is a drug which prolongs the refractory period in the anomalous pathway (like quinidine, procainamide, ajmaline) and/or one which prolongs the ERP in the AV node (like digoxin, beta-blocker, verapamil). Quinidine is used most widely. Tonkin et al[2] reported that it is poorly tolerated by a significant number of patients

TABLE XXV

Preventive Treatment of Tachycardias in the Tel Hashomer Series

Treatment	# Patients
Drugs	
Digoxin	9
Quinidine sulfate	21
Beta-blockers	10
Ajmaline	2
Verapamil	2
Procainamide	2
Valsalva's maneuvers	8
No treatment required	44
	98
Lost to follow-up	11
Total	109

after long-term use. Our experience using it alone or in combination with other drugs in 21 of our 46 patients who received preventive treatment did not corroborate his report: only a few experienced side effects while taking quinidine, and only in the first days of treatment, not after long periods. Quinidine has a much lower incidence of undesirable side effects than procainamide, which is used less frequently (Table XXV).

If one or the other of these drugs is not effective alone, an additional agent which acts on the ERP of the AV node should be added. Our choice is a beta-blocker,[71-75] and we used propranolol together with quinidine or procainamide in 10 of 46 cases with good results (Table XXV). The logical choice, based on wide clinical experience, is digitalis. But, because of its possible role in inducing ventricular fibrillation, discussed previously, digoxin given parenterally is not recommended for treating tachycardias. It is not known, however, whether digoxin has the same effect when administered orally over a long period of time as when it is given intravenously for a short time. Although the preliminary experience of Tonkin et al[2] suggested that refractoriness of the bypass may be less affected by digoxin when given orally, we agree with the authors that it is prudent to avoid its use for the time being. Before introducing this policy, 9 of our 46 patients received 0.25 mg digoxin daily, alone or in combination with quinidine, without developing any side effects or undesirable tachycardias. In a few cases we used oral verapamil and ajmaline, but it is too early to draw any conclusions about the effect of these drugs on the repeated tachycardias of pre-excitation.

A completely different approach to preventive medication is suggested in newborns and infants with pre-excitation and paroxysmal tachycardias. It is the policy of pediatricians to maintain these babies on digitalis, in increasing doses commensurate with their weight gain at least for the first 6-12 months, and to discontinue it if they have remained free of recurrences around the first birthday. With this regimen, Giardina et al[64] reported no recurrences of tachycardias after therapy was stopped in 85% of their cases followed for a long period.

ELECTRICAL TREATMENT

Electrical Cardioversion

Cardioversion can be used as the first treatment in every type of tachycardia seen in the pre-excitation syndrome, usually with good results. However, it is generally employed only after a tachycardia has proved to be resistant to drug treatment,[76-80] and as an emergency treatment. An emergency consists of ventricular fibrillation spontaneously following a rapid pseudoventricular tachycardia, or secondary to iatrogenic factors such as intravenous digoxin during an episode of atrial fibrillation with fast heart rate[36,81,82] or experimental pacing of the atria at fast rates.

Artificial Pacemaker

Tonkin et al[2] claimed that the pacemaker provides a good means of terminating recurrent paroxysmal supraventricular tachycardias. An appropriately timed electrical stimulus is delivered which captures the atrium or the ventricle, blocks the AV node or accessory pathway, and breaks a link in the reentry circuit by rendering the heart chamber refractory to the reentering impulse. The pacing electrode may be placed in the heart in two ways: by the transvenous method, into the right ventricle or coronary sinus (near the left atrium or ventricle), or by the transthoracic method, sutured to the epicardium of any of the cardiac chambers. According to Tonkin, [2] pacing is most effective when the electrodes are as close as possible to the assumed accessory pathway. With the development of a tachycardia, the pacing may be initiated in one of two ways: a magnet applied to the chest wall over an implanted battery-powered demand generator,[83,84,85,86] or a radio frequency device.[87,88] While both methods are equally effective in interrupting a tachycardia,[89-91] the latter carries the potential risk of accidentally initiating an atrial fibrillation if the rate of discharge is too high. Radio frequency pacemakers should be used only in patients in whom the refractory properties of the assumed accessory pathway have been previously investigated (see Chapter VI, Invasive Methods of Investigation, Determination of the effective refractory period of an anomalous AV connection as compared to the normal conduction system). A radio frequency device was used with good results in an infant.[92]

Prior to implantation of a permanent pacemaker it is advisable to try a temporary device, using the same chamber and method of pacing which will be used when the permanent pacemaker is implanted.

SURGERY

The introduction of surgery as a therapeutic method in the treatment of patients with pre-excitation and tachycardia[93-100] was discussed previously (see Chapters I and VI). This section deals with the present status of surgery in pre-excitation.

Surgical Techniques

Two major approaches characterize surgical intervention in pre-excitation: 1) the interruption of anomalous conduction by section of an accessory muscle bundle connecting an atrium with a ventricle; and 2) the interruption of normal conduction by section of the His bundle and production of a complete AV block, always combined with the insertion of an artificial pacemaker.

These two approaches surgically break the ring around which the "circus movement" of an electrical stimulus produces a tachycardia, in the former technique by interrupting the accessory bundle and in the latter by interrupting the normal AV pathway (His bundle). Table XXVI summarizes the techniques and findings in 82 patients (93 operations) operated on for pre-excitation as reported by 16 authors. The primary intention of the surgeon in 71 cases was to find and interrupt an accessory bundle, and in 23 to section the His bundle. Twelve of these 23 were first operations, and 11 were second interventions after failure to interrupt the accessory bundle.

Section of an Accessory Bundle

Since the first successful surgical interruption of an accessory pathway by Cobb et al,[115] more than half of all known operated cases using this approach (50 of 70)[82] were undertaken in the Division of Thoracic Surgery at Duke University in Durham, North Carolina. The following description of the technique is presented in terms of its evolvement at Duke University.

From the beginning it was evident to the surgical team involved in developing this approach that the pathway must be divided at the point where it crosses the atrioventricular ring. At first they approached the ring from the epicardial side and divided the ventricles at their insertion into the annulus fibrosus.[112,115,116] This presented very little difficulty on the right side since the right coronary artery could be easily displaced downwards. However, on the left side the great coronary vein and coronary sinus posed problems. Moreover, injury to the tricuspid valve has much less serious consequences than injury to the mitral valve.[82] Thus, a new technique was developed and now the accessory bundle is divided on the atrial side of the annulus.[31,82,102,117-119] The heart is now approached through a median sternotomy.

On the left side, the accessory bundle is interrupted by incising the atrium just above the annulus of the mitral valve, exposing the fat pad that surrounds the coronary artery and vein. If the tract is located anteriorly on the free wall, the incision is begun at the left trigone (anterior insertion of the aorta into the fibrous skeleton of the heart) and carried counterclockwise to beyond the midportion of the annulus of the free wall. If the pathway is on the posterior free wall, the incision is begun at the right trigone (central fibrous body) and extended clockwise to beyond the midportion. Today, larger and larger incisions are made, based on the real possibility that multiple tracts may be present well away from the dominant one.[82]

On the right side, the pathway is approached through the atrial endocardium and myocardium, with the incision made just above the annulus fibrosus.

In Sealy et al's[82] two most recent patients, the incision was made from the endocardium to the epicardium. With this approach it is possible to make the incision from the anterior aspect of the membranous ventricular septum as it protrudes into the right atrium, and to continue it all the way around to nearly the same point in the posterior aspect of the membranous septum.

A major problem arose in approximately one-quarter of the operated patients at Duke University, in whom ventricular activation occurred posteriorly in the region of the crux.[82] This anomaly may be in either the right or the left side of the ventricular septum, usually within 1 cm of the posterior descending coronary artery. The surgical team stated that the coronary vein should be completely separated from the ventricle in the area of the crux, and all the connections to the muscle at the top of the ventricular septum should be divided. If the earliest area of activation is on the left side of the ventricular septum, the left atrium should be opened and the annulus divided from the right trigone to the lateral midportion of the mitral annulus. If the area is on the right side, the right atrium should be divided from the posterior extent of the membranous ventricular septum to the lateral midpoint of the tricuspid valve.

Division of His Bundle

This second method is used electively as a first operation, or as a second intervention after failure of the first to divide an accessory bundle (Table XXVI). Sealy et al[82] used the elective approach in only one case, and as a second intervention in 6 of 50 cases. They tried to divide the His bundle by suture ligation alone or combined with electrocauterization. Both require insertion of a demand pacemaker. These authors made the interesting observation that although the procedure is theoretically easy to perform, it turns out to be a difficult one (they failed in 3 of their patients).

Results of Surgery

The clinical and electrocardiographic results of operations undertaken for the treatment of tachycardias in the pre-excitation syndrome

TABLE XXVI
Results of Surgery

Study	# Patients operated	# Operations	Kent bundle dissection	His bundle dissection			Delta wave remained unchanged	Delta wave disappeared	Tachycardia present (after 1st op)	Tachycardia disappeared	Death operative	Additional bundle discovered			Comments
				Total	Elective	Secondary						Acc bund	AN conn	HV conn	
1. Sealy et al[82] 1976	50	56	49	7	1	6	15	32	18	29	3	6	3	3	In 3 cases, only retrograde conduction was proven in the acc bundle
2. Iwa[101] 1976	10	10	10	—	—	—	5	5	5	5	1	—	—	—	
3. Karp & Waldo 1975	2	2	2	—	—	—	1	1	1	1	—	—	—	—	Not published, Waldo, personal communication

4. Levitsky[102] 1974	1	1	1	—	—	—	—	1	—	1	—	Quoted by Sealy et al[102] in discussion
5. Knippel et al[103] 1974	5	5	1	4	4	—	4	1	—	3	2	In the addendum an additional operation is indicated without enough clinical details
6. Wellens et al[104] 1974	4	6	2	4	2	2	2	2	1	3	—	Two of the 4 cases were also reported by Meijne et al
7. Coumel et al[105] 1972	1	1	—	1	1	—	1	—	—	1	—	
8. Fontaine et al[106] 1972	1	1	1	—	—	—	—	1	1	—	—	
9. Dunaway et al[107] 1971	1	2	1	1	—	1	1	—	—	1	—	
10. Lindsay et al[108] 1971	1	1	1	—	—	—	1	—	1	—	—	
11. Latour et al[109] 1970	1	1	—	1	1	—	1	—	1	—	—	
12. Cole et al[110] 1970	2	4	2	2	—	2	2	—	—	1	1	
13. Edmonds et al[111] 1969	1	1	—	1	1	—	1	—	—	1	—	
14. Yacoub[112] 1969	1	1	—	1	1	—					1	Quoted by Sealy et al[112]
15. Burchell[113] 1967	1	1	1	—	—	—	1	—	1	—	—	The first surgical intervention intended to section the accessory bundle
16. Dreifus et al[114] 1968	1	1	—	1	1	—	1	—	—	1	—	The first surgical intervention outlined to interrupt the His bundle
Total	83	94	71	23	12	11	36 43.27%	43 53.75%	29 34.93%	47 56 62%	8 9.63%	

must be approached with caution. Not all the operated cases are reported; in Burchell's[120] words, "hearsay evidence suggests that the surgical approach to the treatment of the Wolff-Parkinson-White syndrome has been attempted more often than reported." In addition, the results of the two major techniques are sometimes difficult to compare because authors do not assess their results in the same way, or they do not present enough details.

The findings in papers in which the technique and the results of surgery are described are presented in Table XXVI covering 83 patients and 94 operations. The delta wave disappeared in 43 patients (53.75%) after surgery and remained unchanged or slightly changed in 36 (43.27%) (one patient died before a clear finding could be recorded and 3 others had no delta wave from the beginning). Tachycardia did not recur after the first operation in 47 patients (56.62%), and recurred in 29 (34.43%). Eight patients (9.63%) died during or soon after the surgical procedures, resulting in a total of 37 (44.56%) patients who did not enjoy positive results (both first and second operations are included in this figure).

At present the results have been much more favorable in those cases with pathways in the left or right free wall portion (right or left ventricle) of the AV groove than in those with septal connections. In a partial summary of patients operated on by the Duke University team,[2] the free wall accessory bundle was successfully divided in 20 of 22 patients as judged by disappearance of the delta wave and abolition of tachycardias, while it was successful in only 5 of 14 septal connections.

Additional bypass pathways were discovered in at least 13 operated cases, usually after section of the main accessory bundle; 7 were most probably multiple muscular AV connections, 3 Mahaim bundles, and 3 conducting James' bypass fibers.

Indications and Contraindications for Surgery

There appears to be a general consensus among heart surgeons[82,101] and cardiologists[2,4,120-124] that surgical intervention should be considered only in a very small group of selected patients with intractable episodes of supraventricular tachycardia who are unresponsive to medical and/or electrical therapy, or cases with life threatening arrhythmias, especially atrial fibrillation with an extremely rapid ventricular response.[2] Surgery should be attempted only in those medical centers fully equipped with the necessary equipment for recording before and during the operation, facilities for cardiopulmonary bypass, and personnel specially trained in the electrophysiological techniques for localizing the site of the AV bypass.[2]

The decision to operate on a patient with pre-excitation and attacks of tachycardia should be made in the light of the following considerations: 1) the mortality rate related to surgical intervention approaches 10% as compared to 1-2% mortality in pre-excitation in general; 2) the paroxysms of tachycardia ceased after surgery in only a little more than half of the operated cases (56.62%); 3) the outlook for success following surgery is much less optimistic when the accessory bundle is located septally than when it is on the free wall of the ventricles; 4) even when the most sophisticated investigations show a single AV accessory bundle, multiple connections including Mahaim or James' fibers can be present but silent, and identifiable only after the main bypass bundle has been sectioned;[125] 5) when section of the His bundle is considered (as an elective intervention or a second attempt), it should be clear that, if successful, the reentry mechanism necessary for supraventricular tachycardias will be eliminated but not the AV conduction over the accessory pathway in the event of atrial fibrillation. The documentation of a paroxysmal atrial tachycardia does not exclude the possibility of a fast pseudoventricular tachycardia developing in the same patient at another time (Figure 36d).

There is a category of cases which should not undergo surgery even after taking the above precautions into consideration. This is the group of patients with cardiomyopathy and pre-excitation. In Sealy et al's[82] series of 5 operated cases, this combination was found in 2 of the 3 cases of operative death.

Recently a new surgical technique, cryosurgery, was described by Gallagher et al.[126] Using nitrous oxide as the coolant, a new instrument was designed with which controlled cooling to 0° C could be delivered to an area resulting in cessation of physiological function, with return to normal function on rewarming. In this manner, the result of cooling an area of tissue suspected of containing an accessory pathway could be examined in a reversible manner before applying an irreversible freeze. Once the area to be ablated is located, the instrument has the capacity to produce an ice ball 15 mm in diameter at a temperature of -60° C. This technique will be very useful when an accessory bundle is located superficially in a heart (bridging, for example, the right or left AV grooves) and which previously necessitated the instituting of cardiopulmonary bypass. On the other hand, if the accessory pathway is located close to a coronary artery, cryosurgery is inadvisable. Five successfully treated cases are reported in this paper.[126]

REFERENCES

1. **Mark H, Luna LS:** Treatment of Wolff-Parkinson-White syndrome. *Am Heart J* 83:565, 1972.
2. **Tonkin AM, Gallagher JJ, Wallace AG:** Tachyarrhythmias in Wolff-Parkinson-White syndrome. Treatment and prevention. *JAMA* 235:947, 1976.
3. **Apostolov L:** On a new pharmacologic test in the Wolff-Parkinson-White syndrome. (Ital) *Minerva Med* 58:216, 1967.
4. **Bellet S:** Treatment of arrhythmias, particularly paroxysmal tachycardia, associated with the Wolff-Parkinson-White syndrome. *Am J Cardiol* 17:104, 1966.
5. **Blinder H, Burstein J, Smelin R:** Drug effects in Wolff-Parkinson-White syndrome. *Am Heart J* 44:268, 1952.
6. **Sherf L:** Different arrhythmias seen in the pre-excitation syndrome. Presented at the Sixth Asian-Pacific Congress of Cardiology, Honolulu, Hawaii, October 1976.
7. **Ferrer MI:** Pre-excitation syndrome: mechanisms and treatment. *Cardiovasc Clin* 2:117, 1970.
8. **Takáč M, Rešetár J, Takáčová M, et al:** Contribution to the diagnosis and treatment of paroxysmal tachycardia in Wolff-Parkinson-White syndrome. (Slo) *Vnitr Lek* 13:1090, 1967.
9. **Timio M:** The Wolff-Parkinson-White syndrome. Contribution to the therapeutic study. (Ital) *Minerva Cardioangiol* 15:866, 1967.
10. **Bellet S:** *Clinical Disorders of the Heart Beat,* pp 485-519, Philadelphia: Lea & Febiger, 1971.
11. **Lev M, Leffler WB, Langendorf R, et al:** Anatomic findings in a case of ventricular pre-excitation (WPW) terminating in complete atrioventricular block. *Circulation* 34:718, 1966.
12. **Sakamoto Y, Hiroki T, Kunio S, et al:** Spatial velocity electro-cardiograms in WPW syndrome and complete left bundle branch block. *Jpn J Clin Med* 31:3094, 1973.
13. **James TN, Puech P:** De subitaneis mortibus. IX. Type A Wolff-Parkinson-White syndrome. *Circulation* 50:1264, 1974.
14. **Lev M, Sodi-Pallares D, Friedland C:** A histopathologic study of the atrioventricular communications in a case of WPW with incomplete left bundle branch block. *Am Heart J* 66:399, 1963.
15. **Wellens HJ:** Effect of drugs in the Wolff-Parkinson-White syndrome. *Adv Cardiol* 14:233, 1975.
16. **Dye CL:** Atrial tachycardia in Wolff-Parkinson-White syndrome. Conversion to normal sinus rhythm with lidocaine. *Am J Cardiol* 24:265, 1969.
17. **Gilbert-Queralto J, et al:** Diagnostic and therapeutic significance of ajmaline in the Wolff-Parkinson-White syndrome. (Ger) *Verh Dtsch Ges Kreislaufforsch* 35:231, 1969.
18. **Harris WE, Semler HJ, Griswold HE:** Reversed reciprocating paroxysmal tachycardia controlled by guanethidine in a case of Wolff-Parkinson-White syndrome. *Am Heart J* 67:812, 1964.
19. **Krikler DM:** Verapamil in investigation and treatment of supraventricular tachycardias. (Ital) (Eng Abstr) *Minerva Med* 66:1914, 1975.
20. **Mandel WJ, Laks MM, Obayashi K, et al:** The Wolff-Parkinson-White syndrome: pharmacologic effects of procaine amide. *Am Heart J* 90:744, 1975.
21. **Probst P, Pachinger O:** Influence of propafenon on hemodynamics of the left ventricle and atrioventricular conduction with special reference to the WPW syndrome. (Ger) *Z Kardiol* 65:213, 1976.
22. **Puech P, Latour H, Hertault J, et al:** Injectable ajmaline in paroxysmal tachycardia and the WPW syndrome. Comparison with procaine amide. *Arch Mal Coeur* 57:897, 1964.
23. **Roelandt J, Schamroth L, Hugenholtz P:** Effects of new beta-blocking agent (DL-tiprenolol) on conduction within normal and anomalous atrioventricular pathways in Wolff-Parkinson-White syndrome. *Br Heart J* 34:1272, 1972.
24. **Rosenbaum MB, Chiale PA, Ryba D, et al:** Control of tachyarrhythmias associated

with Wolff-Parkinson-White syndrome by amiodarone hydrochloride. *Am J Cardiol* 34:215, 1974.

25. **Salvador M, Fauvel JM, Lesbre JP, et al:** Prevention of rhythm disorders by a new quinidine salt with sustained action: study of quinidinemia (apropos of 45 cases). *Arch Mal Coeur* 64:712, 1971.
26. **Speckert H, Klepzig H:** Development of a complete left bundle branch block in the WPW syndrome following ajmaline injection. (Ger) *Z Kreislaufforsch* 56:490, 1967.
27. **Spurrell RA, Krikler DM, Sowton E:** Effects of verapamil on electrophysiological properties of anomalous atrioventricular connexion in Wolff-Parkinson-White syndrome. *Br Heart J* 36:256, 1974.
28. **Wellens HJ, Durrer D:** Effect of digitalis on atrioventricular conduction and circus-movement tachycardias in patients with Wolff-Parkinson-White syndrome. *Circulation* 47:1229, 1973.
29. **Wellens HJ, Durrer D:** Effects of procaine amide, quinidine and ajmaline in the Wolff-Parkinson-White syndrome. *Circulation* 50:114, 1974.
30. **Wellens HJ, Lie KI, Bar FW, et al:** Effect of amiodarone in the Wolff-Parkinson-White syndrome. *Am J Cardiol* 38:189, 1976.
31. **Gallagher JJ, Gilbert M, Svenson RH, et al:** Wolff-Parkinson-White syndrome. The problem, evaluation, and surgical correction. *Circulation* 51:767, 1975.
32. **Newman BJ, Donoso E, Friedberg CK:** Arrhythmias in the Wolff-Parkinson-White syndrome. *Prog Cardiovasc Dis* 9:147, 1966.
33. **Mandel WJ, Laks MM, Obayashi K:** Atrioventricular nodal reentry in the Wolff-Parkinson-White syndrome. *Chest* 68:321, 1975.
34. **Wellens HJ:** Contribution of cardiac pacing to our understanding of the Wolff-Parkinson-White syndrome. *Br Heart J* 37:231, 1975.
35. **Castillo-Fenoy A, Goupil A, Offenstadt G, et al:** Syndrome de Wolff-Parkinson-White et mort subite. *Ann Med Interne* 124:871, 1973.
36. **Dreifus LS, Haiat R, Watanabe Y, et al:** Ventricular fibrillation. A possible mechanism of sudden death in patients and Wolff-Parkinson-White syndrome. *Circulation* 43:520, 1971.
37. **Cloetens W, de Mey D:** Evolution particuliere d'un syndrome de Wolff-Parkinson-White. Influence pharmacologigue de la procaïne-amide. *Acta Cardiol* 8:632, 1953.
38. **van Hees CA:** L'effet de la procaïne-amide sur le syndrome de Wolff-Parkinson-White. *Acta Cardiol* 8:639, 1953.
39. **Hoffman I, Abernathy RS, Haedicke TA:** Effect of procaine amide on anomalous conduction and paroxysmal tachycardia in a case resembling the Wolff-Parkinson-White syndrome. *Am Heart J* 44:154, 1952.
40. **Lavenne F, Tyberghein J, Sonnet J:** Action de la procaïne amide dans le syndrome de Wolff-Parkinson et White. *Acta Cardiol* 8:384, 1953.
41. **Zapata Diaz J, Cabrera CE, Rodriguez MI, et al:** Acción de la procainamida (pronestyl) sobre el corazón. Sindrome de Wolff-Parkinson-White (WPW) y procainamida. *Arch Inst Cardiol Mex* 23:87, 1953.
42. **Bobba P, Baldrighi V, Tronconi L, et al:** Injectable ajmaline in the Wolff-Parkinson-White syndrome. Compared with procainamide. (Ital) *Atti Soc Ital Cardiol* 1:54, 1966.
43. **Bucur V, Enescu R, Siraulescu R, et al:** Di-mono-chloroacetyl-ajmaline in the treatment of disorders of cardiac rhythm. (Rum) *Med Intern* (Bucur) 21:47, 1969.
44. **Duclos F:** Efficacy of ajmaline (gylurritmal) in the treatment of a paroxysmal auricular fibrillation lasting for 3 weeks in a case of WPW. (Sp) *Rev Esp Cardiol* 22:240, 1969.
45. **Lombardi M, Masini G:** Utility and importance of ajmaline in the Wolff-Parkinson-White syndrome. *Atti Soc Ital Cardiol* 1:50, 1966.
46. **Masior J, Platek D, Zgorniak M:** The effect of ajmaline on the Wolff-Parkinson-White syndrome. (Pol) *Kardiol Pol* 8:185, 1965.
47. **Nicita-Mauro V:** Considerations on the use of ajmaline in Wolff-Parkinson-White syndrome. (Ital) *Boll Soc Ital Cardiol* 15:123, 1970.
48. **Tronconi L:** Use of ajmaline by venous route and of procaine amide in the treatment of Wolff-Parkinson-White syndrome. (Ital) *Minerva Cardioangiol* 14:228, 1966.

49. **Okel BB:** The Wolff-Parkinson-White syndrome, report of a case with fatal arrhythmia and autopsy findings of myocarditis, inter-atrial lipomatous hypertrophy, and prominent right moderator band. *Am Heart J* 75:673, 1968.
50. **Josephson ME, Kastor JA, Kitchen JG III:** Lidocaine in Wolff-Parkinson-White syndrome with atrial fibrillation. *Ann Intern Med* 84:44, 1976.
51. **Morris JT:** Case report. Supraventricular tachycardia of Wolff-Parkinson-White syndrome converted to sinus rhythm by intravenous lidocaine. *J Med Assoc State Ala* 42:271, 1972.
52. **Rosen KM, Barwolf C, Ehsani A, et al:** Effects of lidocaine and propranolol on the normal and anomalous pathways with pre-excitation. *Am J Cardiol* 30:801, 1972.
53. **Chaudron JM, Lebacq E:** Demonstration of a major pre-excitation syndrome during treatment of auricular flutter using intravenous injection of verapamil. (Eng Abstr) *Acta Cardiol* (Bruxelles) 30:137, 1975.
54. **Bhargava AN, Chandra K, Dhanda PC:** Effects of digitalis and quinidine in a patient with short P-R interval. *J Assoc Physicians India* 13:563, 1965.
55. **Malaguti R, Soffritti E:** Reflex nervous stimulations and digitalis in the Wolff-Parkinson-White syndrome. (Ital) *Arcisped S Anna Ferrara* 17:981, 1964.
56. **Nakamoto K:** Simultaneous block of the normal A-V pathway and abnormal AV bypass tract by digitalis intoxication in a patient with type A WPW syndrome. *Jpn Circ J* 32:861, 1968.
57. **Bracchetti D, Binetti G, D'Osualdo F, et al:** Contraindications to the use of digitalis in atrial fibrillation in the course of Wolff-Parkinson-White syndrome. (Eng Abstr) *Boll Soc Ital Cardiol* 19:181, 1974.
58. **Dyk T, Kozlowski W:** On the problems of administration of digitalis in Wolff-Parkinson-White syndrome. (Pol). *Pol Tyg Lek* 23:1280, 1968.
59. **Swiderski J, Lees MH, Nadas AS:** The Wolff-Parkinson-White syndrome in infancy and childhood. *Br Heart J* 24:561, 1962.
60. **Engle MA:** Wolff-Parkinson-White syndrome in infants and children. *Am J Dis Child* 84:692, 1952.
61. **Gleckler WJ, Lay JVM:** Wolff-Parkinson-White syndrome and paroxysmal tachycardia in infancy. *JAMA* 150:683, 1952.
62. **Kaye HH, Reid DS, Tynan M:** Studies in a newborn infant with supraventricular tachycardia and Wolff-Parkinson-White syndrome. *Br Heart J* 37:332, 1975.
63. **Kreidberg MB, Dushan TA:** Paroxysmal auricular tachycardia associated with Wolff-Parkinson-White syndrome in a newborn infant. *J Pediatr* 43:92, 1953.
64. **Giardina AC, Ehlers KH, Engle MA:** Wolff-Parkinson-White syndrome in infants and children. A long-term follow-up study. *Br Heart J* 34:839, 1972.
65. **Przybylski J, Chiale PA, Quinteiro RA, et al:** The occurrence of phase-block in the anomalous bundle of patients with Wolff-Parkinson-White syndrome. *Eur J Cardiol* 3:267, 1975.
66. **Cragnolino HJ:** Guanethidine in paroxysmal tachycardia. *Lancet* 1:606, 1965.
67. **Cragnolino HJ, Podio C, Gagliardi J, et al:** Guanethidine in the Wolff-Parkinson-White syndrome. Prevention and treatment of the paroxysmal tachycardia crisis. (Sp) *Prensa Med Argent* 52:2759, 1965.
68. **Georgopoulos A, Tountas K:** The influence of the antihistaminic drug antazoline on the Wolff-Parkinson-White syndrome. (Gr) *Hellen Iatr* 32:1235, 1963.
69. **Montoyo JV, Llamas A, Guasi C:** Antazoline and auricular fibrillation in the Wolff-Parkinson-White syndrome. (Sp) *Rev Esp Cardiol* 29:55, 1976.
70. **Maisuradze MZ, Mamaladze GT:** On the use of prednisolone in the treatment of Wolff-Parkinson-White syndrome. *Ter Arkh* 37:116, 1965.
71. **Ballarino M, Bruno L, Prevosti L:** Propranolol and ventricular pre-excitation (Wolff-Parkinson-White syndrome). (Ital) *Atti Soc Ital Cardiol* 2:161, 1968.
72. **Demiroglu C, Göskel MF, Yavuz A, et al:** Effect of propranolol on the Wolff-Parkinson-White syndrome. (Turk) *Tip Fak Med* (Istanbul) 30:333, 1967.
73. **Golden RL:** Prevention of paroxysmal tachycardia of Wolff-Parkinson-White syndrome with combined propranolol and quinidine therapy. *Psychosomatics* 11:585,

1970.
74. **Moene RJ, Roos JP:** Practolol in attacks of tachycardia in Wolff-Parkinson-White syndrome. (Eng Abstr) *Ned Tijdschr Geneeskd* 116:877, 1972.
75. **Ogan H, Ağbaba S, Ercan N, et al:** Wolff-Parkinson-White syndrome not remitting after treatment with propranolol (Inderal). *Turk Tip Cem Mec* 34.246, 1968.
76. **Ahlinder S, Granath A, Holmer S, et al:** Wolff-Parkinson-White syndrome with external heart massage and defibrillation. (Sw) *Nord Med* 70:1336, 1963.
77. **Battistini G, Riva P, Acito P, et al:** Our preliminary experience with treatment of atrial fibrillation and fibrioolflutter with external electric shock. (Ital) *G Clin Med* 48:1011, 1967.
78. **Castellanos A Jr, Johnson D, Mas I, et al:** Electrical conversion of paroxysmal atrial fibrillation in the Wolff-Parkinson-White (pre-excitation) syndrome. *Am J Cardiol* 17:91, 1966.
79. **Lapi PR, Santini L, Masotti G, et al:** The treatment of paroxysmal auricular fibrillation in the Wolff-Parkinson-White syndrome with electric cardioversion. (Ital). *Boll Soc Ital Cardiol* 13:146, 1968.
80. **Meyer AD, Greenberg HB:** Cardioversion of recurrent postoperative supraventricular tachycardia in Wolff-Parkinson-White syndrome. *Am J Cardiol* 18:904, 1966.
81. **Dreifus LS, Wellens HJ, Watanabe Y, et al:** Sinus bradycardia and atrial fibrillation associated with the Wolff-Parkinson-White syndrome. *Am J Cardiol* 38:149, 1976.
82. **Sealy WC, Gallagher JJ, Wallace AG:** The surgical treatment of Wolff-Parkinson-White syndrome: evolution of improved methods for identification and interruption of the Kent bundle. *Ann Thorac Surg* 22:443, 1976.
83. **Preston TA, Kirsh MM:** Permanent pacing of the left atrium for treatment of WPW tachycardia. *Circulation* 42:1073, 1970.
84. **Ravara L, Kaku S:** A case of paroxysmal tachycardia treated by atrial stimulation. *Rev Port Ter Med* 6:42, 1972.
85. **Ryan GF, Easly RM Jr, Zaroff LI, et al:** Paradoxical use of a demand pacemaker in treatment of supraventricular tachycardia due to the Wolff-Parkinson-White syndrome. Observation on termination of reciprocal rhythm. *Circulation* 38:1037, 1968.
86. **Wirtzfeld A, Kiefhaber S, Baedeker W:** Pacemaker therapy in the WPW syndrome. (Ger) *Verh Dtsch Ges Kreislaufforsch* 39:322, 1973.
87. **Fantini F, Camilli L, Grassi G:** Radio-frequency stimulation in the therapy of paroxysmal tachycardia crises in Wolff-Parkinson-White syndrome. *Minerva Med* 64:2281, 1973.
88. **Fantini F, Camilli L, Grassi G, et al:** Stimulation with radio-frequency in the therapy of paroxysmal tachycardia crises in the Wolff-Parkinson-White syndrome. *Boll Soc Ital Cardiol* 18:517, 1973.
89. **Gallagher JJ:** Letter: Reply to S Goldstein. *Circulation* 53:205, 1976.
90. **Goldstein S:** Letter: Pacing the WPW patient. *Circulation* 53:204, 1976.
91. **Mandel WJ, Laks MM, Yamaguchi I, et al:** Recurrent reciprocating tachycardias in the Wolff-Parkinson-White syndrome. Control by the use of a scanning pacemaker. *Chest* 69:769, 1976.
92. **Dreifus LS, Arriaga J, Watanabe J, et al:** Recurrent Wolff-Parkinson-White tachycardia in an infant. Successful treatment by a radio-frequency pacemaker. *Am J Cardiol* 28:586, 1971.
93. **Akakura I:** Surgical treatment of WPW (Wolff-Parkinson-White) syndrome. (Jpn) *Jpn J Thorac Surg* 22:765, 1969.
94. **Iwa T, Misaki T, Sugimoto K, et al:** Surgical approach in Wolff-Parkinson-White syndrome. (Jpn) *Respir Circ* (Tokyo) 20:901, 1972.
95. **Iwa T, Sugimoto K, Misaki T, et al:** Surgery of Wolff-Parkinson-White syndrome. (Jpn) *J Jpn Assoc Thorac Surg* 20:568, 1972.
96. **Iwa T, Todo K, Misaki T, et al:** Surgical treatment of Wolff-Parkinson-White syndrome. (Jpn) *Jpn J Clin Med* 30:1779, 1972.
97. **Meijne NG, Mellink HM, Van Dam RT, et al:** Surgical treatment of the Wolff-Parkinson-White syndrome. (Dut) *Ned Tijdschr Geneeskd* 115:90, 1971.

98. **Nakamura K, Okada M, Kishmoto A, et al:** Surgical interruption of Kent's bundle in WPW syndrome—postoperative recurrence. (Jpn) (Eng Abstr) *Jpn J Thorac Surg* 26:883, 1973.
99. **Tarao M, Hirose M, Murase K, et al:** Surgical treatment of Wolff-Parkinson-White (WPW) syndrome associated with mitral stenosis. *J Jpn Assoc Thorac Surg* 23:1366, 1975.
100. **Wellens HJ, Schuilenberg RM, Durrer D:** Indications for surgical treatment of the Wolff-Parkinson-White syndrome. (Dut) *Ned Tijdschr Geneeskd* 115:89, 1971.
101. **Iwa T:** Surgical experiences with the Wolff-Parkinson-White syndrome. *J Cardiovasc Surg* 17:549, 1976.
102. **Sealy WC, Wallace AG:** Surgical treatment of Wolff-Parkinson-White syndrome. *J Thorac Cardiovasc Surg* 68:757, 1974.
103. **Knippel M, Pioselli D, Rovelli F, et al:** Tachicardiac ribelle nella sindrome de preccitazione. Trattamento chirurgico di cinque casi. *G Ital Cardiol* 4:657, 1974.
104. **Wellens HJ, Janse MJ, Van Dam R, et al:** Epicardial mapping and surgical treatment in Wolff-Parkinson-White syndrome type A. *Am Heart J* 88:69, 1974.
105. **Coumel P, Waynberger M, Fabiato A, et al:** Wolff-Parkinson-White syndrome: problems in evaluation of multiple accessory pathways and surgical therapy. *Circulation* 45:1216, 1972.
106. **Fontaine G, Guiraudon G, Bonnet M, et al:** Kent's bundle section in a case of A-B Wolff-Parkinson-White syndrome. II. Epicardial cartographies. *Arch Mal Coeur* 65:925, 1972.
107. **Dunaway MC, King SB, Hatcher CR, et al:** Disabling supraventricular tachycardia of Wolff-Parkinson-White syndrome type A, controlled by surgical A-V block and a demand pacemaker after epicardial mapping studies. *Circulation* 45:522, 1972.
108. **Lindsay AE, Nelson RM, Abildskov JA, et al:** Attempted surgical division of the pre-excitation pathway in the Wolff-Parkinson-White syndrome *Am J Cardiol* 28:581, 1971.
109. **Latour H, Puech P, Grolleau R, et al:** Surgical treatment of severe attacks of paroxysmal tachycardia in Wolff-Parkinson-White syndrome and its limitations. *Arch Mal Coeur* 63:977, 1970.
110. **Cole JS, Wills RE, Winterscheid LC, et al:** The Wolff-Parkinson-White syndrome problems in evaluation and surgical therapy. *Circulation* 42:111, 1970.
111. **Edmonds JH Jr, Ellison RG, Crews TL:** Surgically induced atrioventricular block as treatment for recurrent atrial tachycardia in Wolff-Parkinson-White syndrome. *Circulation* 39 (suppl 105):11, 1969.
112. **Sealy WC, Hatller BG, Blumenschein SD, et al:** Surgical treatment of Wolff-Parkinson-White syndrome. *Ann Thorac Surg* 8:1, 1969.
113. **Burchell HB, Frye RL, Anderson MW, et al:** Atrioventricular and ventriculoatrial excitation in Wolff-Parkinson-White syndrome (type B). Temporary ablation at surgery. *Circulation* 36:663, 1967.
114. **Dreifus LS, Nichols H, Morse D, et al:** Control of recurrent tachycardia of Wolff-Parkinson-White syndrome by surgical ligature of the A-V bundle. *Circulation* 38:1030, 1968.
115. **Cobb FR, Blumenschein SD, Sealy, WC, et al:** Successful surgical interruption of the bundle of Kent in a patient with Wolff-Parkinson-White syndrome. *Circulation* 38:1018, 1968.
116. **Sealy WC:** Surgical treatment of Wolff-Parkinson-White syndrome. *Bull Soc Int Chir* 29:252, 1970.
117. **Gallagher JJ, Svenson RH, Sealy WC, et al:** The Wolff-Parkinson-White syndrome and the preexcitation dysrhythmias. *Med Clin North Amer* 60:101, 1976.
118. **Sealy WC, Wallace AJ, Ramming KP, et al:** An improved operation for the definitive treatment of the Wolff-Parkinson-White syndrome. *Ann Thorac Surg* 17:107, 1974.
119. **Wallace AG, Sealy WC, Gallagher JJ, et al:** Surgical correction of anomalous left ventricular pre-excitation: Wolff-Parkinson-White (type A) *Circulation* 49:206, 1974.
120. **Burchell HB:** Surgical approach to the treatment of ventricular pre-excitation. *Adv*

Intern Med 16:43, 1970.

121. Surgery in Wolff-Parkinson-White syndrome. *Brit Med J* 4:547, 1974.
122. **Holzmann M:** New diagnostic and therapeutic developments in Wolff-Parkinson-White syndrome. *Schweiz Med Wochenschr* 101:494, 1971.
123. **James TN:** The Wolff-Parkinson-White syndrome. *Ann Intern Med* 71:399, 1969.
124. **King SB, Logue RB:** Surgery for WPW syndrome. *Circulation* 49:1020, 1974.
125. **Gallagher JJ, Sealy WC, Kasell J, et al:** Multiple accessory pathways in patients with the Wolff-Parkinson-White syndrome. *Circulation* 54:571, 1976.
126. **Gallagher JJ, Sealy WC, Anderson RW, et al:** Cryosurgical ablation of accessory atrioventricular connections. A method for correction of the pre-excitation syndrome. *Circulation* 55:471, 1977.

IX

Problems of Diagnosis: Some Illustrative Cases

PRE-EXCITATION PATTERNS MIMICKING OR OBSCURING OTHER ECG ABNORMALITIES

According to Wolff[1] and Ruskin et al,[2] diagnostic errors are made in more than one-third of the patients with the pre-excitation syndrome. Not only does the anomalous ECG mimic[3-9] or obscure[10–23] other pathologic entities, but the grossly abnormal appearance of the ECG tracing in itself frequently leads to an erroneous diagnosis of heart disease.[24–30]

Pitfalls in the Diagnosis of Myocardial Infarction

The following case report illustrates this problem:

A 52-year-old man, suffering from hypertension for 15 years and from diabetes mellitus for 4 years, came to the hospital after 2 days of increasingly severe precordial pain. He complained of severe pain in the anterior chest area, which radiated into the neck and left arm. Blood pressure fell from 180/120 to 100/80. In the ECG (Figure 12a) the QS pattern in leads II, III and aVF suggested an infarct in the diaphragmatic myocardium. A delta-like wave raised the possibility of pre-excitation, but this was excluded on the basis of a normal P-R interval (0.14′). The following day (Figure 12b) deep negative T waves appeared in V_2-V_6, which were interpreted as pre-excitation syndrome and ischemia of the anterior and septal wall. The severity of the clinical course and the nature of the ECG changes led to a diagnosis of acute myocardial infarct. The ECG remained unchanged during the patient's 3 weeks in the hospital (Figure 12c). On reexamination 2 years later, the signs of pre-excitation and the assumed diaphragmatic myocardial infarction had disappeared (Figure 12d). However, the typical sign of an old anterior septal infarct was now present in precisely the same place in which the negative T waves had made their appearance the day after admission 2 years earlier.

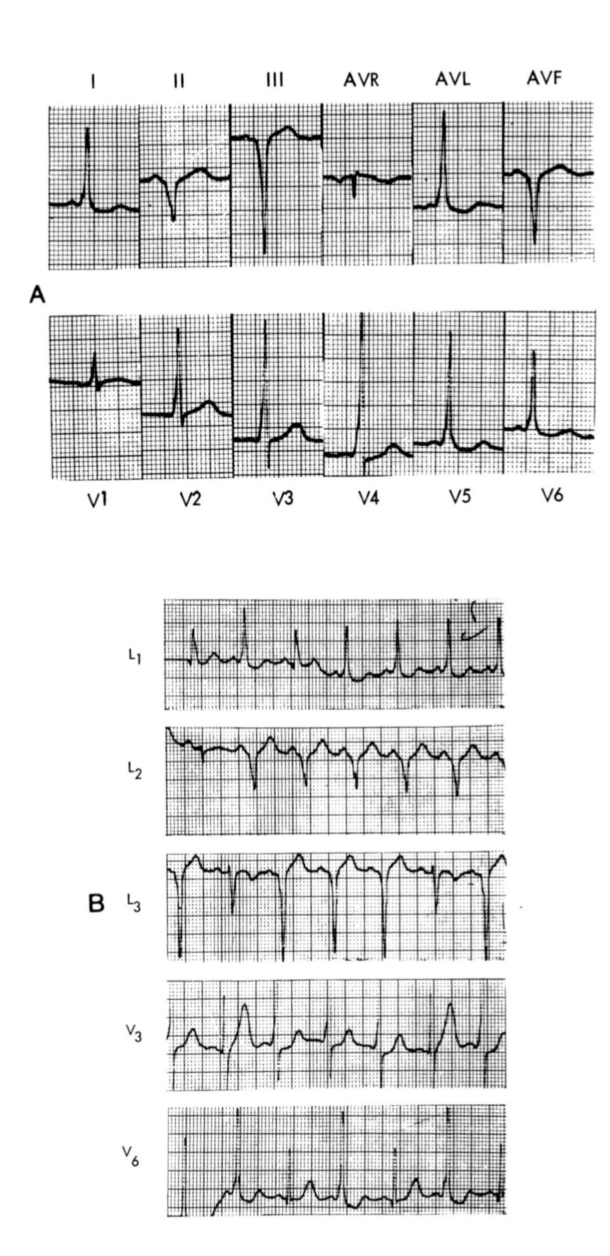

Figure 51. *Pre-excitation mimicking left ventricular hypertrophy in a 59-year-old man.* **a.** *Pre-excitation conduction.* **b.** *Intermittent pre-excitation. Note high R complexes in lead V_6 and deep S waves in lead V_3.*

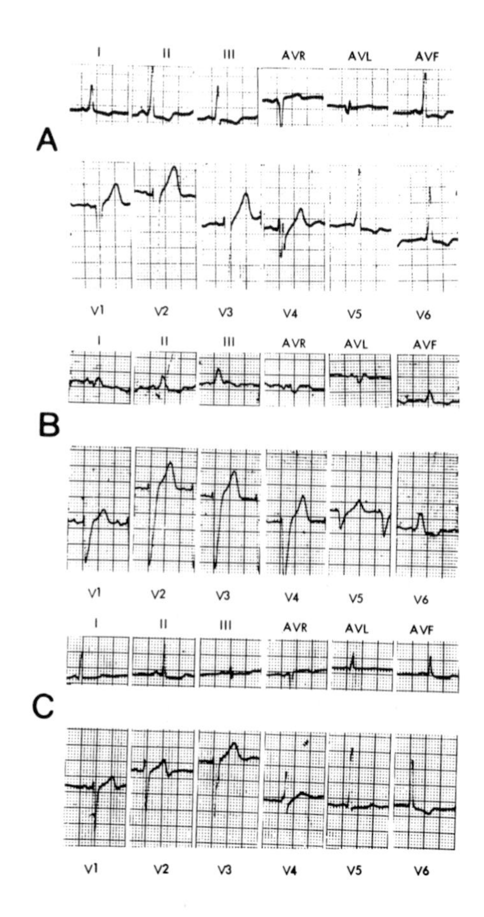

Figure 52. *Pre-excitation and additional left bundle branch block.* **a.** *Pre-excitation type BI.* **b.** *In addition to the short P-R interval and delta waves, note the appearance of broad QRS complexes and a configuration resembling left bundle branch block.* **c.** *Normal AV conduction. Left heart strain pattern.*

This case illustrates the dual problem in the diagnosis of myocardial infarction in the presence of the pre-excitation syndrome. Q waves in leads II, III and aVF can be an indication of either pre-excitation or an infarct in the diaphragmatic myocardium, and the typical QRS pattern of the pre-excitation syndrome, which may remain unchanged during the acute infarction, may mask the underlying acute condition.

Diaphragmatic myocardial infarction is the most common diagnostic error in cases of pre-excitation,[31-33] accounting for almost half of the misdiagnoses (see Table XIII and Figure 8, 14, 15, 51, 53). However,

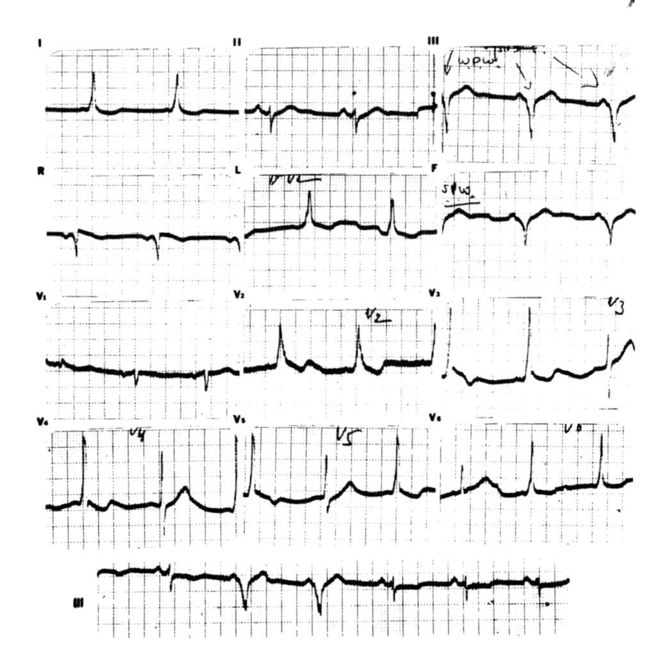

Figure 53. *Pre-excitation mimicking ventricular premature beats in an 18-year-old man. Pre-excitation type BS. The beats of pre-excitation (2, 3) on the bottom (lead III) during intermittency were erroneously interpreted as "ventricular bigeminy".*

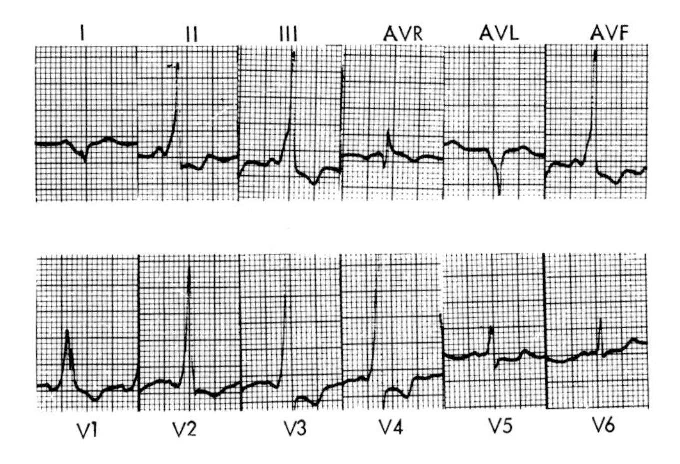

Figure 54. *Pre-excitation mimicking high lateral myocardial infarction in a 21-year-old man. Note QS patterns in leads I and aVL.*

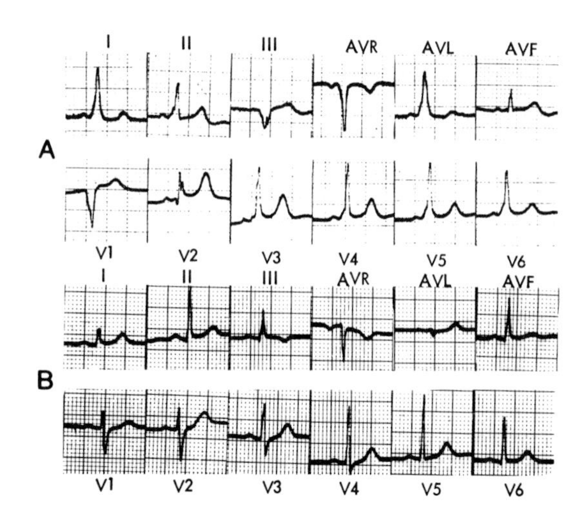

Figure 55. *Pre-excitation mimicking recent anteroseptal myocardial infarction in a 28-year-old woman.* **a.** *Pre-excitation. Note QS in lead V_1 and qR with an elevation of the ST-T segment in lead V_2.* **b.** *Normal AV conduction, normal ECG tracing.*

other pre-excitation patterns in ECG tracings may mimic an old or recent infarction in other locations of the heart. In Figure 5, for example, a QS pattern in leads I and aVL (Figure 5b) suggests the presence of an old high lateral myocardial infarction with ischemic changes in the anteroseptal wall (negative T waves in leads V_1-V_4). A deep Q in leads I and aVL (Figure 54), followed by R and deep T waves in the precordial leads, is another example of pre-excitation mimicking high lateral myocardial infarction. On the other hand, QS in V_1 and QR in V_2, together with an elevation of the ST-T segment (V_2), may arouse suspicion of a recent anteroseptal myocardial infarction[34] (Figure 55). In all these cases, the age of the patients, their complaints (or lack thereof), laboratory findings and especially ECG tracings obtained during intermittency when all these strange infarction–like patterns disappear, make a diagnosis of infarction improbable.

Pitfalls in the Diagnosis of Ventricular Hypertrophy

Pre-excitation has also been seen to mimic or obscure a right or left ventricular hypertrophy:

A 33-year-old woman began experiencing attacks of palpitations 3 years prior to hospitalization, having them 3 to 4 times a year. During one such paroxysmal tachycardia she was hospitalized and a diagnosis of pure *mitral stenosis was established. According to the hospital summary report (August 1954), a diagnosis of "silent" old posterior myocardial infarction due to arteriosclerotic coronary thrombosis was also "proven" (Figure 8b). Because of the presence of pure mitral stenosis, the patterns in V_1 and aVR strongly suggested early right ventricular hypertrophy (incomplete RBBB) (Figure 8b and 8c). Clear delta waves appeared (Figure 8c) which were not present in the previous tracing (Figure 8b), and the pre-excitation syndrome was considered. Later, during a period of intermittency, a normally conducted ECG tracing was obtained (Figure 8a), and both the patterns of "old posterior myocardial infarction" and "early right ventricular hypertrophy" disappeared. Thus, both these diagnoses were found to be erroneous.*

A similar disappearance, this time of a "left hypertrophy," with normalization of the pre-excitation pattern in the ECG tracing is illustrated in Figure 51:

A 59-year-old man suffered from long-standing hypertension. His first ECG (Figure 51) was considered to be compatible with pre-excitation syndrome: an old diaphragmatic myocardial infarction and left heart hypertrophy (high voltage in V_5, V_6 and aVL, and biphasic T waves with a slight ST-T depression). The next day an ECG showing normal conduction was obtained: rs pattern replaced the QS in II and III, voiding the previous diagnosis of old inferior infarction, and more important, the high voltage in leads V_5 and V_6 disappeared as well as all the pathological ST-T changes, and a completely normal QRST pattern became evident in these leads (Figure 51b, lead V_6). The presence of high voltage in the left chest leads as well as an increased S wave in V_1 (Figure 56, lead V_1) may thus represent features of the pre-excitation syndrome which mimic left ventricular hypertrophy, particularly when accompanying clinical signs such as hypertension are present.

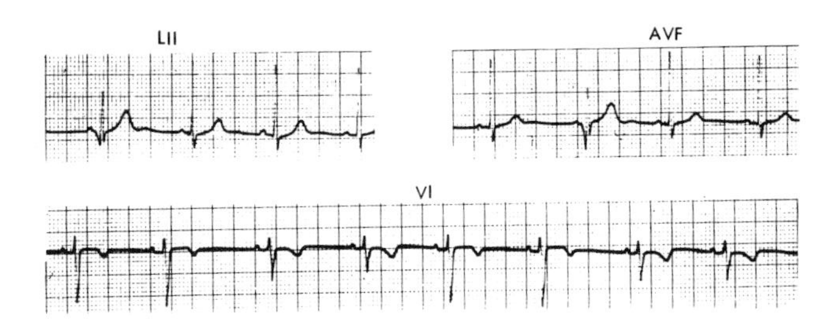

Figure 56. *Intermittent pre-excitation. Note deep q waves in pre-excitation beats in leads II and III, and differences in the depth of S waves in lead V_1.*

In other circumstances the ECG pattern during pre-excitation can obscure left ventricular hypertrophy (Figure 10):

A 10-year-old boy suffered from extreme rheumatic aortic insufficiency. The chest X-ray film revealed an enlarged left ventricle confirmed by the ECG tracing (Figure 10a), which showed high voltage (SV_1 + RV_6 50mm—Sokolov's criterion for left ventricular hypertrophy). Later, however, the patient exhibited pre-excitation pattern (Figure 10b), the high R in V_6 disappeared, and the deep S wave in V_1 was replaced by an upright R of the type A pre-excitation. This late appearance of the anomalous pre-excitation conduction obscured a previously confirmed ECG finding of left ventricular hypertrophy. This patient also suffered from attacks of supraventricular tachycardia and died at the age of 13, during an episode of pulmonary edema (Figure 10c).

Right and Left Bundle Branch Block in the Presence of Pre-excitation

All the cases of pre-excitation published before Wolff, Parkinson and White's classic paper[35] considered the QRS pattern to represent a right or left bundle branch block (BBB). Even Wolff, Parkinson and White who in 1930 defined pre-excitation as a nosological independent entity, called it short P-R with BBB. These two disorders share common features and lead to the confusion in diagnosis.[36-39] For instance, Figure 54, showing a tracing of a pre-excitation pattern, deep S waves in I, aVL, V_5 and V_6, and RR pattern in V_1, and Figure 57 in which the pre-excitation tracing lacks the "q" waves in I, aVL, V_5 and V_6 and a QS

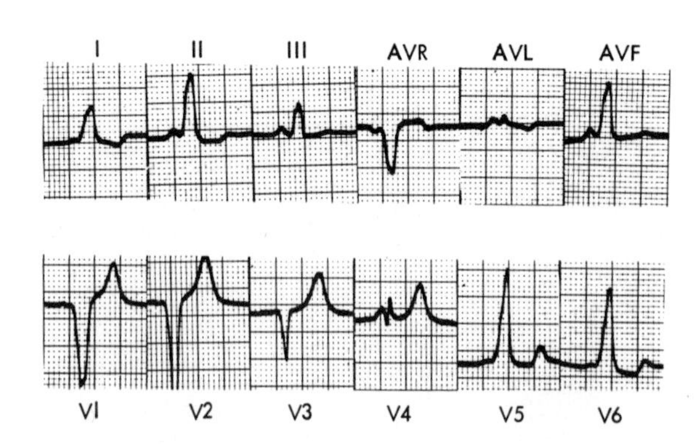

Figure 57. *Pre-excitation mimicking left bundle branch block in a 47-year-old man. Note lack of q waves in leads I, aVL and V_6 and QS pattern in lead V_1.*

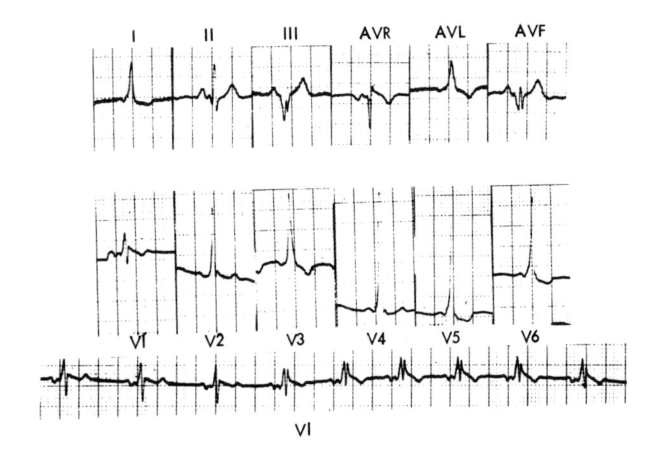

Figure 58. *Pre-excitation and right bundle branch block in a 50-year-old man. Pre-excitation type AS. In the bottom tracing (lead V_1) direct evidence of an additional right bundle branch block pattern is observed (intermittent RBBB).*

pattern in V_1-V_3, may represent right and left BBB, respectively, in addition to the anomalous conduction, or they may be pre-excitation patterns alone. In some cases the interpretation is easier because features characteristic of RBBB or LBBB appear in tracings of patients previously known to be suffering from pre-excitation: these cases may be regarded as showing two conduction disturbances and not just one.[40] For example, in Figure 52a, pre-excitation, type B; in Figure 52b there appears what resembles an additional LBBB with an "M" pattern in I and aVL, and broad low QRS patterns in V_6; and in Figure 52c there is normal conduction showing only nonspecific ST-T changes in leads I, aVL and V_6.

A similar later appearance of a BBB, this time RBBB, against the background of a previously documented pre-excitation case is seen in Figure 28a, pre-excitation, type B; in Figure 28b there is an RR^1 pattern in V_1; and in Figure 28c there is intermittent normal and pre-excitation conduction, without signs of RBBB (see also Figure 58).

The exact diagnosis of pre-excitation and additional BBB is facilitated by spatial vectorcardiogram (see Chapter V).

Right Bundle Branch Block, Left Axis Deviation and Coronary Insufficiency

The detection in ECG tracings of a RBBB and a left axis deviation

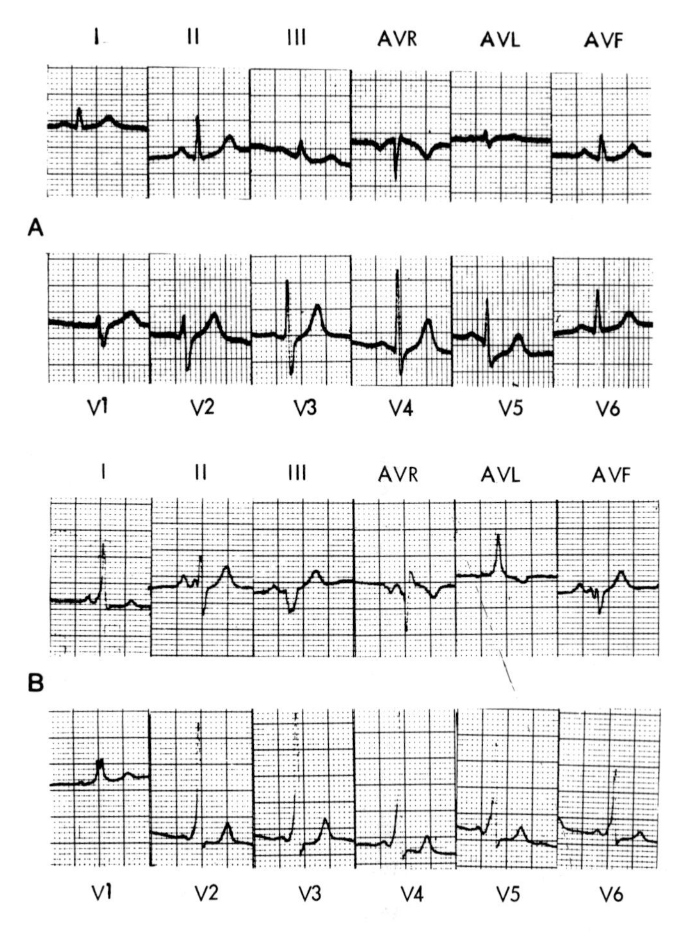

Figure 59. *Pre-excitation mimicking right bundle branch block with left axis deviation in a 33-year-old man.* **a.** *Normal AV conduction, normal ECG tracing.* **b.** *Pre-excitation. Note the appearance of left axis deviation and a pattern in lead* V_1 *resembling right bundle branch block.*

was considered in the last decade to be the expression of a partial multifascicular BBB, which in 10% of the cases [41] was a precursor to a complete AV block (due to the complete inability of both bundle branches to conduct). The prognosis in such cases was considered "unstable." Some of the pre-excitation patterns are also able to mimic this "multifascicular block"[42] (Figure 59):

A 48-year-old man had no cardiac complaints or signs of pathological heart condition. Routine ECG (Figure 59b) and examination prior to *surgery for hydrocele showed a RBBB and LAD. The question was raised whether to use a temporary artificial pacemaker during the operation to prevent a complete AV block (note the 30° left axis deviation and RBBB pattern in Figure 59b). However, a second ECG showed a completely normal pattern and the diagnosis of intermittent pre-excitation mimicking multifascicular block was established. During pre-excitation conduction there was a clear depression of the ST-T section in leads V_3 V_4, a finding usually identified as coronary insufficiency. When later normal conduction replaced the pre-excitation pattern, this ST-T depression disappeared together with the pseudo-multifascicular block.*

This kind of ST-T depression, often associated with negative or biphasic T waves, is not a rare occurrence in the pre-excitation syndrome and is often mistaken as a sign of additional coronary heart disease. Further cardiac investigations can clarify this point. Figure 60 shows a marked ST-T depression in a 24-year-old woman without any complaints or signs of cardiac disease. Figures 35a and 35b show a 26-year-old man suffering from attacks of palpitation, which were documented as simple sinus tachycardia. No trace of the "coronary insufficiency" seen in Figure 35a could be detected during the intermittent normal conduction. In both these patients an ergometry test showed good work capacity with no chest pains or any other discomfort. The differential diagnosis of this kind of pseudocoronary heart disease is simple when it involves young women, but becomes more difficult in elderly persons.

Pre-excitation Mimicking Arrhythmias: Ventricular Premature Beats and Ventricular Tachycardia

In rare cases the intermittent appearance of pre-excitation beats have been thought by some physicians to represent ventricular premature beats (Figure 28c). Such patients have been sent to our clinic for antiarrhythmia treatment. In another instance the patient arrived with a tentative diagnosis of ventricular bigeminy (Figure 53, lead III, bottom). In a third patient (Figure 11) suffering from a long-standing hypertensive cardiovascular disease, we found left ventricular hypertrophy, coronary insufficiency and first degree AV block (P-R 0.22) in the normally conducted ECG tracing (Figure 11a); signs of ischemia (V_2-V_4) and pseudo high lateral myocardial infarction appeared during pre-excitation (Figure 11b). In this case, however, both true ventricular bigeminy (Figure 11c, V_1) and bigeminy due to intermittent pre-excitation (Figure 11c, V_4) were also observed on different occasions. Thus, caution is required before establishing a diagnosis of ventricular premature beats in cases of pre-excitation.

It is sometimes difficult to differentiate true ventricular tachycardia from the pseudoventricular tachycardia often encountered in patients with pre-excitation,[43–44] especially when the rhythm of the tachycardia is regular (Figure 36d). This is also illustrated in the following case:

An elderly patient with atherosclerotic heart disease showed intermittency (Figure 48a) and bouts of true ventricular tachycardia (Figure 48b) with the help of ambulatory monitoring. He reported vertigo and a generally uncomfortable feeling, which is usually not observed in cases with the "pseudoventricular tachycardia" of pre-excitation. It was our belief that the true ventricular tachycardia detected by the ambulatory monitoring was probably secondary to his coronary condition and not to the pre-excitation syndrome.

DIAGNOSTIC CLUES FOR DISCOVERING ACUTE MYOCARDIAL INFARCTION IN THE PRESENCE OF PRE-EXCITATION

The detection of different cogenital heart diseases (especially Ebstein's anomaly) and rheumatic valvular diseases is not problematic, because of the characteristic murmurs and clinical picture. The main difficulty is in diagnosing myocarditis and allied disorders, and acute myocardial infarction.

The correct assessment of the ECG in the diagnosis of acute myocardial infarction in the presence of anomalous AV conduction has often been stressed[45] and is outlined here.[46-58] As spelled out by Scherf and Cohen,[59] the question is whether the ECG changes are part of the picture of pre-excitation uncomplicated by myocardial infarction, or whether they are a picture of myocardial disease in one who previously exhibited the pre-excitation syndrome. Brackbill et al[60] published an excellent review on this subject in 1974, with the following conclusions:

(1) Serial T wave changes alone cannot be relied upon for diagnosing myocardial infarction, since the pre-excitation itself may produce striking and often unstable ST-T wave abnormalities in the absence of myocardial infarction. The authors refer here to Tamagna et al,[61] who experimentally produced variable degrees of pre-excitation in cats. Only a slight change in the degree of right ventricular pre-excitation in the control state produced striking changes in both the QRS complexes and T wave changes. When the influence of experimental acute myocardial infarction on the ST-T wave changes of pre-excitation was studied, they found that T wave changes due to pre-excitation could not be reliably differentiated from T wave changes due to underlying myocardial damage. We found, however, that the appearance of "dynamic" T wave

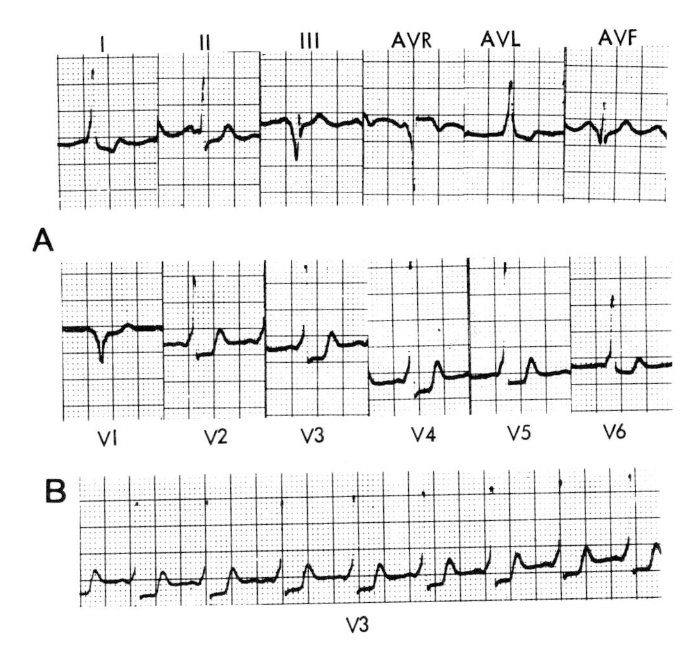

Figure 60. *Pre-excitation mimicking coronary insufficiency in a 24-year-old woman.* **a.** *Pre-excitation type BS. Note the depression of the ST-T segment in leads $V_{2\text{-}5}$.* **b.** *Long strip of lead V_3 to emphasize the ST-T depression.*

changes in the presence of the pre-excitation syndrome, when accompanied by characteristic clinical signs of myocardial infarction, may be of great help in establishing the diagnosis of myocardial infarction. This was seen in one of our patients (Figure 12a, b, c, d), and similar cases have been reported.[62]

(2) Brackbill et al[60] considered the characteristic elevation of the ST-T segment to be much more reliable in the diagnosis of acute myocardial infarction with pre-excitation, although it is quite rare.[56] An illustrative and convincing case was presented by him [60] (and by other investigators[55] as well). On the other hand, they stress that even the marked ST-T elevation of acute myocardial infarction may be completely absent during anomalous AV conduction.

(3) In the presence of pre-excitation, left ventricular extrasystoles which may unmask an infarcted area must satisfy two conditions to be of diagnostic value: the morphology must be either a qR or qRs but not a QS configuration (which can be produced by the type B pre-excitation by itself), and the qR or qRs must not be recorded in aVR where the qRs is expected in any event to be predominantly negative. Schamroth and

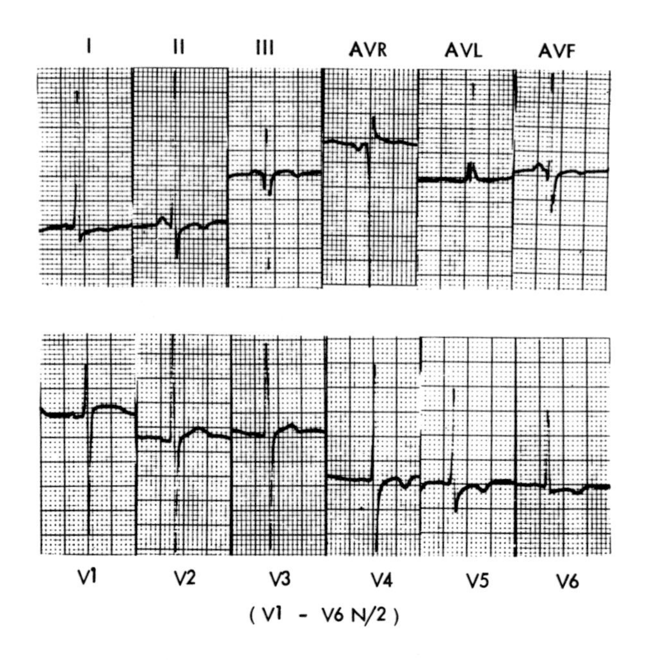

Figure 61. *Atypical pre-excitation in a 22-year-old man. Note short P-R intervals but not sufficiently convincing delta waves in leads I and V_5 (see case report).*

Lapinsky[63] demonstrated such changes.

(4) Brackbill[60] and others have stressed the importance of inducing normal AV conduction before making a diagnosis of myocardial infarction. Very often normal conduction returns spontaneously and the diagnosis becomes obvious. If this does not occur, some authors suggest attempting to induce them; a simple manipulation such as increased respiratory excursions[64] or upright posture can sometimes normalize the QRS complex. Or, if that fails, some investigators[65,66] have used drugs with success, including amyl nitrite inhalation,[57] ajmaline,[25,67-75] quinidine[48,57] and atropine.[47] (This approach is sometimes also important in demonstrating the presence or absence of old myocardial infarction.)

According to the criteria established by the World Health Organization, the diagnosis of a definite myocardial infarction can be made today even in the presence of an equivocal ECG tracing, if the enzymes and the clinical picture are characteristic for infarction. With our capability of receiving precise data not only on enzymes (each one appearing at a definite period during the acute stage), but also specific isoenzymes, the diagnosis of acute myocardial infarction in the presence of pre-excitation has now been made relatively easy.

ATYPICAL CASES

In the great majority of cases it is easy to diagnose a patient with pre-excitation from a standard 12-lead ECG tracing, even when there are unusual features such as a normal P-R interval or a narrow QRS complex. There are, however, instances where cardiologists do not agree on the interpretation of the record.[76-78] Several cases from the Tel Hashomer Heart Institute, which were not included in the study, follow:

A 22-year-old man experienced several paroxysms of palpitations in recent years, but the arrhythmias were never registered in ECG tracings. Some physicians considered a 2/4 grade systolic murmur as being functional, others suspected obstructive cardiomyopathy. The case was summarized as one of pre-excitation syndrome and possible subaortic stenosis. In the ECG tracing (Figure 61), very high voltage was found in lead I, aVL and precordial leads and a negative T wave in leads II, III, aVF, V_4-V_6. A slurring at the beginning of the QRS complex in leads I and V_5 and a rR pattern in aVL were considered as delta waves; however, the delta was not sufficiently convincing to allow a definite diagnosis of pre-excitation. Nevertheless, because of the short P-R interval (0.10-0.12 second) we considered this case an atypical form (or forme fruste) of pre-excitation.

A 59-year-old woman suffered anginal pains for 2 years before presentation. She suffered from moderate hypertension, and palpitations were assumed in the last year from her objective report. The ECG tracing (Figure 62) showed a short P-R (0.08 second), a slurring at the beginning of leads III, aVF and V_6 and negative T waves in most limb and precordial leads. The slurring was considered to represent delta waves, and pre-excitation was diagnosed. Like the previous case, it was our opinion that the delta wave form did not permit a positive diagnosis of pre-excitation, and we therefore labeled it another atypical case of pre-excitation.

A 35-year-old woman suffered for years from palpitations, and was considered "an atypical case of pre-excitation" because of a borderline P-R interval (0.12-0.13 second), unconvincing delta waves in leads II, III and aVF, an rsr_1 pattern in V_1 and a qR pattern in V_5 and V_6. In other tracings a negative P wave appeared without a concomitant change in the QRS complex (Figures 63a and 63b). The case was considered atypical for another reason: an ECG tracing registered during what the patient described as a paroxysm of palpitations turned out at one time to be only a sinus tachycardia (Figure 63c), and at another time ventricular tachycardia. However, it seemed to us that this was a pseudoventricular tachycardia

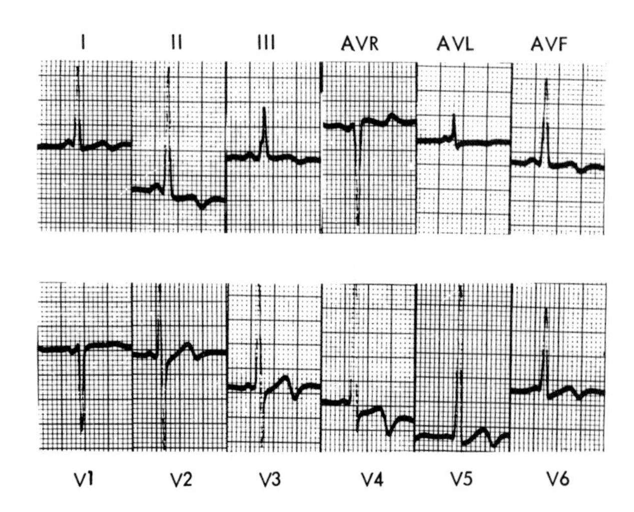

Figure 62. *Atypical pre-excitation in a 59-year-old woman. ECG tracing similar in some aspects to Figure 61: short P-R interval, unclear delta waves (see case report).*

(atrial fibrillation with broad QRS complexes, Figure 63d), seen exclusively in pre-excitation. Thus, this case can be labeled pre-excitation with changing P waves, paroxysmal sinus tachycardial and fast atrial fibrillation showing broad QRS complexes. The clue to its diagnosis was the pseudoventricular tachycardia which is almost pathognomonic for this disorder.

These problems of atypical versus typical pre-excitation are of particular importance to the general practitioner working outside the larger cardiology centers. At present such cases with troublesome tachycardias should be transferred to hospitals where noninvasive or invasive intracardiac investigations can be performed, and a more exact diagnosis can be reached.

The Short P-R Normal QRS Syndrome

Cases which do not show a delta wave at the beginning of the QRS complexes do not, by definition, belong to the pre-excitation syndrome, but are considered by us to be unusual cases of the disorder.[4,79-85]

Massumi and Vera[86] investigated a 62-year-old woman with short P-R interval and normal QRS complex who suffered numerous episodes of tachycardia at rates above 180/min (case 10). Normal QRS complexes were found in control beats and in all tracings taken during tachycardia, but atrial stimulation at rates up to 150/min uncovered impressive delta

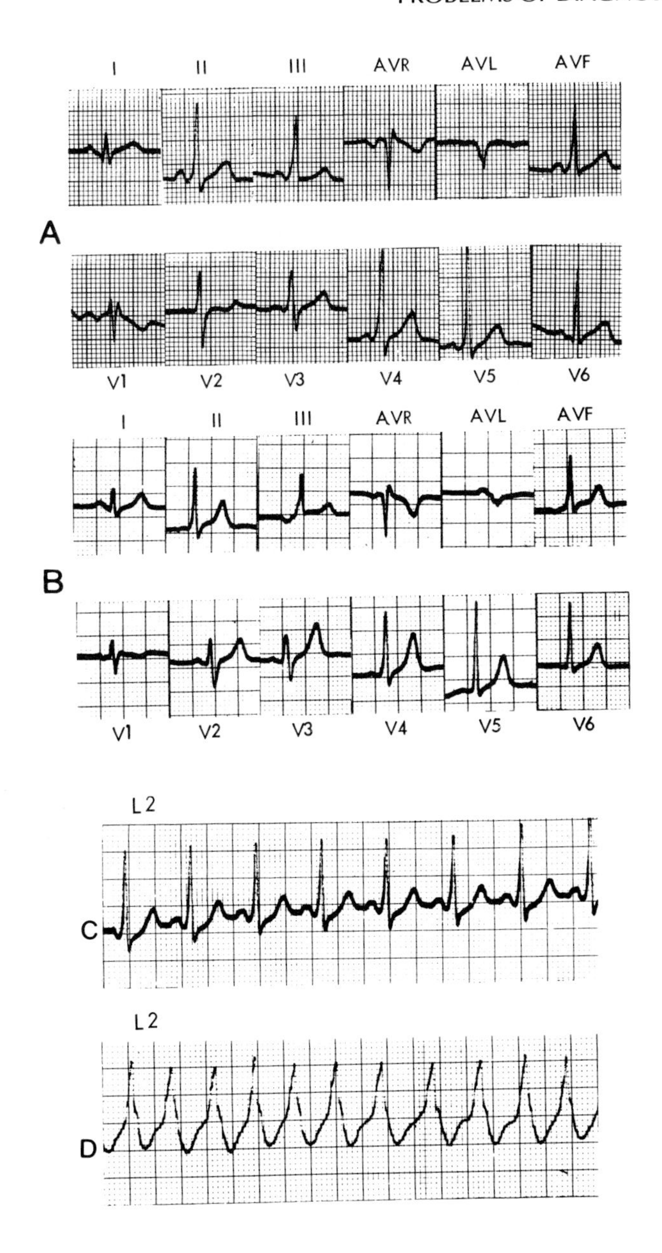

Figure 63. *Atypical pre-excitation in a 35-year-old woman.* **a.** *Pre-excitation conduction. Note delta waves in leads II, III, aVF and $V_{4\text{-}5}$.* **b.** *Similar tracing with P wave changes. Note negative P waves in leads II, III, aVF.* **c.** *Sinus tachycardia.* **d.** *Pseudoventricular tachycardia (atrial fibrillation with broad QRS complexes (see case report).*

waves.[86, Figure 14] At the same time the His deflection moved further into the QRS so that, in the extreme situation of total excitation through the accessory pathway, the His deflection was buried in the ventricular potential. Intracardiac registration studies and atrial pacing have taught us that this kind of electrophysiologic behavior is the "hallmark" of pre-excitation due to an accessory bundle (see Chapter VI).

Brechenmacher et al[87] described a 65-year-old man who suffered from attacks of paroxysmal tachycardia for 13 years. The ECG showed a sinus rhythm with a short P-R interval (0.12 second) and a narrow QRS complex. He was diagnosed as a case of Lown-Ganong-Levine syndrome. The ECG tracings recorded during attacks of tachycardia[87, Figure 6,7] were considered to represent atrial fibrillation with a functional RBBB and a ventricular rate of 200, or an atrial flutter of 300, with broad QRS complexes. In our opinion these were tracings of pseudoventricular tachycardias seen often in pre-excitation type A (where the pattern in V_1 is always similar to RBBB). This resemblance is striking when Brechenmacher et al's tracings[87, Figure 6,7] are compared to our Figure 37b recorded during a paroxysm of tachycardia in a case of pre-excitation (Figure 37a). There is even an obvious similarity between Brechenmacher et al's case 1 and 2[87, Figure 1] in the precordial leads; but the tachycardia in case 1 was labeled atrial fibrillation with "an aspect of a functional RBBB, probably with major pre-excitation," and no mention of pre-excitation was made in case 2. This discrepancy in the interpretations of almost identical ECG tracings arose from the results of pathological investigations: in case 1 the authors found an accessory bundle outside the normal conduction system (Kent bundle), believed to produce the ECG patterns during atrial fibrillation; in case 2 they found an atrial-His bundle connection which could be classified as an accessory bundle partly included in the normal conduction system (inlet connections, see Chapter IV). A similar case was reported by Lev et al,[88] where no accessory bundle outside the normal conduction system (Kent bundle) was found, and only "Mahaim and James' " fibers were described. Nevertheless, this patient as well as Brechenmacher's exhibited bouts of atrial fibrillation with broad QRS complexes (pseudoventricular tachycardia) resembling LBBB (type B in Rosenbaum's classification). Finally, the short P-R normal QRS tracing presented by Brechenmacher had a P-R of 0.12 second, an interval considered normal by our criteria (Chapter II). We consider this to be a case of intermittent pre-excitation rather than of Lown-Ganong-Levine syndrome.

A 26-year-old woman, mother of 4, suffered from palpitations for 8 years. No special findings were revealed by physical examination or X-ray studies. The ECG showed a short P-R interval (0.10-0.11 second) with

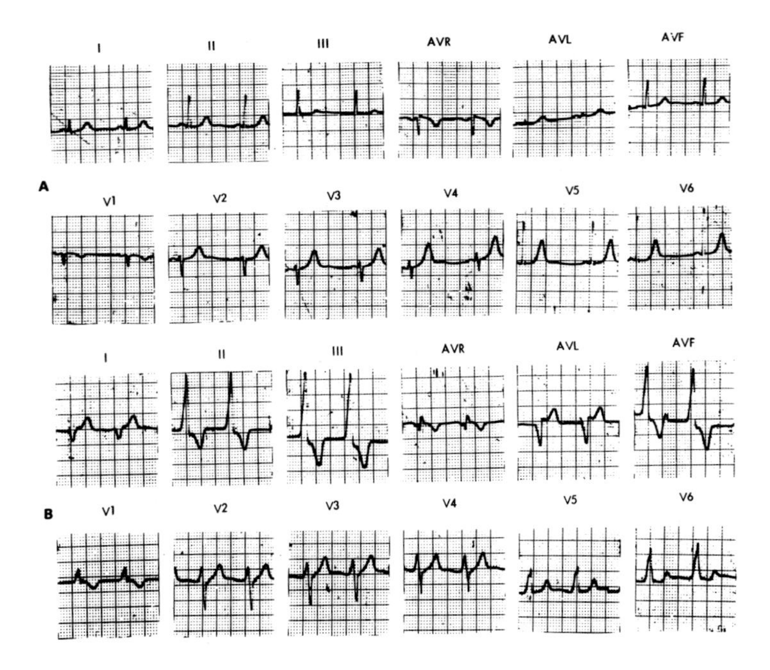

Figure 64. *The short P-R, normal QRS syndromes.* **a.** *Normal ECG tracing. Note the short P-R interval (0.10—0.11 second).* **b.** *Junctional rhythm, with retrograde conducted P waves (rate 95). Note the QRS pattern resembling pre-excitation type AI (see case report).*

normal QRS complexes during control sinus rhythm (Figure 64a). An ECG tracing taken during an attack of palpitations showed the following (Figure 64b): rate 95, rhythm junctional with retrograde conducted P waves, broad QRS complexes (0.11-0.12 second) resembling pre-excitation type A with clear delta waves in leads II, III, aVF, V_1, V_5 *and* V_6.

This group of short P-R normal QRS complexes was included here also because of the ability of bypass fibers of James[89] to reproduce, in combination with other unusual features (such as conducting Mahaim fibers or asynchronization of the normal order of ventricular depolarization), a typical ECG pattern of pre-excitation which is indistinguishable in the regular ECG tracing from patterns assumed to be produced by an accessory bundle outside the conduction system.

There are many cases of short P-R and normal QRS complexes, both with and without attacks of tachycardia. Intracardiac investigations in such cases throw a different light on these ECG findings. [90] The data of Caracta et al[79] suggested the following possible explanations for the

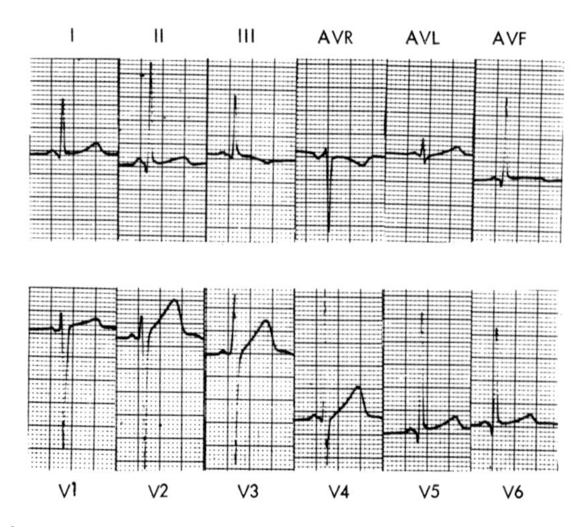

Figure 65. *Short P-R interval, aberrant (?) QRS complexes in an 18-year-old man. Note short P-R interval, deep S waves in lead V_1 and high R deflections in $V_{5,6}$.*

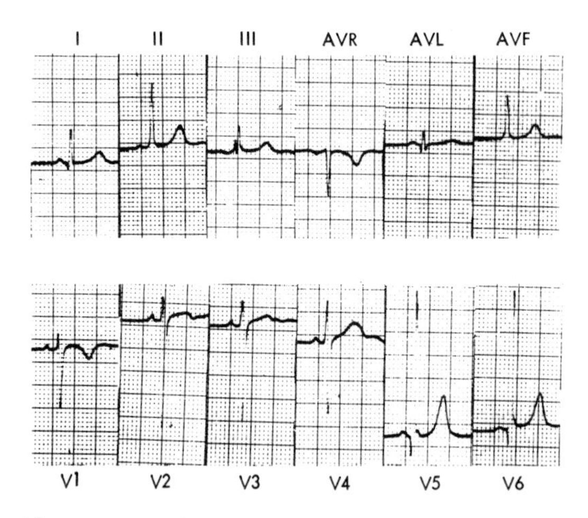

Figure 66. *Short P-R, aberrant (?) QRS complexes in a 24-year-old man. Note short P-R intervals, deep Q waves in leads $V_{5,6}$ with high R deflections.*

short P-R intervals: 1) total or partial bypass of the AV node; 2) an anatomically small AV node; 3) a short or rapidly conducting intranodal pathway; or 4) isorhythmic AV dissociation. The first three may be involved in the formation of some forms of pre-excitation (see Chapter VI).

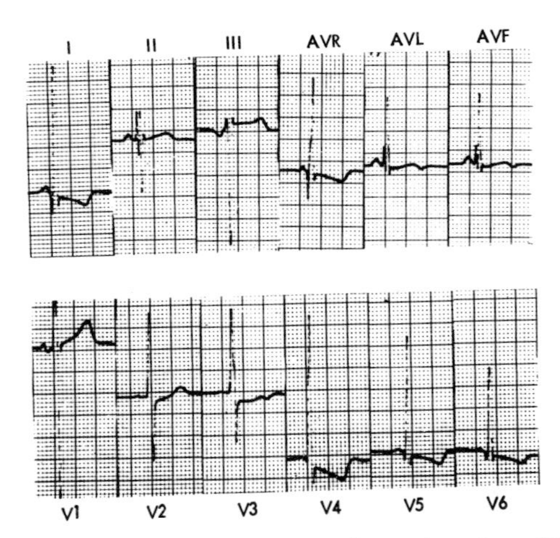

Figure 67. *Short P-R, aberrant (?) QRS complexes in a 17-year-old boy. Short P-R intervals. Note deep Q waves in lead I and very high voltage in leads I, V_1, V_5; negative T waves in I, $V_{4\text{-}6}$.*

There were 6 patients with unusual ECG tracings in the Tel Hashomer files. They exhibited short P-R intervals followed by q waves and QRS complexes resembling LVH (deep S waves in V_1, high R's in V_5, V_6, Figure 65, 66, 67) or combined ventricular hypertrophy (additional high R waves in V_1, Figure 68, 69, 70). Interestingly, all but one were young men without any other cardiopathy. Two of the 6 cases complained of palpitations but no paroxysmal tachycardia was ever documented by ECG. These cases were included in this chapter, since we suspect that the strange QRS patterns following short P-R intervals may represent some unusual form of QRS aberrations in the absence of any other indication of ventricular hypertrophy. They are not considered cases of pre-excitation because the delta wave is absent, but we mention them as a possible additional form of anomalous conduction.

When studying the short P-R normal QRS syndromes we are often confronted with a diagnosis of Lown-Ganong-Levine syndrome. Lown, Ganong and Levine [91] published a collection of cases in 1952 with a short P-R interval arbitrarily chosen to be 0.12 or 0.13 second, seen mostly in middle-aged women (two-thirds of the cases) with no organic heart disease, exhibiting snapping first apical heart sound and (in 11%) attacks of paroxysmal tachycardia. To the best of our knowledge, no other series of cases has been reported with these criteria. Mandel et al[90] published a series of 3 cases under the heading of "Lown-Ganong-Levine Syndrome," 2 men and one woman, all with recurrent tachycardias but none with a snapping first heart sound. Most of the other papers dealing with

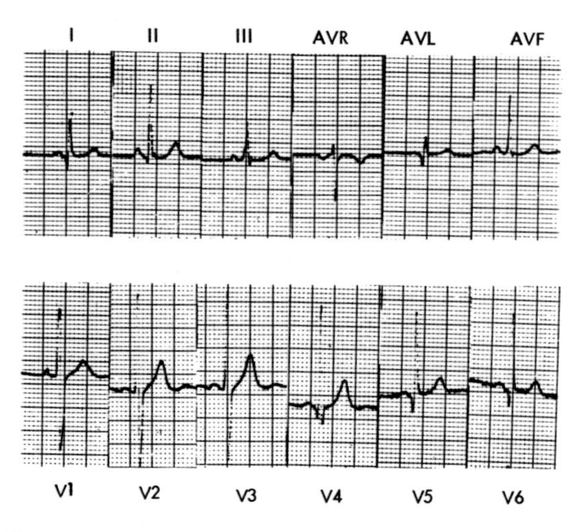

Figure 68. *Short P-R, aberrant (?) QRS complexes in a 37-year-old man. Note short P-R interval, deep Q waves in leads I, aVL, $V_{5,6}$ and R-S in V_1.*

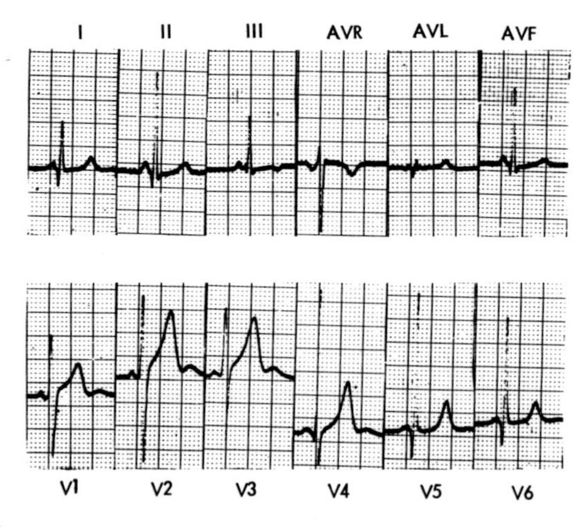

Figure 69. *Short P-R, aberrant (?) QRS complexes in a 17-year-old boy. Note short P-R interval, deep q waves in leads I, II, $V_{5,6}$, high voltage in lead V_5 and R-S in V_1.*

this syndrome were sporadic reports showing none of the features described by the authors, except for a short P-R interval.

Since publication of the original paper, [91] the term Lown-Ganong-Levine syndrome has been used loosely for a large number of patients

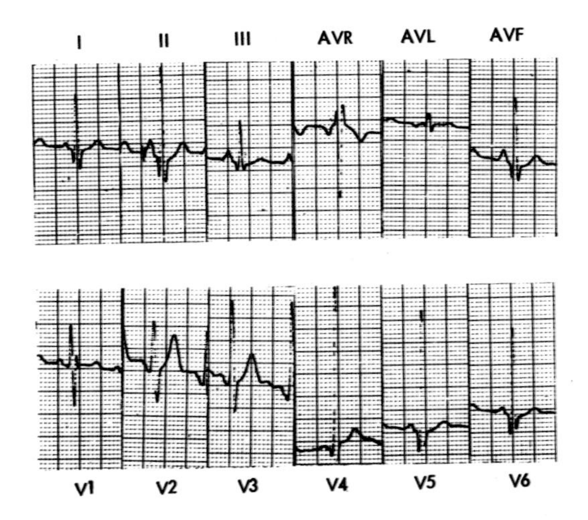

Figure 70. *Short P-R, aberrant (?) QRS complexes in a 24-year-old man. Note short P-R interval, deep q waves in leads I, II, aVF, $V_{5,6}$, high R complexes in lead V_5 and R-S in V_1.*

with short P-R intervals. For some authors it constitutes a separate and independent clinical entity (Bellet [92]); for others (Massumi[83]) it is only one extreme of a wide spectrum of cases, with classical pre-excitation cases located at the other end. For most, it is a synonym for all kinds of short P-R intervals.[87] Caracta et al,[79] summarizing the electrophysiological investigations of 18 patients with short P-R intervals, reported that 15 of the 18 cases were men (4:1 male-female ratio), with an incidence of palpitations of 27%. Thus, the question arises whether a separate clinical and ECG entity, corresponding to the criteria put forth by Lown, Ganong and Levine, exists at all. We can also wonder whether the originally suggested P-R interval of 0.12-0.13 second really represents a "short P-R interval." The important point is, like with the WPW syndrome, a term is used loosely, coming to mean different things to different investigators, not fulfilling the criteria of the original paper, and easily misleading subsequent investigators or readers. We suggest, therefore, that the term LGL be eliminated and replaced by the descriptive name of "short P-R normal QRS syndromes."

We have already considered the possibility that some intermittent pre-excitation beats may be misdiagnosed as ventricular premature beats (VPB). The opposite possibility also exists, namely the diagnosis of true VPB as pre-excitation. In Figure 47, VPB is seen with characteristic delta waves. To avoid such an erroneous diagnosis we accept Katz and Pick's[93] criteria of at least 3 consecutive beats in order to make a diagnosis of pre-excitation.

REFERENCES

1. **Wolff L:** Wolff-Parkinson-White syndrome: Historical and clinical features. *Prog Cardiovasc Dis* 2:677, 1960.
2. **Ruskin JM, Akhtar M, Damato AN, et al:** Abnormal Q waves in Wolff-Parkinson-White syndrome. Incidence and clinical significance. *JAMA* 235:2727, 1976.
3. **Campbell M:** Inversion of T waves after long paroxysms of tachycardia. *Br Heart J* 4:49, 1942.
4. **Castellanos A, Aranda J, Gutierrez R, et al:** Effects of pacing site on QRS morphology in Wolff-Parkinson-White syndrome with special reference to "pseudo-tachycardia-dependent block in accessory pathway" and "atrial gap." *Br Heart J* 38:363, 1976.
5. **Ghosh MB, Chia BL:** The Wolff-Parkinson-White syndrome simulating myocardial infarction. *Singapore Med J* 12:183, 1971.
6. **Kariv I:** Wolff-Parkinson-White syndrome simulating myocardial infarction. *Am Heart J* 55:406, 1958.
7. **Martinez JM, Basili RM, Suãrez LD:** Myocardial lateral pseudonecrosis in a case of Wolff Parkinson White syndrome. *Medicina* (B Aires) 32:646, 1972.
8. **Narayanan GB, Balasubramanian V, Prasad B, et al:** Pre-excitation syndrome simulating ischaemic heart disease—a diagnostic problem in the services. *J Assoc Physicians India* 22:751, 1974.
9. **Trifunovíc S, Kovác M:** Electrocardiographic imitation of myocardial infarct in acquired Wolff-Parkinson-White syndrome. (Ser) *Med Pregl* 28:223, 1975.
10. **Anselmino A, Milone PA, Nazzi V:** Electrocardiography and clinical relations between Wolff-Parkinson-White syndrome and coronary disease. (Ital) *Minerva Cardioangiol* 11:677, 1963.
11. **Apostolov L:** Myocardial infarction and Wolff-Parkinson-White syndrome. *Folia Med* (Plovdiv) 6:113, 1964.
12. **Baer CG:** Die Maskierung Elektrokardiographischer Infarktzeichen durch ein WPW-Syndrom. *Arch Kreislaufforsch* 45:301, 1956.
13. **Croce L, Noseda V:** Wolff-Parkinson-White syndrome and myocardial infarct. (Ital) *Minerva Med* 55:158, 1964.
14. **Gorelov NS, Agranovich RI:** Wolff-Parkinson-White syndrome and myocardial infarct. (Rus) *Kazan Med Zh* 40:46, 1963.
15. **Grazi S, Bonazzi O, Romano S, et al:** Intermittent ventricular pre-excitation during acute myocardial infarct. *Cardiol Prat* 25:313, 1974.
16. **Hilmer W:** About differential diagnosis of the negative T-waves. Myocardial infarction or anomaly of the excitation. *Cardiology* (Basel) 49:305, 1966.
17. **Kiss T, Tóth J:** Intermittent WPW syndrome in myocardial infarcts. (Hun) *Orv Hetil* 113:395, 1972.
18. **Ogigaya N, Kodama K, Hisabori H, et al:** Proceedings: acute myocardial infarct associated with Wolff-Parkinson-White syndrome, A case study. *Jpn Circ J* 39:841, 1975.
19. **Pashkov VV:** Diagnosis of myocardial infarct in patients with Wolff-Parkinson-White syndrome. (Rus) *Kardiologiia* 9:45, 1969.
20. **Preston TA, Kirsh MM:** Permanent pacing of the left atrium for treatment of WPW tachycardia. *Circulation* 42:1073, 1970.
21. **Tkaczewski W, Lada J:** The WPW auriculoventricular block. Masked ischemia. (Pol) *Kardiol Pol* 6:291, 1963.
22. **Vancini B, Artuso P, Marangolo A, et al:** Clinical electrocardiographic and vectorcardiographic aspects of Wolff-Parkinson-White syndrome associated with myocardial infarct. (Ital) *Boll Soc Ital Cardiol* 13:562, 1968.
23. **Videla JG:** El diagnóstico del infarcto de miocardio en presencia del sindrome de Wolff-Parkinson-White. *Rev Argent Cardiol* 24:18, 1957.
24. **Agnoletto A, Foa V:** Diagnostic errors in the Wolff-Parkinson-White syndrome. Analysis of two cases. (Ital) *Acta Gerontol* (Milano) 14:95, 1964.
25. **Anfossi F, Garraggi F, Rossi PL:** Importance of ajmaline in the diagnosis of the

Wolff-Parkinson-White syndrome. (Ital) *Minerva Cardioangiol* 15:408, 1967.
26. **Hua G, Witz F:** Electrocardiographic diagnosis of the Wolff-Parkinson-White syndrome. *Ann Cardiol Angeiol* (Paris) 21:327, 1972.
27. **Okel BB:** The Wolff-Parkinson-White syndrome, report of a case with fatal arrhythmia and autopsy findings of myocarditis, interatrial lipomatous hypertrophy, and prominent right moderator band. *Am Heart J* 75:673, 1968.
28. **Parashar S:** Misdiagnosed intermittent Wolff-Parkinson-White syndrome-Group B (case report). *J Assoc Physicians India* 20:257, 1972.
29. **Taylor JS:** Electrocardiographic diagnosis of the Wolff-Parkinson-White syndrome with abnormal QRS complexes and paroxysmal tachycardia. *J Arkansas Med Soc* 67:297, 1971.
30. **Thurmann M:** The Wolff-Parkinson-White syndrome differential diagnosis. *Missouri Med* 63:114, 1966.
31. **Di Bello R, Gonzales Puig R, Rivero C, et al:** False diagnoses of myocardial infarct in Wolff-Parkinson-White syndrome. (Sp) *Torax* 14:108, 1965.
32. **Fekete AM:** Wolff-Parkinson-White (pre-excitation) syndrome simulating posterior infarction. *Virginia Med Monthly* 91:535, 1964.
33. **Wasserburger RH, White DH, Lindsay ER:** Noninfarctional QS II, III, aVF complexes as seen in Wolff-Parkinson-White syndrome and left bundle branch block. *Am Heart J* 64:617, 1962.
34. **Kulbertus HE, Collignon PG:** Ventricular pre-excitation simulating anteroseptal infarction. *Dis Chest* 56:461, 1969.
35. **Wilson FN:** A case in which the vagus influenced the form of the ventricular complex of the electrocardiogram. *Arch Intern Med* 16:1008, 1915.
36. **Apostolov L:** Wolff-Parkinson-White syndrome and branch block. (Ital) *Minerva Cardioangiol* 15:20, 1967.
37. **Bardan V:** Wolff-Parkinson-White syndrome associated with branch block. (Rum) *Med Intern* (Bucur) 17:1231, 1965.
38. **Homola D, Srnova V:** Right bundle branch block in Wolff-Parkinson-White syndrome. (Cz) *Vnitr Lek* 13:1197, 1967.
39. **Richard S, Pardy L:** The occurrence of type B Wolff-Parkinson-White conduction in the presence of right bundle branch block. *J Mt Sinai Hosp NY* 36:96, 1969.
40. **Bauer GE:** Wolff-Parkinson-White complexes alternating with bundle branch block. *Am Heart J* 54:452, 1957.
41. **Lasser RP, Haft JI, Friedberg CK:** Relationship of right-bundle-branch block and marked left axis deviation (with left parietal or peri-infarction block) to complete heart block and syncope. *Circulation* 37:429, 1968.
42. **Sobrino JA, Mate I, Munoz JE, et al:** Disappearance of right bundle branch block with left anterior hemiblock when associated with a type B pre-excitation syndrome. *Am Heart J* 87:497, 1974.
43. **Blanchot P, Warin JR:** Auricular fibrillation and Wolff-Parkinson-White syndrome simulating ventricular tachycardia. (Fr) *Bord Med* 4:1999, 1971.
44. **Marriott HJ, Rogers HM:** Mimics of ventricular tachycardia associated with the W-P-W syndrome. *J Electrocardiol* 2:77, 1969.
45. **Wolff L:** Diagnostic clues in the Wolff-Parkinson-White syndrome. *NEJM* 261; 637, 1959.
46. **Angle WD:** Myocardial infarction in the Wolff-Parkinson-White syndrome. A method of vector analysis of ECG changes. *Am Heart J* 56:36, 1958.
47. **Goldman RI:** WPW syndrome and myocardial infarction. *Dis Chest* 44:620, 1963.
48. **Kistin AD, Robb GP:** Modification of the electrocardiogram of myocardial infarction by anomalous atrioventricular excitation (Wolff-Parkinson-White syndrome). *Am Heart J* 37:249, 1949.
49. **Moll A:** Myocardial infarction associated with Wolff-Parkinson-White syndrome. *Tagl Prax* 16:3, 1975.
50. **de Monchaux RJ:** A case of Wolff-Parkinson-White syndrome complicated by myocardial infarction. *Med J Aust* 1:544, 1965.

51. **Rinzler SH, Travell J:** The electrocardiographic diagnosis of acute myocardial infarction in the presence of the Wolff-Parkinson-White syndrome. *Am J Med* 3:106, 1947.
52. **Sodi-Pallares D, Cisneros F, Medrano GA, et al:** Electrocardiographic diagnosis of myocardial infarction in the presence of bundle branch block (right and left), ventricular premature beats and Wolff-Parkinson-White syndrome. *Prog Cardiovasc Dis* 6:107, 1963.
53. **Spritz N, Cohen BD, Frimcter GW, et al:** Electrocardiographic interrelationship of the pre-excitation (Wolff-Parkinson-White) syndrome and myocardial infarction. *Am Heart J* 56:715, 1958.
54. **Stein I, Wroblewski F:** Myocardial infarction in Wolff-Parkinson-White syndrome. *Am Heart J* 42:624, 1951.
55. **Thomas JR:** Electrocardiogram of the month. Sequential ECG changes of myocardial infarction in WPW syndrome. *Dis Chest* 53:217, 1968.
56. **Verani MS, Baron H, Maia IG:** Myocardial infarction associated with Wolff-Parkinson-White syndrome. *Am Heart J* 83:684, 1972.
57. **Wolff L, Richman JL:** The diagnosis of myocardial infarction in patients with anomalous atrioventricular excitation (Wolff-Parkinson-White syndrome). *Am Heart J* 45:545, 1953.
58. **Zoll PM, Sacks DR:** Myocardial infarction superimposed on short P-R prolonged QRS complex. A case report. *Am Heart J* 30:527, 1945.
59. **Scherf D, Cohen J:** *The Atrioventricular Node and Selected Cardiac Arrhythmias,* pp 373-447, New York-London: Grune & Stratton, 1964.
60. **Brackbill TA, Dove JT, Murphy GW, et al:** The diagnosis of myocardial infarction in the Wolff-Parkinson-White syndrome. *Chest* 65:493, 1974.
61. **Ticzon AR, Damato AN, Caracta AR, et al:** Intraventricular septal motion during preexcitation and normal conduction in Wolff-Parkinson-White syndrome. *Am J Cardiol* 37:840, 1976.
62. **Borkenstein E:** Wolff-Parkinson-White syndrome . . . hypoxia . . . infarct? *Wien Med Wochenschr* 115:301. 1965.
63. **Schamroth L, Lapinsky GB:** The Wolff-Parkinson-White syndrome associated with myocardial infarction and right bundle branch block. *J Electrocardiol* 5:299, 1972.
64. **Lamb LE, Dermksian G, Sarnoff CA:** Significant cardiac arrhythmias induced by common respiratory maneuvers. *Am J Cardiol* 2:563, 1958.
65. **Mojkowska H:** Effect of some drugs on the electrocardiographic curve in Wolff-Parkinson-White syndrome. (Pol) *Pol Arch Med Wewn* 40:437, 1966.
66. **Szczeklik J, Pyzik Z:** Effect of propranolol on the electrocardiogram in Wolff-Parkinson-White disease. (Pol) *Pol Tyg Lek* 26:302, 1971.
67. **Accatino G, Brocchi G:** Use of ajmaline in the diagnosis of myocardial necrosis in Wolff-Parkinson-White syndrome. *Minerva Cardioangiol* 21:46, 1973.
68. **Correale E, Corsini G:** Value of ajmaline in the clinical diagnosis of Wolff-Parkinson-White syndrome. (Ital) *Rass Int Clin Ter* 50:989, 1970.
69. **Gilbert-Queralto J, et al:** Diagnostic and therapeutic significance of ajmaline in the Wolff-Parkinson-White syndrome. (Ger) *Verh Dtsch Ges Kreislaufforsch* 35:231, 1969.
70. **Homola D, Srnová V:** The use of ajmaline for the normalisation of the WPW syndrome (preliminary report). *Scr Med Fac Med Brunensis* 38:1, 1965.
71. **Homola D, Srnová V:** The diagnosis of combined heart conduction disorders with ajmaline. (Ger) *Z Kreislaufforsch* 58:89, 1969.
72. **Kaltenbach M:** Ajmaline test in the WPW syndrome. *Dtsch Med Wochenschr* 100:2553, 1975.
73. **Lombardi M, Masini G:** Usefulness and importance of ajmaline in the Wolff-Parkinson-White syndrome. (Ital) *Minerva Med* 57:1296, 1966.
74. **Soler Soler J, Casellas Bernat A, Trilla Sánchez E:** Action of ajmaline in the Wolff-Parkinson-White syndrome. Diagnostic value. (Sp) *Arch Inst Cardiol Mex* 36:68, 1966.
75. **Warembourg H, Pauchant M, Ducloux G, et al:** Wolff-Parkinson-White syndrome. I. Generalities. Electrocardiographic and complementary methods. Ajmaline test. (Fr) *Lille Med* 14:979, 1969.

76. **Gorelov NS:** On several atypical forms of Wolff-Parkinson-White syndrome. (Rus) *Kardiologiia* 9:126, 1969.
77. **Homola D:** Atypical forms of the Wolff-Parkinson-White syndrome and kindred ECG pictures. *Cor Vasa* 5:288, 1963.
78. **Marti Garcia JL, Candel Delgrado JM, Guijarro Morales A, et al:** Atypical aspects of the Wolff-Parkinson-White syndrome (WPW). *Rev Clin Esp* 136:563, 1975.
79. **Caracta AR, Damato AN, Gallagher JJ, et al:** Electrophysiologic studies in the syndrome of short P-R interval, normal QRS complex. *Am J Cardiol* 31:245, 1973.
80. **Castellanos A Jr, Castillo CA, Agha AS, et al:** His bundle electrograms in patients with short P-R intervals, narrow QRS complexes, and paroxysmal tachycardias. *Circulation* 43:667, 1971.
81. **Littmann D:** Aberrant atrioventricular conduction in a patient with paroxysmal tachycardia, a short P-R interval and a normal QRS complex. *Am J Med* 2:126, 1947.
82. **Masini G, Lombardi M:** The short PR syndrome. Occult Wolff-Parkinson-White syndrome or an independent clinical entity? (Ital) *Boll Soc Ital Cardiol* 11:660, 1966.
83. **Massumi RA, Vera Z, Ertem G:** "The Wolff-Parkinson-White syndrome and other types of short P-Q interval: related or unrelated entities?" in *Advances in Electrocardiography* (Schlant RC, Hurst JW, Eds), p 275, New York: Grune & Stratton, 1972.
84. **Pawluk W:** Shortened P-Q wave with unchanged ventricular complex (Lown-Ganong-Levine syndrome). (Pol) *Pol Tyg Lek* 19:1461, 1964.
85. **Soederstroem N:** Observations on the significance of shortened P-Q intervals in the electrocardiogram. *Cardiologia* 7:1, 1943.
86. **Massumi RA, Vera Z:** Patterns and mechanisms of QRS normalization in patients with Wolff-Parkinson-White syndrome. *Am J Cardiol* 28:541, 1971.
87. **Brechenmacher C, Laham J, Iris L, et al:** Etude histologique des voies anormales de la conduction dans un syndrome de Wolff-Parkinson-White et dans un syndrome de Lown-Ganong-Levine. *Arch Mal Coeur* 67:507, 1974.
88. **Lev M, Fox SM, Greenfield JC, et al:** Mahaim and James' fibers as a basis for a unique variety of ventricular pre-excitation. *AM J Cardiol* 36:880, 1975.
89. **Ferrer MI:** New concepts relating to the pre-excitation syndrome. *JAMA* 201:1938, 1967.
90. **Mandel WJ, Danzig R, Hayakawa H:** Lown-Ganong-Levine syndrome. A study using His bundle electrograms. *Circulation* 44:696, 1971.
91. **Lown B, Ganong WF, Levine SA:** The syndrome of short P-R interval normal QRS complex and paroxysmal rapid heart action. *Circulation* 5:593, 1952.
92. **Bellet S:** *Clinical Disorders of the Heart Beat,* pp 485-519, Philadelphia: Lea & Febiger, 1971.
93. **Katz LN, Pick A:** *Clinical Electrocardiography: The Arrhythmias,* pp 100-104, Philadelphia: Lea & Febiger, 1956.

PART TWO

THEORIES

X

A Summary of the Findings and Theories of the Pre-excitation Syndrome

FINDINGS

Electrophysiological Phenomena

Enhanced Activity:

1) a high incidence of premature beats, most of which are supraventricular in origin.

2) a high incidence of P wave changes, usually at a rate higher than the predominant basic rhythm (active ectopic rhythm).

3) different forms of supraventricular tachycardia (50% of cases) including reentry tachycardias using the normal and accessory pathways, and reentry tachycardias inside the AV node.

4) a high incidence of atrial fibrillation or flutter, mostly with broad and distorted QRS complexes.

5) cases with paroxysmal, nonexertional sinus tachycardia.

6) cases with active AV dissociation in which a subsidiary pacemaker is firing at a rate faster than the sinus node.

7) cases of wandering pacemakers.

Suppressed Activity and Conduction:

1) cases with sinus bradycardia and a high incidence of sinus arrhythmia in all age groups; both groups show junctional escape beats, usually during pre-excitation conduction.

2) cases with first degree AV block during normal conduction and two cases during pre-excitation conduction.

3) cases with second degree AV block with Wenckebach's phenomenon, or 2:1 conduction during pre-excitation and sometimes during normal conduction.

4) cases with pseudocomplete AV block with QRS complexes similar to classical pre-excitation; a few cases where complete AV block developed with normal complexes except for ventricular captures in which the pattern was characteristic of pre-excitation.

5) coexistence of right or left bundle branch block.

Other Phenomena:

1) changes in P wave configuration and/or polarity.

2) concomitant P and QRS changes during both normal and anomalous conduction.

3) correlation of the P-R interval during normal conduction and during pre-excitation.

4) intermittency seen in more than 25% of cases.

5) complete disappearance of pre-excitation, mainly in infants and young children, but sometimes also in adults.

6) electrophysiologic evidence of assumed structural AV connection in patients without pre-excitation (concealed pre-excitation).

Anatomical and Structural Findings

1) an accessory bundle completely outside the normal conduction system, connecting the atria with the ventricles at different locations around the AV groove or connecting the atrial with the ventricular septum.

2) more than one accessory bundle, completely outside the normal conduction system.

3) different topographic locations of these accessory bundles: subepicardial, intramyocardial, or starting in one of these locations and terminating in another.

4) varying cell composition of the bundles: working myocardial cells, Purkinje-like fibers, slender transitional cells resembling AV node fibers, and P cells simulating the structural appearance of a sinus node.

5) other anomalous AV bypass tracts: atrio-His connections, Mahaim paraspecific fibers, with and without an additional accessory tract completely outside the normal conduction system.

6) absence of any anomalous structural AV connection.

7) an accessory connection outside the normal conduction system in cases which never exhibited pre-excitation.

Histopathological Findings

1) findings in approximately one-third of all studied cases, involving part or all of the normal conduction system: AV node, His bundle, bundle branches, internodal tracts, sinus node and sometimes even an anomalous AV connection.

2) consisting of fibrosis, cell infiltration, fat infiltration, cell degeneration and cell atrophy; usually more than one feature present.

3) these findings were in some cases the only pathological involvement of the heart, and in others they accompanied AV connections completely outside or partly mediated through the normal conduction system.

4) the histopathological involvement of the conduction system was sometimes attributed to a known cardiac disease such as myocarditis or Chagas' disease, and in some cases the etiology was unknown.

5) some findings were characteristic for known cardiac diseases: myocardial infarction, congenital heart disease, rheumatic valvular involvement, *etc.*

Clinical Findings

1) a preponderance of males to females, 2:1.

2) a different clinical picture in newborns and infants than in adults.

3) different responses of tachycardias to drugs in patients showing similar clinical and ECG features.

4) poor results in surgery, despite evidence by intracardiac preoperative investigations and cardiac mapping during surgery, indicating the exact location of an accessory bundle completely outside the normal conduction system.

5) familial occurrence of the syndrome (sometimes accompanying cardiomyopathies).

6) a reported increased incidence of pre-excitation in patients with hyperthyroidism, psychiatric disturbances, nervous system diseases, acute rheumatic fever, acute myocardial infarction, cardiomyopathies and Ebstein's anomaly.

7) predominance of the right ventricular electrical forces in type A pre-excitation despite the known anatomical preponderance of the left ventricular mass.

8) normalization of conduction with exercise in some cases, and appearance of pre-excitation conduction in others.

9) the unpredictability of appearance and disappearance of ECG anomalies, including tachycardias.

SUMMARY OF THE MOST POPULAR THEORIES

The theories on pre-excitation fall into two categories: those based on the presence of an accessory anomalous muscular AV bundle, located completely or partly outside the normal AV conduction system, and those based on a physiopathological rather than structural background.[1-29]

Structural Theories

"Kent Bundle" Theory (accessory bundles completely outside the normal AV conduction system)

The first to propose this theory were Holzmann and Scherf[8] and Wolferth and Wood.[30] One of the original authors, Scherf, together with Cohen[31] summarized it in 1964 as follows: "In one carefully studied observation the P waves remained unchanged during normal sinus rhythm and during interspersed abnormal beats exhibiting typical delta waves." This was seen at rest as well as after exercise and, therefore, Holzmann and Scherf concluded that the sinus node was always the pacemaker in both patterns. A shortened atrioventricular conduction time associated with an abnormal complex, followed by beats with normal atrioventricular conduction and normal ventricular complexes can, according to the authors, be best explained by assuming that some impulses are conducted over an anomalous, accessory atrioventricular connection, circumventing the delaying AV node. The authors mention as a possibility particularly, but not invariably, the right lateral atrioventricular bundle described by Kent in 1895 (Kent bundle). The authors were fully aware of the fact that the bundle of Kent was not proved to exist in the human heart. ". . . Such an accessory atrioventricular connection would fully explain the electrocardiographic characteristics of the pre-excitation syndrome. The short P-R interval is due to the direct pathway from the atrium to the ventricle avoiding the delay of conduction in the AV system." The premature excitation of a certain portion of the ventricle results in a delta wave which was stressed by the authors as a characteristic finding and the abnormal speed of the excitation causes the broadening of the ventricular complex.

Other Accessory AV Bypasses (partly included in the normal AV conduction system)

In 1966 Lev[11] arrived at the conclusion that "bundles of Kent" are in some way related to the "pre-excitation syndrome," but are not indispensable for its production and that other structural bypasses may also be involved. At that time it also became clear that in addition to classic pre-excitation, several variants had come to be recognized. These included a short P-R interval (less than 0.12 second) with normal QRS; *ie,* a QRS with normal duration and no delta wave, and a normal or even prolonged P-R interval coupled with a prolonged QRS containing a delta wave.[32]

Based on this, Ferrer[33,34] proposed a new concept of the anatomic background to explain the various forms of anomalous AV conduction: "1) In the classic form of the WPW syndrome, with short P-R, long QRS and delta wave, the bypass or short circuiting may occur in one of two ways—excitation may move across a bundle of Kent, thus avoiding the specific conduction tissues completely or almost completely, bringing pre-excitation or early depolarization to the ipsilateral ventricle and somewhat later depolarization to the contralateral ventricle, and resulting in a short P-R and a long QRS; or excitation may travel over two separate bypass tracts, namely, the AV bypass of James, which would produce the short P-R, and then over Mahaim fibers, which would produce the delta wave of early ventricular invasion and the prolonged QRS. 2) With a short P-R and normal QRS, only the James' bypass or paranodal fibers would exist or be utilized. With no other anomalous pathway functioning, a normal ventricular event and QRS would follow. 3) When a delta wave, indicating pre-excitation of the ventricles, and a prolonged QRS exist with a normal P-R interval (an unusual variant of the WPW syndrome), it is likely that Mahaim fibers alone, and not a bundle of Kent, are being used to bypass normal intraventricular conduction. The excitation enters the upper margin of the AV node from the atria, is susceptible to the normal AV nodal delay and hence produces a normal and not a short P-R interval, and moves into the common bundle of His. There it short-circuits into the ventricular septum over Mahaim fibers, thus bypassing the lower branches of the bundle of His. If the AV nodal function is impaired by drugs or disease, then the P-R interval will of course be longer than normal."

Theory of Postnatal Remnants of AV Nodal and His Bundle Tissues

The third and most recent attempt to explain the mechanisms of pre-excitation and associated arrhythmias and tachycardias was put forward by James[9]: "By the end of the sixth or eighth week of gestation the heart of the human fetus contains a clearly identifiable AV node and His bundle and the general structure of the entire heart is very similar to that of the adult. This has led to the erroneous concept that the AV node and His bundle are completed at that time and that there are no further alterations in their structure. Actually, there are rather extensive postnatal structural changes which lead to the final anatomic configurations of the adult AV node and His bundle. Between the second month of gestation and full term delivery the AV node and His bundle do not change much in appearance, and are loosely organized structures which freely connect much of the atrial septum directly with the ventricular septum. The margins of these conduction centers are grossly irregular, and the smoothing off process which molds and shapes them to the adult form is a critically important activity beginning at birth and normally not terminating for several months or even years. When the molding and shaping process is more protracted, an irregular shape of the His bundle, with outpocketings and loop connections, is observed up through the first two decades of life. These irregularities may be of considerable significance in the genesis of reentrant arrhythmias with and without associated pre-excitation of the ventricles; the potential significance of this in the WPW syndrome is illustrated by the splintering and multiple division of the AV node and His bundle, which are well shown in some carefully studied cases. Conversely, the evolutionary normal destruction of these irregularities, and the transformation of the node and the bundle into relatively smoothly outlined structures, may account for "spontaneous" termination of some paroxysmal tachycardias of childhood. These same postnatal changes may also be one explanation why the WPW syndrome occurs in infancy or childhood, and later in life sometimes simply disappears."

Pathophysiological Theories

Over the years many of the pathological investigations in hearts of patients suffering from the pre-excitation syndrome during life failed to turn up any anomalous structural AV connections. Some investigators concluded therefore that some pathophysiological changes occurring in the structurally normal AV conduction system may produce pre-excitation without the presence of accessory anomalous structural AV connections and bundles.

"Excitable Center" Theory

This theory was first postulated by Holzmann and Scherf[8]; "An excitable center may exist in the ventricles which can explain the abnormal QRS complex." In order to explain the short P-R interval, these authors had to assume that an atrial contraction as a result of a sinus impulse, through mechanical action, stimulated this excitable center, thereby initiating anticipation of a ventricular complex and consequently resulting in shortening of the P-R interval. This would explain the short P-R interval, the wide QRS complex and the occurrence of paroxysmal tachycardia would be understandable. ". . . at a later date the possibility was considered that a premature impulse may be formed in the ventricles at an abnormal site because of the effect of electrotonus from the excitation of the atria" (Sodi-Pallares[28,35]).

"Accelerated Conduction" Theory

A second pathophysiological theory was postulated by Prinzmetal et al[22] in 1952. Their hypothesis required that four conditions be fulfilled: "1) There must normally be a delay of the auricular impulse at the AV node; 2) In WPW aberration the normal delay takes place except at that part of the node which has been altered. This altered portion must allow part of the impulse to pass through without the normal delay; 3) The abnormal impulse must pass down the bundle of His; and 4) The AV node and bundle must be constituted in such a way that specific cells or areas of the node 'supply' specific areas of the 'ventricular myocardium'."

In further elaboration of the second point, the authors theorized that: "this abolition of normal delay can take place in either of two ways. The first possibility is that there exists some kind of alteration of a portion of the node which permits the impulse to traverse it more rapidly than normal. The other possibility is that there is no increase in the velocity of the impulse but that it enters the node lower than does the normal impulse. Thus, it would have less distance to travel within the node and would enter the ventricle prematurely."

"Synchronized Sino-Ventricular Conduction" Theory

In 1966 Sherf and James[26] postulated this theory based on recently described internodal and interatrial conduction pathways, longitudinal, dissociated conduction in the AV node and His-Purkinje axis, and a predisposition for the terminal station of every electrical stimulus leaving the sinus node. Two phenomena of pre-excitation influenced the development of this theory: 1) the fact that many fibers originating from the posterior internodal tract bypass the bulk of the AV node in order to enter its lower pole (thus fulfilling the second possibility of "accelerated conduction" postulated by Prinzmetal et al[22]); and 2) the electrophysiological unrest in the atria which expresses itself by an increased number of supraventricular premature beats and P wave changes. Considering these two features the authors argued that "if a beat originates within the posterior internodal tract (PIT) and arrives at a lateral margin of the AV node abnormally early, the spread of excitation within the AV node would be other than usual and arrival of the excitation front in the His bundle would be distorted. Presuming selective and separate conduction within the His bundle, evidence for which has been presented, the activation of any of several direct connections with the IV septum could account for the delta wave and later activation of the right and left bundle branches would produce the normal late portion of QRS."

REFERENCES

1. **Apostolov L:** The value of electrocardiography in studying the pathogenesis of the Wolff-Parkinson-White syndrome. (Rus) *Kardiologiia* 8:93, 1968.
2. **Cagán S, Cagánová A, Králóvá G:** Contribution to the etiology and pathogenesis of the Wolff-Parkinson-White syndrome. (Slo) *Cas Lek Cesk* 107:1564, 1968.
3. **Castorina S:** On the pathogenesis of Wolff-Parkinson-White syndrome. (Ital) *Riv Med Aeronaut Spaz* 31:88, 1968.
4. **Caucino L:** Pathogenetic considerations on a case of Wolff-Parkinson-White syndrome. (Ital) *Minerva Med* 57:1902, 1966.
5. **Feldmann H, Koch E:** Zum Syndrom der Verbreiterung der QRS-Gruppe auf Kosten von PQ. Grundsaetzliches zur Theorie des sog Kentschen Buendels. Ein Beitrag zur Frage der Entstehung der verlaengerten PQ-Zeit. *Z Kreislaufforsch* 35:512, 1943.
6. **Gorelov NS:** Pathogenesis of Wolff-Parkinson-White syndrome. (Rus) *Ter Arkh* 43:57, 1970.
7. **Gorelov NS:** Etiology of Wolff-Parkinson-White syndrome. (Rus) *Ter Arkh* 43:115, 1971.
8. **Holzmann M, Scherf D:** Über Elektrokardiogramme mit verkürtzter Vorhof-Kammer-Distanz und positiven P-Zacken. *Z Klin Med* 121:404, 1932.
9. **James TN:** The Wolff-Parkinson-White syndrome: evolving concepts of its pathogenesis. *Prog Cardiovasc Dis* 13:159, 1970.
10. **Koike T, Sugiyama S, Okajima T, et al:** Proceedings: effects of vagus nerve stimulation in Wolff-Parkinson-White syndrome. *Jpn Circ J* 39:881, 1975.
11. **Lev M:** "The Pre-excitation Syndrome: Anatomic Considerations of Anomalous A-V Pathways," in *Mechanisms and Therapy of Cardiac Arrythmias* (Dreifus LS, Koff WS, Eds), pp 665-670, New York: Grune & Stratton, 1966.
12. **Lombardi M, Masini G:** Clinical and electrocardiographic physiopathogenetic aspects of ventricular pre-excitation. (Ital) *Boll Soc Ital Cardiol* 11:1430, 1966.
13. **Mahaim I:** Le syndrome de Wolff-Parkinson-White et sa pathogénie. *Helv Med Acta* 8:483, 1941.
14. **Mahaim I:** Kent's fibers and the A-V paraspecific conduction through the upper connections of the bundle of His-Tawara. *Am Heart J* 33:651, 1947.
15. **Mahaim I:** Les "fibres de Kent" et la conduction a-v paraspécifique par les connections superiéures du faisceau de His-Tawara. *Arch Inst Cardiol Mex* 18:46, 1948.
16. **Mahaim I, Winston MR:** Recherches d'anatomie comparée et de pathologie expérimentale sur les connexions hautes du faisceau de His-Tawara. *Cardiologia* 5:189, 1941.
17. **Sherf L, James TN:** A new electrocardiographic concept: synchronized sinoventricular conduction. *Dis Chest* 55:127, 1969.
18. **Masini G, Lombardi M:** Old and new problems on the subject of the pathogenesis of the Wolff-Parkinson-White syndrome. (Ital) *Riv Crit Clin Med* 65:596, 1965.
19. **Mathur KS, Elhence GP, Wanal PK, et al:** Accelerated A-V conduction; an unusual manifestation of digitalis toxicity. *J Assoc Physicians India* 19:283, 1971.
20. **Mauck HP Jr, Hockman CH, Hoff EC:** ECG changes after cerebral stimulation. I. Anomalous atrioventricular excitation elicited by electrical stimulation of the mesencephalic reticular formation. *Am Heart J* 68:98, 1964.
21. **Portillo Acosta B:** New electrovectorcardiographic concepts on the Wolff-Parkinson-White syndrome. (Sp) *Rev Venez Sanid Asist Soc* 33:369, 1968.
22. **Prinzmetal M, Kennamer R, Corday E, et al:** *Accelerated Conduction. The Wolff-Parkinson-White Syndrome and Related Conditions*, New York: Grune & Stratton, 1952.
23. **Ritchie WT:** The action of the vagus on the human heart. *Q J Med* 6:47, 1912.
24. **Rozenblit J, Reich J:** The Wolff-Parkinson-White syndrome, the result of simple atrio-ventricular dissociation. (Pol) *Kardiol Pol* 7:317, 1964.
25. **Santini M, Milazzotto F, Dini P, et al:** Supernormal conduction in the ventricular pre-excitation syndrome (report of a case). *G Ital Cardiol* 4:84, 1974.
26. **Sherf L, James TN:** A new look at some old questions in clinical electrocardiography. *Henry Ford Hosp Med Bull* 14:265, 1966.

27. **Sodi-Pallares D, Bisteni A, Hedrano G:** Estudios sobre el sindrome de Wolff-Parkinson-White. I. La activácion ventricular en el sindrome de Wolff-Parkinson-White. *Arch Inst Cardiol Mex* 25:676, 1955.
28. **Sodi-Pallares D, Galder RM:** *New Bases of Electrocardiography,* St. Louis: Mosby, 1956.
29. **Stepanov MA, Goldberg GA, Panova AG, et al:** The genesis of electrocardiographic changes in the Wolff-Parkinson-White syndrome. (Rus) *Kardiologiia* 10:104, 1970.
30. **Wolferth CC, Wood FC:** Further observations on the mechanism of production of a short P-R interval in association with prolongation of the QRS complex. *Am Heart J* 22:450, 1941.
31. **Scherf D, Cohen J:** *The Atrioventricular Node and Selected Cardiac Arrhythmias,* pp 373-447, New York-London: Grune & Stratton, 1964.
32. **Antonioli G, Pozzar C, Abrahamsohn R:** Clerc-Levy-Cristesco syndrome (clinical contribution). (Ital) *Arcisped S Anna Ferrara* 18:1145, 1965.
33. **Ferrer MI:** New concepts relating to the pre-excitation syndrome. *JAMA* 201:1938, 1967.
34. **Ferrer IM:** *Pre-excitation,* Mt. Kisco, New York: Futura, 1976.
35. **Sodi-Pallares D:** Fusion beats and anomalous atrioventricular excitation. Anomalous atrioventricular excitation: panel discussion. *Ann NY Acad Sci* 65:845, 1957.

XI

Concept of Interacting Structural and Functional Factors

Dual or Multiple AV Conduction Pathways

The fundamental pathophysiological characteristic of the pre-excitation syndrome is the existence of two pathways capable of transmitting impulses from the atria to the ventricles at two different velocities. According to this view, put forth by Massumi,[1] it matters little whether the anomalous (or accessory) bundle courses through the AV node, around the AV node, or is situated peripherally on the outside of the conduction system, running like a bridge between the walls of the atrium and ventricle. The mere existence of unequal velocities of conduction is responsible for the most important features of the pre-excitation syndrome: the abnormal QRS and the paroxysmal tachycardia.

Considering the electrophysiological, pathological and surgical findings, we can describe four basic double (or multiple) parallel AV conduction pathways: 1) a direct accessory atrioventricular bypass, completely outside the normal conduction system (labeled by us RAB, LAB or SAB, Table XVI, and by Rossi[2] "Kent bundle"); 2) an atrioventricular bypass tract, partly mediated through the normal conduction system, formed by an anomalous atrial-His bundle connection (AH), combined with His bundle fibers leading to the ventricular septum (HV connections in our nomenclature, Table XV, and Mahaim paraspecific fibers in Rossi's[2]); 3) atrioventricular bypass tracts through remnants of sequestered conduction tissues formed during the normal postnatal development of the AV node and His bundle. These connections may or may not be partially mediated through the normal conduction system; 4) an atrioventricular bypass pathway completely included in the normal conduction system induced by an early conduction through James' AV nodal bypass fibers and producing, in consequence, an asynchronization of the normal electrical depolarization fronts in the AV node, His bundle and ventricles.

In addition to these four basic forms, other combinations of anomalous AV bypass pathways may exist. For instance, an early conducting AV nodal bypass tract (James' fibers) may act together with a group of Mahaim paraspecific fibers;[3] an atrial-His bundle connection may produce an asynchronization of the downstream normal conduction system by longitudinal dissociated conduction;[4] Mahaim (NV or HV connections) fibers may act independently, without any other AV nodal bypass mechanism, but still bypassing part of the normal conduction system (His bundle, bundle branches, normal Purkinje network). There is extensive documentation that more than one single AV bypass pathway may be present in one patient: two or more accessory bundles completely outside the normal conduction system were reported in some cases,[5] while combinations of such accessory connections with an additional atrial-His bundle[4] or with His bundle–ventricular septum (Mahaim paraspecific fibers) were described in others.[6,7]

Most of the accessory bundles referred to in the first mentioned pathway above, and the atrial-His bundle connections in the second, are the results of faults in the normal embryological development of the heart. They pass through holes and clefts left over in the fibrous septa which normally completely divide the atria from the ventricles (except for the location where the His bundle penetrates through it). These congenital faults can be compared to similar well known congenital defects in the heart, such as Ebstein's anomaly or ventricular septal defect (VSD). The presence of Mahaim paraspecific fibers and remnants of AV nodal or His bundle tissue are the results of abnormal or delayed postnatal dynamic structural changes. Mahaim fibers are often seen in newborns, but disappear later; AV nodal or His bundle tissue, which is present during the final molding of the functional area of the conduction system, is normally absorbed during the first year of life. These postnatal developmental failures can be compared to a lack of closure of a ductus arteriosus (patent ductus arteriosus—PDA) or a foramen ovale after birth. The two parallel conducting pathways are potentially present in every normal conduction system.

These findings of an impressive number of parallel double or multiple AV conduction pathways in human hearts raise some interesting questions:—If there are direct structural bypass tracts which completely avoid the AV node with its characteristic slowing down of the electrical conduction velocity, why don't such cases always display the pattern of pre-excitation? What can account for the observed intermittency in many patients?

—If there are, as recently reported,[2] direct structural accessory AV bundles in human hearts, why don't these anomalous connections ever express themselves in the classic QRS pattern of pre-excitation, despite innumerable ECG tracings and long-term monitoring ("concealed pre-excitation")?

—If every normal heart contains AV nodal bypass fibers (James' fibers) that can short-circuit the AV nodal electrical delay, why don't normal human ECG tracings always display short P-R intervals, or during intracardiac electrophysiological studies, short P-H conduction times?

The logical answer to these puzzling questions can only be that the mere anatomical presence of additional atrioventricular conduction pathways is not sufficient to induce the pre-excitation phenomenon. It appears that they may, rather, serve as channels for its production, but have to be triggered first by some additional factors which will transform them from "silent partners" into "active speakers."

Electrical Instability in the Heart in Pre-excitation

Massumi's statement,[1] which forms the foundation of our concept, says specifically that not only must there be two parallel atrioventricular connections (one normal, one anomalous) present in a heart in order to induce pre-excitation, there must also be two different conduction velocities. In other words, factors must be present which produce a faster conductivity in one AV pathway or a slower one in the other, and the electrical stimulus descending from the atria will use the faster conducting pathway (the one with the shorter ERP) (see Chapter VI).

The rate of conductivity of the electrical stimulus in any cardiac tissue is a function of the cell composition of the structure and of the status of the surrounding biochemical extracellular environment. The cell composition of different accessory or other kinds of AV nodal bypass structures was found to vary from case to case. Thus, one can envision a very fast conduction if the accessory AV connection is composed of Purkinje or Purkinje-like cells (2-4 meters per second); a slow conduction of 0.8 to 1 meter per second if the connection is formed exclusively or mainly of working myocardial cells;[8,9] slower yet (0.05 meters per second) if it is composed of short and slender transitional cells (normally found in the AV node[4]), or of clusters of P and transitional cells[10] (encountered mostly in the sinus node).

Thus, the cellular structure of the bypass tract is one of the two major factors establishing the basic electrical conduction facility in the anomalous pathway. But, since the cellular composition of an accessory pathway, and consequently its basic rate of conductivity, will always be the same, the phenomenon of intermittency remains unexplained. Therefore, we appeal to the second major factor influencing the rate of electrical conductivity—the status of the biochemical extracellular environment in the area of a given cardiac structure. A change in this milieu can alter the functional behavior of the cells: it can enhance conduction velocities in accessory bundles or slow the electrical delay in the normal AV node; it can influence the translocation of the primary pacemaking center of the heart, inside the sinus node or outside it, into an ectopic focus. Of the various extracellular factors known to influence the electrophysiological behavior of cells, the most important seems to be a disequilibrium between the ordinary sympathetic tonus (catecholamine distribution and/or amount of production) and the vagal forms (acetylcholine distribution and/or amount of production). Other factors capable of changing the cellular electrophysiological properties, directly or mediated through the autonomic nervous system of the heart, are disturbed electrolyte relations (K, Na, Ca, Mg, etc) and metabolic concentrations (such as lactic acid) in pathological quantities around a given cardiac structure.

Such autonomic, electrolytic or metabolic changes around a potential double or multiple atrioventricular conduction system may constitute the triggering element for producing pre-excitation. This may be accomplished by a temporarily accelerated conduction through accessory bundles, Mahaim fibers and other bypass channels, or by a slowed electrical conduction in the AV node or other structures with a similar cell composition and cell behavior. When no accessory or other structural AV bypass pathways can be demonstrated in the heart of a pre-excitation case, it is postulated that the double atrioventricular pathway is located inside the normal conduction system. In such a case a change in the extracellular environment around the sinus node may produce a migration of the normal pacemaker from its regular location in the node to an ectopic center near to or in the posterior internodal tract. Normally, the electrical stimulus leaves the sinus node in a fixed order, using first the anterior internodal tract (Bachman's bundle), then the middle internodal tract and finally the posterior internodal tract (the longest of the three). This being the sequence of atrial depolarization, the stimulus descending through the PIT arrives late at the lower pole of the AV node to find it already in a refractory period induced by the stimulus which arrived earlier on the AIT and MIT (Diagram VIII). If, however, the primary pacemaker is dislocated by functional factors into the PIT, then the order of atrial depolarization is changed and the PIT will be the first to conduct the electrical stimulus to the AV node. It will arrive through the AV nodal bypass fibers to its lower pole, avoiding the bulk

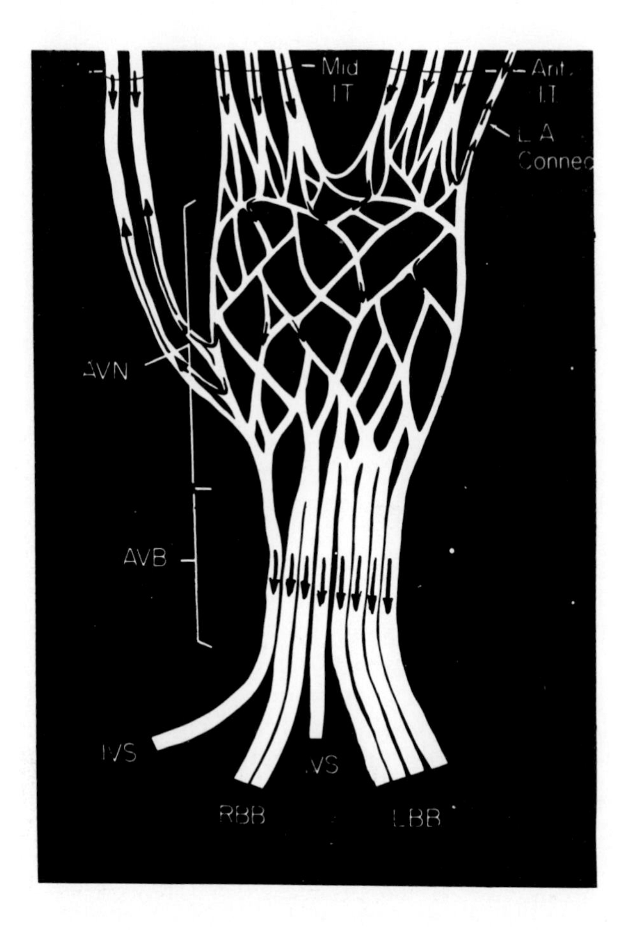

Diagram VIII. *General schema of the AV node and His bundle, including the inputs of the former and the output of the latter. Abbreviations here and in subsequent similar figures are: Post IT: posterior internodal tract; Mid IT: middle internodal tract; Ant IT: anterior internodal tract; LA: left atrium; AVN: AV node; AVB: His (AV) bundle; IVS labels direct connections to the interventricular septum; RBB and LBB are right and left bundle branches. Thin black arrows indicate the direction of spread of excitation as we consider it normally occurs. The major and earliest excitation of the AV node occurs via the middle and anterior internodal tracts, especially the latter, which intermingle at the crest of the node. This excitation at the crest enters the node and is filtered through to the His bundle as a uniform front, which may or may not be perfectly straight (shown so schematically here), but which we consider must be consistent in each normal cardiac cycle. Because excitation arrives earliest from the anterior and middle internodal tracts, this not only penetrates the AV node but also enters adjacent atrial connections with the node which are not yet excited, such as those to the posterior internodal tract and those along the septal endocardium to the left atrium. Entry into these latter connections effectively cancels late arriving impulses from those directions.*

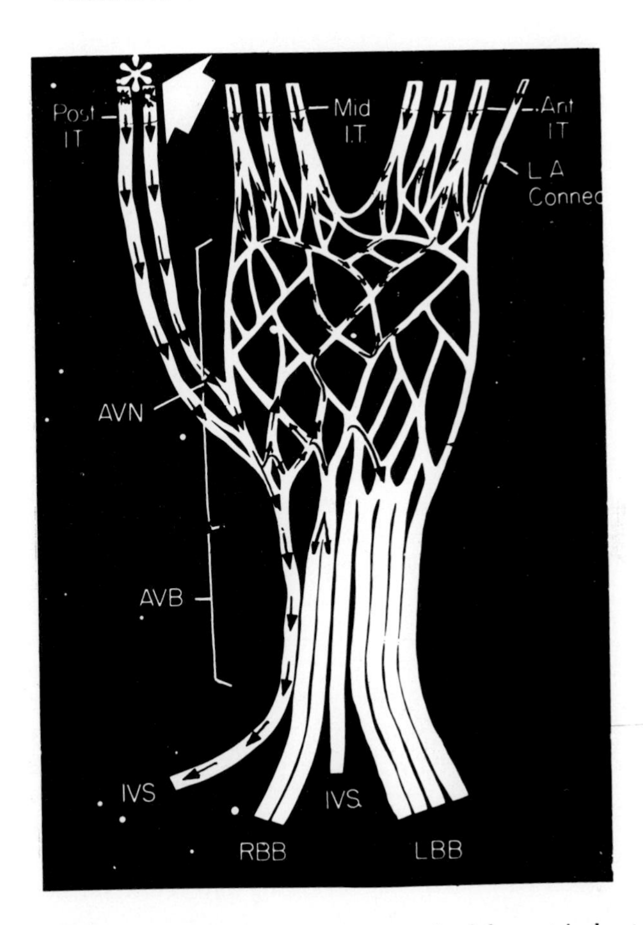

Diagram IX. *Our hypothesis concerning genesis of the ventricular pre-excitation QRS and its probable relationship to enhanced supraventricular ectopic pacemaking (and paroxysmal tachycardias). If a beat originates within the posterior internodal tract (arrow and asterisk) and arrives at a lateral margin of the AV node abnormally early, the spread of excitation within the AV node would be other than usual, and arrival of the excitation front in the His bundle would be distorted. Presuming selective and separate conduction within the His bundle, the activation of any of several direct connections with the IV septum could account for the delta wave and later activation of the right and left bundle branches would produce the normal late portion of QRS.*

of the AV nodal tissue, and produce a short P-R interval (Diagram IX). It can be assumed that this early arrival of the electrical stimulus at the lower part of the AV node produces an asynchronization of the normal depolarization front in the AV node and His bundle, and starts depolarizing ventricular myocardium at an unusual site, inducing pre-excitation.

This pathological mechanism of pre-excitation, based on interaction between normal and/or anomalous structural atrioventricular connections and electrophysiological instabilities in the heart can, in our opinion, explain the induction of all forms of pre-excitation encountered. However, the pre-excitation syndrome does not express itself solely by the premature depolarization of a ventricular myocardial section, but also by a broad range of arrhythmias. These include premature beats, tachycardias, atrioventricular dissociations, bradycardia, escape beats and different forms of AV block. These arrhythmias—expressions of enhanced electrophysiological activity on the one hand and suppressed activity on the other—raise two major questions:

—Are they produced by a common cause, which is also responsible for the anatomical faults encountered in most cases?

—Or are they forerunners of pre-excitation, due to electrical instabilities in the heart not connected to the existence of the different anatomical channels, but able to trigger the anomalous AV pathway and induce pre-excitation?

There are, at present, no clear-cut answers to these questions. The many factors involved in the production of this syndrome form a kind of kaleidoscope, subject to change every second, giving the unpredictable and strange features observed in the syndrome. These factors include the different accessory or other anomalous AV connections, running subendocardially, subepicardially or in the depth of the myocardium, connecting the atria to the ventricles at different locations and composed of a great variety of myocardial cells with different morphological and physiological behaviors, and the changing amounts of sympathetic and/or vagal hormones surrounding them or the normal conduction system. In this complex structural-physiological conglomerate, it is difficult to sort out the precise sequence of events.

REFERENCES

1. **Massumi RA, Vera Z, Mason DT:** The Wolff-Parkinson-White syndrome. A new look at an old problem. *Mod Concepts Cardiovasc Dis* 42:41, 1973.
2. **Rossi L:** A histological survey of pre-excitation syndrome and related arrhythmias. *G Ital Cardiol* 5:817, 1975.
3. **Lev M, Fox SM, Greenfield JC, et al:** Mahaim and James' fibers as a basis for a unique variety of ventricular pre-excitation. *Am J Cardiol* 36:880, 1975.
4. **Brechenmacher C, Coumel P, Fauchier JP, et al:** De subitaneis mortibus. XXII. Intractable paroxysmal tachycardias which proved fatal in type A Wolff-Parkinson-White syndrome. *Circulation* 55:408, 1977.

5. **Verduyn Lunel AA:** Significance of annulus fibrosus of heart in relation to AV conduction and ventricular activation in cases of Wolff-Parkinson-White syndrome. *Br Heart J* 34:1263, 1972.
6. **Lev M, Gibson S, Miller RA:** Ebstein's disease with Wolff-Parkinson-White syndrome: report of a case with a histopathologic study of possible conduction pathways. *Am Heart J* 49:724, 1955.
7. **Lev M, Sodi-Pallares D, Friedland C:** A histopathologic study of the atrioventricular communications in a case of WPW with incomplete left bundle branch block. *Am Heart J* 66:399, 1963.
8. **Ohnell RF:** Pre-excitation, a cardiac abnormality. *Acta Med Scand* (suppl 152), 1944.
9. **Segers M, Sanabria T, Lequime J, et al:** Le syndrome de Wolff-Parkinson-White. Mise en evidence d'une connexion A-V septale directe. *Acta Cardiol* 2:21, 1947.
10. **James TN, Puech P:** De subitaneis mortibus. IX. Type A Wolff-Parkinson-White syndrome. *Circulation* 50:1264, 1974.

XII

Explanation of Some Phenomena of the Pre-excitation Syndrome by the Proposed Concepts

Congenital and Postnatal Pre-excitation

If we assume that the same faulty development is responsible for both the anomalous structural AV tracts and the electrical instabilities in the heart, then some cases may be congenital, and others may be postnatal in origin. This could work in a number of ways:

1) The electrical instability could be a primary physiological dysfunction of the normal conduction system developing parallel with the anatomical anomaly; 2) The different structural anomalous AV connections are sometimes built by cells possessing potential pacemaker properties (P cells or Purkinje cells), and their mere presence increases the potential for electrical instabilities. Such a specialized cell composition may occur in accessory bundles, outside the normal conduction system, in remnants of the developing AV node and His bundle, and in the Mahaim paraspecific fibers; 3) Through holes in an imperfectly developed fibrotic septa separating the atria from the ventricles, there are not only muscular bridges which connect the upper and lower cardiac chambers but also chains of nerves which accompany the accessory myocardial bundles and which may influence their electrophysiological status.

Acquired Pre-excitation

The theory that the electrical instability of the heart is not connected to the structural anomaly but is, rather, a secondary factor in pre-excitation can explain the hypothesis of acquired pre-excitation.[1] The existence of such an entity has been a matter of controversy for decades.

Support for it comes from many examples in which pre-excitation appeared during an acute disease or disappeared after the basic pathological factor was eliminated. Others, however, point out the intermittent nature of pre-excitation and attribute the appearance or disappearance of the syndrome to sheer chance at the time of examination. In their monolythic approach whereby the syndrome is due to the presence of a Kent bundle only, the development of such a bundle later in life was unacceptable. If, however, the broader concept of interacting structural and electrophysiological anomalies is accepted, then cardiac diseases or disorders can influence the autonomic equilibrium in the heart. Based on the histopathological findings, it is our opinion that there is more evidence for the "two-step" development of pre-excitation than for a single congenital or postnatal developmental failure.

The vegetative equilibrium may change from cardiac diseases, systemic diseases, or from disorders influencing heart activity. The first group includes inflammatory (bacterial, parasitic, viral or allergic) involvement of the heart such as myocarditis, acute rheumatic fever and similar conditions. Histopathological findings in different parts of the conduction system showed cell infiltrations, which may be a direct indication of an acute inflammatory process, or fibrosis, which may be connected to old scars of myocarditis. In one case no connecting AV bundle was detected,[2] and the only pathological finding was the characteristic histological features of acute Chagas' disease. (In this case the mechanism of pre-excitation probably consisted of the splitting of the normal conduction system.) Most of the publications stressing the appearance of pre-excitation during an acute cardiac disease focus on myocarditis or acute rheumatic fever. It is easy to accept the view that such diffuse disturbances in the myocardium and around nerves may produce serious changes in the vegetative equilibrium of the heart, enhancing or suppressing different electrophysiological activities and triggering the appearance of pre-excitation.

Acute myocardial infarction is another pathological situation, and the appearance of pre-excitation during or soon after was thought to be an acquired form of the disorder.[3-5] The increased catecholamine level of the blood, and sometimes increased acetylcholine concentrations around vital structures such as the sinus node or AV node, may also produce an enhanced activity in specific myocardial cells (P cells, Purkinje cells), or their depressions, and result in premature beats or first, second and complete AV block, respectively. This could be an ideal situation for the formation of a double AV conduction system with different conduction velocities. In cases where an accessory bundle completely outside the normal conduction system has existed silently from birth, this changed autonomic status could activate it by inducing a faster conduction of the electrical stimulus through it.[6,7] When, however, such a bundle was not found[8,9] or demonstrated by intracardiac investigation after a myocardial infarction,[5] we have to assume a splitting of the normal conduction system by a functional disturbance in the sinus node area, which produces an earlier activation of the PIT (and consequently James' AV nodal bypass fibers).[5] An alternative explanation in such cases is destruction of the AIT and the MIT by a right atrial infarction, forcing the electrical stimulus to reach the AV node solely through PIT.

Degenerative changes in or around the normal[2] or accessory pathways[10] were demonstrated by pathological studies, and can also change the autonomic equilibrium of the hearts. In some cases a very advanced degenerative disease was found, the etiological background of which remained an enigma even after the pathological studies of the heart.[11] Other cardiac conditions which can produce a disequilibrium of the autonomic system in the heart are hemodynamic disturbances seen in many cases of congenital heart disease.

Changed intracardiac catecholamine-acetylcholine relations are also involved in generalized or systemic diseases. It is known that cerebral disorders can produce premature beats or inversion of the T waves in ECG tracings by changing the normal distribution of catecholamines in the heart. This may also be the background of many reports (Chapter I) of a connection between cerebral or psychiatric disorders and the pre-excitation syndrome. The interaction of the thyroid hormone with catecholamine levels is known, and this hormonal disorder may therefore also disturb the normal sympathetic-vagal relationship in the heart.

Genesis of Tachycardias in Pre-excitation

The single factor inducing all forms of tachycardia seen in the pre-excitation syndrome is the premature beat. In "circus movement" tachycardia, an accessory pathway completely outside the normal conduction system is used in combination with the normal AV conduction system; the accessory pathway is blocked by a premature beat so that the next stimulus descends through the AV node–His bundle and returns to the atria by the again excitable accessory bundle. Regarding the reentry tachycardia inside the AV node, after a local factor (increased vagal tonus or pathological environmental change around the AV node) has produced an increased AV nodal delay or a Wenckebach's period, it is again the premature beat which exploits this timely asynchronization of conduction in the AV node, permitting echo beats to appear. The perpetuation of the echo beats results in an intra–AV nodal reeentry tachycardia. Atrial fibrillation is produced when a premature beat falls in the vulnerable period of atrial repolarization. It takes two forms: the broad QRS (pseudoventricular tachycardia) when the electrical stimulus descends to the ventricles through the accessory bundle, or the regular atrial fibrillation when the stimulus uses the normal AV nodal-His-Purkinje axis (or a combination of both, Figure 37c). Finally, if we consider the possibility of an automatic ectopic (or sinus node) center firing at high rates from one single location as the cause of tachycardia (as postulated by Scherf and Cohen[12] and Sherf and James[13]), then such a tachycardia can *de facto* be regarded as a long series of premature beats.

In summary, it can be stated that the premature beat is responsible for starting most (if not all) tachycardias seen in pre-excitation. This premature beat is the result of enhanced electrophysiological activity in potential pacemaker cells located in ectopic centers. The enhanced activity permits the phase 4-depolarization to arrive at the depolarization level of the cell and produce a premature beat. It makes no difference whether this enhanced sympatheticomimetic activity takes place inside the accessory bundle, is due to an excess of coffee or smoking or both, or is produced by pathological conditions associated with heart or systemic diseases. The end result in all these instances is a premature beat which is able to induce one of the many types of tachycardia seen in pre-excitation.

The increased incidence of premature beats observed in the pre-excitation syndrome (as compared to healthy and hospitalized populations) is most probably due to a great number of additional pacemaker cells dispersed in the different anomalous structural AV connections. These premature beats induce a higher incidence of tachycardias in pre-excitation cases than in patients without an additional AV junctional pathway;[14] this is because in the pre-excitation syndrome, in addition to the usual opportunities for the starting of a tachycardia, there is also the dual or multiple AV conduction system over which a "circus movement" tachycardia can take place. Our inability to predict which patient will suffer from tachycardias and when an attack of paroxysmal tachycardia will appear is due to the continuous dynamic changes in the electrophysiological behavior of the syndrome.

Concomitant P and QRS Changes

P wave changes appear in a large number of pre-excitation cases (see Chapter II). They represent changes in the ordinary atrial depolarization order and vector, and are usually the result of one of two electrical developments: a dislocation of the driving pacemaker from the sinus node into an ectopic supraventricular center (enhanced electrophysiological activity), or a functional block in one or more of the internodal and/or interatrial conduction pathways (depressed activity). Both are produced by a disequilibrium of the vegetative cardiac conduction system. Our speculation that there may be a greater incidence of P wave changes in pre-excitation can be partly substantiated by the fact that some of the anomalous AV pathways contain an increased number of specialized cells (Purkinje cells, P cells, transitional cells), which means more potential electrophysiological disturbances (enhanced or depressed activities). Or, alternatively, coexisting with the structural congenital anomalies, there may be functional inborn disturbances of the normal electrical stimulus production and conduction system.

In most cases P wave changes were accompanied by concomitant QRS changes, either in the form of intermittency (to normal or to pre-excitation patterns) or the appearance of two or more different forms of pre-excitation QRS patterns. In rare cases a completely aberrant QRS complex, resembling neither normal nor pre-excitation in the patient, has been observed (Figure 19d).

We shall consider two possible explanations of these phenomena. When an accessory pathway is present completely outside the normal conduction system or partly incorporated in it, a change in the location of the driving pacemaker, bringing it nearer or farther away from the atrial end of the additional AV connection, may induce the appearance or disappearance, respectively, of pre-excitation, together with changes in P wave morphology. This would be the spontaneous equivalent of a similar phenomenon observed during experimental or diagnostic electrical pacing of the atria in patients with pre-excitation. (Remember the case which was considered to be only a short P-R normal QRS syndrome but turned out to be pre-excitation when, during pacing, the typical QRS of the disorder appeared.[15])

In cases where the presence of one accessory structural AV connection was not documented by intracardiac or pathological studies, it must be assumed that the double electrical AV conduction developed inside the normal conduction system. A dislocation of the sinus node pacemaker (due to a spontaneous change in the autonomic homeostasis in the area) into the posterior internodal tract (PIT) will transform this pathway into the primary atrial conduction channel, and induce the activation of James' AV nodal bypass fibers and pre-excitation. When the ectopic center on the PIT is near the AV node, retrograde atrial depolarization may be observed, expressed in the ECG tracing as a negative P wave together with pre-excitation. When, on the other hand, the ectopic stimulus originates near the sinus node, but still within the PIT, the vector of the atrial depolarization will still closely resemble that of the sinus rhythm, and also the P waves. Finally, when the ectopic center is

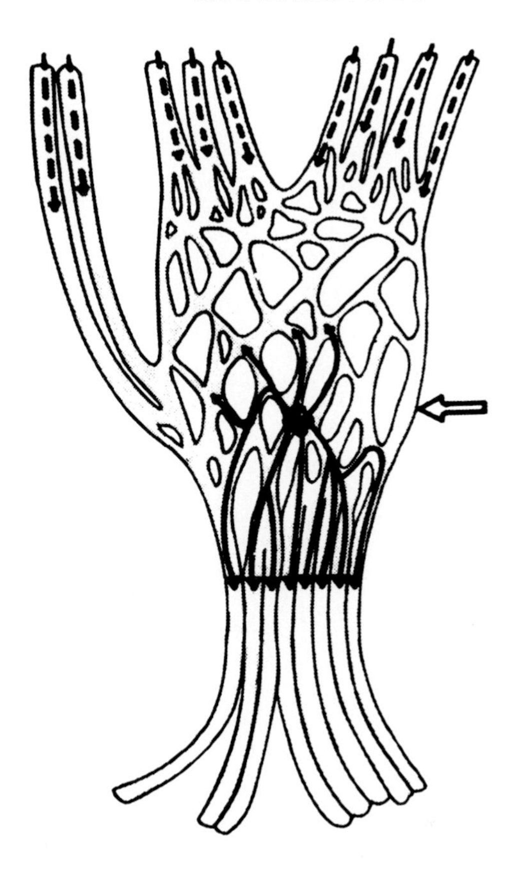

Diagram X. *Location of an ectopic AV junctional focus on the axis of the normal AV nodal depolarization front. In this case it will not disrupt the order of the usual sequence of depolarization distal to the focus, and no aberration will be seen in the QRS complexes.*

located somewhere between these two extreme points, *eg*, within the lower portion of PIT, varying degrees of alterations in P wave form will be seen, usually with additional QRS changes (Figure 3). If the electrophysiological changes produce a migration of the pacemaker into the junctional area (and not into the PIT), junctional rhythms appear, with normal QRS configuration when it originates on the central axis of the normal AV nodal-His bundle conduction (Diagram X), or with aberration when its origin is eccentrically located in the junctional area (Diagram XI).

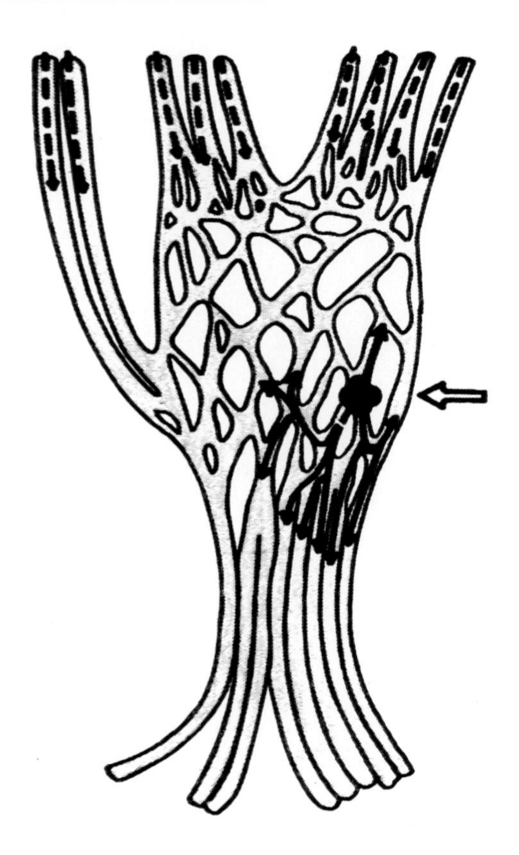

Diagram XI. *A similar junctional ectopic focus, but placed eccentrically or peripherally in the junctional region. It will produce an asynchronization of the ordinary depolarization front both in the AV node and distal to it, resulting in a QRS complex with aberration (see Diagram X).*

Correlation of the P-R Interval during Normal and Anomalous Conduction in a Single Patient

The high correlation between normal and anomalous AV conduction times in the same patient (see Chapter II) is difficult to explain in terms of a structural bridge stretching over the AV sulcus, circumventing the AV node. However, it becomes clearer in light of our broader concept of interplaying electrophysiological and structural anomalies.

The AV and sinus nodes are very sensitive to increased vagal tonus.

Experimental stimulation of the right vagus produces a sinus standstill, and stimulation of the left produces a prolonged P-R interval, followed by a second degree and then complete AV block. The mechanism of this sensibility to acetylcholine is not understood at present but is most probably due to specific electrophysiological properties of some special cells —the P cells, and/or the slender transitional cells, which are the main components of the sinus and AV nodes. The prolonged P-R interval is usually the result of an increased vagus tonus, secondary to pathological development around the AV node. This was observed in cases with acute diaphragmatic myocardial infarction or acute rheumatic fever. A prolonged P-R interval during regular conduction in a patient with a normal P-R interval (0.12-0.20 second) during pre-excitation implies a similarity in cell composition of the AV node and the AV anomalous bypass, responsible for the production of pre-excitation. When an accessory bundle completely outside the normal AV conduction system is suspected, it is assumed to be composed mainly of P cells and/or transitional cells.[11] In such a case, autonomic changes which would influence the electrophysiological status of the AV node may in the same way influence the accessory AV bundle. The same may be said for structures connecting the atria with the ventricle which are partly incorporated in the normal AV conduction system (atrial-His bundle connection or remnants of the AV nodal tissue in the early postnatal period).

Consider, on the other hand, cases where no structural AV connection exists, and the mechanism of pre-excitation is attributed to disturbed electrophysiological events inside the normal conduction system. The bypass fibers of James', which conduct early the electrical impulse towards the AV node and ventricles, enter the AV node at its lower pole but still have to travel downwards through a part of its special cell population. Thus, it is not surprising that if a pathological slowing down of electrical conduction exists in the AV node, for any reason whatsoever, it will also be reflected proportionately during pre-excitation conduction when the stimulus is passing through a part of the AV node.

It may also be that in all adult cases where an increased P-R interval during pre-excitation is observed, a cell population correlation of one kind or another exists between the anomalous AV connections and the AV node.

First Degree, Second Degree, and Complete AV Block

Only two cases are known where P-R conduction time was longer than 0.20 second during pre-excitation, and in one of them the P-R interval during normal conduction was 0.40 second. In our opinion such cases represent only extreme situations of what was discussed in the preceding section. When the block is of a Wenckebach type, AV nodal tissue is involved in the mechanism of pre-excitation in one way or another (James' bypass fibers or AV accessory connections composed of similar cells as the AV node), since they are only able to produce the gradually prolonged P-R intervals and finally the dropped beat.

An AV block of the 2:1 or 3:1 conducting type seen together with the pre-excitation implies active conduction by direct His bundle connections to the ventricular septum (Mahaim fibers). If the second degree block is located in the AV node, producing a 2:1 conduction, the para-specific Mahaim fibers will transfer every second stimulus filtrated through the AV node and produce the 2:1 block with pre-excitation. In such instances, however, the P-R interval will be normal during pre-excitation.

When a block is seen with a short rather than normal P-R interval, one can assume a structural background similar to that described by Lev et al[10] (case 29 in Table XVI): an accessory bundle connecting the right atrium with the right ventricle (RAB), composed of Purkinje-like cells. There was fatty degeneration and infiltration with mononuclear cells in all the parts of the conduction system (including the AV node), and degeneration and fibrosis were also detected in the RAB. Conduction over the RAB should be fast in such a case, producing a short P-R due to the Purkinje cell composition. If the degenerative and/or infiltrative process produces a complete block of the normal conduction system, then a partial lesion of the RAB, as was observed in this case, will produce a 2:1 block with pre-excitation and short P-R. (This is only speculative since no such ECG changes were described in this case.) There is no pathological material available on second degree AV block with pre-excitation.

In all the cases of pre-excitation in which a complete AV block developed, the QRS complexes during the block were of idioventricular or idiojunctional origin without the characteristic pattern of pre-excitation. All the anomalous AV connections have to be blocked. In some cases the successful interruption of the normal AV conduction (His bundle) by surgery was followed by the appearance of an idioventricular rhythm; we can assume that the anomalous AV pathway was connected to the AV nodal area (His-ventricular septum connections or an asynchronization of the normal conduction system). In most cases, however, despite the interruption of the normal AV conduction by surgery, the pre-excitation pattern persisted (although the tachycardia disappeared); we may assume that an accessory AV connection, completely outside the normal conduction system, remained intact and continued to transfer electrical stimuli from the atria to the ventricles. In cases with a complete

AV block and an idioventricular rhythm driving the ventricles, but where an occasional ventricular capture with pre-excitation pattern can be detected, all the above mentioned possibilities exist. The mere appearance of a ventricular capture in a case of complete AV block indicates that the block is functional rather than structural in nature.

The only case seen with complete AV block and QRS patterns characteristic of pre-excitation (labeled "pseudo–pre-excitation" by us) will be discussed in Chapter XIII.

Bradycardia-Tachycardia Syndrome

This combination appears strange, since every theory described thus far located the origin of the pre-excitation far from the area of the sinus node. The bradycardia-tachycardia syndrome (sick sinus syndrome) may be produced by histopathological involvements of the sinus node (fibrosis, degeneration, infiltrates secondary to a reduced blood supply, inflammatory process or unknown etiological causes), or may be the result of a functional disturbance of the normal conduction system (in some cases the AV node is also involved). Since both these syndromes are quite rare, their coexistence[16,17] seems to indicate a common background unknown at present. It may be due to a disease of the sinus node in the presence of pre-excitation, or the former may produce pre-excitation by forcing the primary pacemaker into the PIT and James' AV nodal bypass fibers. However, one cannot ignore the possibility that the two syndromes may appear with a structurally normal sinus node. In such instances it must be reemphasized that the pre-excitation syndrome as well as the bradycardia-tachycardia syndrome may both be the result of functional and electrophysiological diseases or disorders of a structurally normal conduction system. No scientific proof for such a disorder is available at present.

REFERENCES

1. **Zamfir C, Turcu E, Giora G, et al:** On the organic form of shortening the P-R interval of the electrocardiogram with deformity of the ventricular complex. (Rus) *Kardiologiia* 3:71, 1963.
2. **Lev M, Kennamer R, Prinzmetal M, et al:** A histopathologic study of the atrioventricular communication in two hearts with the Wolff-Parkinson-White syndrome. *Circulation* 24:41, 1961.
3. **Fabre H:** Wolff-Parkinson-White syndrome after infarct. Stability during 15 years; disappearance 2 years ago. *Arch Mal Coeur* 63:301, 1970.
4. **Goel BG, Han J:** Manifestations of the Wolff-Parkinson-White syndrome after myocardial infarction. *Am Heart J* 87:633, 1974.

5. **Mathew G, Raftery E:** Accelerated atrioventricular conduction after myocardial infarction, a study using His bundle electrograms. *Br Heart J* 35:985, 1973.
6. **Levine HD, Burge JC Jr:** Septal infarction with complete heart block and intermittent anomalous atrioventricular excitation (Wolff-Parkinson-White syndrome); histologic demonstration of a right lateral bundle. *Am Heart J* 36:431, 1948.
7. **Schumann G, Jansen HH, Anschütz F:** On the pathogenesis of the WPW syndrome. (Ger) *Virchows Arch (Pathol Anat)* 349:48, 1970.
8. **Plavsic C, Marie D, Zimolo A:** Infarctus du myocarde et syndrome de Wolff-Parkinson-White. *Acta Cardiol* 11:190, 1956.
9. **Prinzmetal M, Kennamer R, Corday E, et al:** "Accelerated Conduction," in *The Wolff-Parkinson-White Syndrome and Related Conditions*, New York: Grune & Stratton, 1952.
10. **Lev M, Sodi-Pallares D, Friedland C:** A histopathologic study of the atrioventricular communications in a case of WPW with incomplete left bundle branch block. *Am Heart J* 66:399, 1963.
11. **James TN, Puech P:** De subitaneis mortibus. IX. Type A Wolff-Parkinson-White syndrome. *Circulation* 50:1264, 1974.
12. **Scherf D, Cohen J:** *The Atrioventricular Node and Selected Cardiac Arrhythmias*, pp 373-447, New York-London: Grune & Stratton, 1964.
13. **Sherf L, James TN:** A new electrocardiographic concept: synchronized sinoventricular conduction. *Dis Chest* 55:127, 1969.
14. **Wellens HJ, Durrer D:** The role of an accessory atrioventricular pathway in reciprocal tachycardia. Observations in patients with and without the Wolff-Parkinson-White syndrome. *Circulation* 52:58, 1975.
15. **Massumi RA, Vera Z:** Patterns and mechanisms of QRS normalization in patients with Wolff-Parkinson-White syndrome. *Am J Cardiol* 28:541, 1971.
16. **Masoni A, Pradella A, Tomasi AM, et al:** WPW anomaly with paradoxical pathogenesis due to sino-auricular block. (Ital) *Minerva Cardioangiol* 14:20, 1966.
17. **Mochizuki S, Kiriyama T, Kakusui K, et al:** A case of Wolff-Parkinson-White (WPW) syndrome accompanied with atrioventricular conduction disturbance and sick sinus syndrome. *J Jpn Soc Intern Med* 64:1264, 1975.

XIII

Various QRS Patterns in the Different Forms of Pre-excitation

In the historical survey of the pre-excitation syndrome presented in Chapter I, it was noted that in addition to the short P-R interval, the early investigators were also impressed by the strange deformation of the QRS complexes, which they considered to be BBB.[1,2,3] This interpretation was soon discarded[4,5] and the broad, bizarre ventricular complexes were considered specific for the pre-excitation syndrome. In this last chapter we will close the circle of the mysterious phenomena encountered in this fascinating electrical disorder of the heart by discussing again the nature and pathological mechanism of these QRS deformations in the light of our proposed concept. These deformations raise two important questions

First, is there an explanation for the genesis of the characteristic QRS deformation in pre-excitation other than the "fusion beat" theory?[5]

The answer to this question appears to be affirmative. The simple concept of "fusion beat" has been repeatedly questioned by many electrocardiographers. It has been suggested[4,6] that a specific ventricular center is also able to produce all the changes which characterize the pre-excitation QRS pattern. Our case of complete AV block (Figure 6a and 6b) illustrates this point: termed "pseudo-pre-excitation" to distinguish it from the classic form, the case demonstrates that stimuli arising in a certain area of the ventricular myocardium can reproduce the typical QRS seen in pre-excitation. Because of the complete AV block, it is presumed that the rapid late part of the ventricular depolarization cannot be produced by a coinciding supraventricular stimulus. The logical conclusion is that somewhere in the nonatrial myocardium there must be a center capable of producing QRS complexes identical to those seen in the pre-excitation syndrome; and, that the entire depolarization of the ventricle from beginning to end must be anomalous, without the participation of depolarization produced by atrial stimuli descending through the usual conduction pathways. Where would this center be? On the basis of experiments in dogs, Sodi-Pallares[7] concluded that the initial slurring delta wave of the R seen in pre-excitation corresponds to early activation of certain portions of the upper ventricular septum near the tricuspid or the aortic valve. It may be argued that there is a different mechanism of depolarization of the ventricles which produces the specific form of the QRS in cases of complete AV block and in cases with the true pre-excitation syndrome: an idioventricular beat in the former, and a fusion beat in the latter. In at least one example of classic pre-excitation,[8] it was shown that the entire ventricular depolarization must have been anomalous without the participation of any stimuli arriving from the atria (see Chapter III, Other Arrhythmias). In this case, during normal atrioventricular conduction the P-R interval had a duration of nearly 0.40 second, while during pre-excitation the QRS vectors were generated only 0.16 second after the P wave. Ventricular conduction was interpreted as being entirely due to the pre-excitation wave, because the whole P-S duration during the anomalous conduction ended before the normal P-R interval elapsed. Rather than being the exception, this case may represent what happens in all cases of pre-excitation with a stimulus from a specific area in the intraventricular septum leading to anomalous depolarization of both ventricles. This belief is strengthened by the appearance of ventricular premature beats in a patient without pre-excitation (Figure 47), which show a characteristic pre-excitation QRS pattern in both ECG and VCG tracings.

Nevertheless, the general consensus today appears to be that the characteristic QRS complex encountered in this syndrome is a "fusion beat" between a prematurely activated sector of ventricular myocardium and the normal depolarization of the rest of the ventricles. Under certain conditions all the depolarization of the ventricles is anomalous (*eg*, atrial pacing in the neighborhood of an assumed accessory AV bundle), while under other conditions it is completely normal—intermittency (spontaneous or experimentally produced). All other cases are different stages of "fusion" between the anomalous and normal depolarization of the ventricles. This approach is based mainly on the results of intracardiac registrations and pacing in patients with pre-excitation. It is our opinion that these findings are all either indirect or assumed evidence of the existence of ventricular fusion. There is at present no direct way to study the part of the atrial stimulus which descends directly through the His bundle and actually takes part in depolarization of the ventricles.

Second, can the bizarre QRS complexes be explained in any other way except by the "fusion beat" hypothesis?

The answer to this second question is also affirmative. Theories about the mechanism of the QRS deformation in pre-excitation changed during the years. Wolff, Parkinson and White[9] regarded it as true BBB resulting from some abnormal vagal effect, which simultaneously accelerated AV conduction. Numerous authors were of the opinion that there is no true BBB, in particular, Holzman and Scherf[4] and Wolferth and Wood.[10] They found that the P-T (or P-J) interval was within the limit of normal ventricular conduction, contrary to the rule in true BBB. Furthermore, they found the same P-J interval in both normal and abnormal complexes, where intermittency was present. Hunter et al[5] also rejected the assumption of a BBB in pre-excitation. They regarded the special form of the QRS in pre-excitation as representing a delta wave (produced by an early ventricular depolarization) simply annexed to the normal QRS: "A nodal beat might be superimposed upon a normal P-QRST complex, so that the P-R interval was shortened by the early ventricular complex of the nodal beat which was then succeeded by the normal ventricular complex, the two in juxtaposition giving a broad wave simulating branch block."

By 1948 Wolff and White[11] arrived at a more sophisticated view: namely, that although ventricular depolarization is initiated prematurely through an accessory conducting pathway, the impulse traversing the normal auriculoventricular connections ultimately arrives at the ventricular muscle, not yet reached by anomalous excitation, and depolarization of the entire ventricular mass is completed. The result of the two ventricular depolarization vectors will produce the QRS deformation. Assuming this to be true, the P-R interval will be shortened to the same extent that the QRS interval is lenthened. Thus, in records showing both anomalous and normal complexes, the P-J intervals of the two types should be the same. This was not always the case, however. Table XXVII presents 24 consecutive cases of pre-excitation with intermittency from our series, and Table XXVIII summarizes 4 series from the literature plus the Tel Hashomer series where such measurements were given. In most cases (54 of 84) the P-J interval was not identical during normal and anomalous conduction and therefore doubt is cast on the concept of a simple "fusion beat."

The "fusion beat" theory also claims support from the "fact" that the characteristic pattern of the QRS has some features of the normal QRS complexes and some of the anomalous forms. Actually, in some cases the differences in form during pre-excitation are so great that the authors considered all the muscle mass to have undergone anomalous depolarization. This is especially true in type A cases, where the R in V_1

TABLE XXVII
Electrocardiographic Values in 24 Cases during Pre-excitation and Normal Conduction in the Tel Hashomer Series

Case #	During Pre-excitation Conduction				During Normal Conduction				Diverse
	P-R	QRS	P-J	Axis	P-R	QRS	P-J	Axis	
8	0.12	0.12	0.24	−60°	—	0.08	—	+20°	Normal during SVT
	0.14	0.08	0.22	−20°					
9	0.16	0.12	0.28	−90°	0.20	0.10	0.30	+60°	Normal tracing I degree HB
10	0.14	0.12	0.26	−60°	0.18	0.08	0.26	−20°	
11	0.08	0.12	0.20	0°	0.10	0.12	0.22	0°	
13	0.12	0.12	0.24	+60°	—	0.08	—	undetermined	Normal during SVT
16	0.06	0.12	0.18	+60°	0.14	0.08	0.22	+90°	
17	0.10	0.10	0.20	0°	0.14	0.08	0.22	+60°	
28	0.12	0.10	0.22	−40°	0.16	0.10	0.26	+60°	
33	0.08	0.12	0.20	−40°	—	0.08	—	+60°	Normal during SVT
	0.08	0.16	0.24	−70°					
35	0.12	0.14	0.26	−40°	0.16	0.10	0.26	−10°	
37	0.12	0.08	0.20	0°	0.15	0.08	0.23	+60°	
	0.12	0.12	0.24	−25°					
38	0.10	0.14	0.24	+60°	0.15	0.08	0.24	+80°	
39	0.10	0.14	0.24	+70°	0.14	0.10	0.24	+70°	
41	0.10	0.14	0.24	+90°	0.20	0.08	0.28	undetermined	Normal I degree HB
45	0.12	0.12	0.24	−35°	0.16	0.08	0.24	−10°	
48	0.12	0.12	0.24	+60°	0.14	0.08	0.22	+30°	Trans LBBB
	0.10	0.16	0.26	+90°	—		—		
51	0.10	0.10	0.20	+35°	—	0.08	—	+60°	Normal during SVT
57	0.08	0.12	0.20	0°	0.10	0.12	0.22	+10°	
62	0.12	0.10	0.22	−50°	0.16	0.08	0.24	+30°	
	0.10	0.16	0.26	−70°					
75	0.10	0.12	0.22	+90°	0.16	0.08	0.24	−10°	
77	0.12	0.14	0.26	+110°	0.16	0.10	0.26	0°	
84	0.08	0.12	0.20	+90°	—	0.08	—	+60°	Normal during SVT
89	0.09	0.14	0.23	+30°	0.12	0.06	0.18		
95	0.10	0.12	0.22	+60°	0.16	0.08	0.24		

and V_4R is always tall and bears no resemblance to the rS of normal conduction, and in cases with QS in leads II, III and aVF where during normal conduction R's are observed in all or part of the leads.

The lack of fixed times between the P-J intervals during normal and pre-excitation conduction, and the complete dissimilarity in the

TABLE XXVIII
Pre-excitation with Intermittency in 4 Studies from the Literature and the Tel Hashomer Series

Series	# Cases	# Cases with Data of Normal and Abnormal P-S Interval in the Same Patient	P-S Interval during Normal and Anomalous Conduction in the Same Patient		Case #
			Identical P-S Intervals	Nonidentical P-S Intervals	
Averill[12]	107	20	5	15	4, 14, 17, 18, 22, 48, 49, 51, 53, 55, 75, 61, 104, 19, 6
Wolff[11]	41	14	5	9	7, 9, 16, 20, 23, 26, 34, 36, 39
Schiebler[13]	28	5	0	5	13, 16, 18, 23, 24
Littman[13]	9	4	1	3	1, 2, 7
Tel Hashomer series	85*	17	6	11	2, 4, 6, 7, 8, 11, 14, 16, 18, 19, 20
Total	270	60	17 (28%)	43 (72%)	

*Early series

QRS patterns led us to conclude that even if the electrical stimulus descending through the AV node–His axis does make some contribution to the depolarization of the ventricles, it does not influence the QRS pattern in pre-excitation. The pre-excitation component (expressed by the vector of the delta wave) is, rather, the factor which dictates what direction the rest of the QRS ventricular depolarization form will take (see Chapter V, Vectorcardiography).

A logical explanation of the order of ventricular depolarization during pre-excitation was hypothesized by Grant et al[8] in 1958: in the majority of cases a stimulus from the atrium crosses the atrioventricular ring into the intraventricular septum, giving rise to the pre-excitation event. This stimulus may enter at any of a number of different points in the septum. If it enters near the diaphragmatic margin of the septum, the pre-excitation wave will spread leftward and superiorly, producing a leftward and superiorly directed delta wave. With this direction the wave will soon reach the territory of the inferior division of the left bundle; if it reaches these fibers before the normal atrioventricular impulse is de-

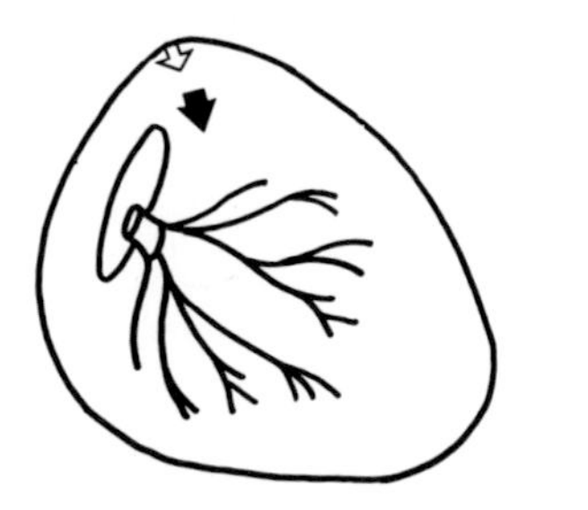
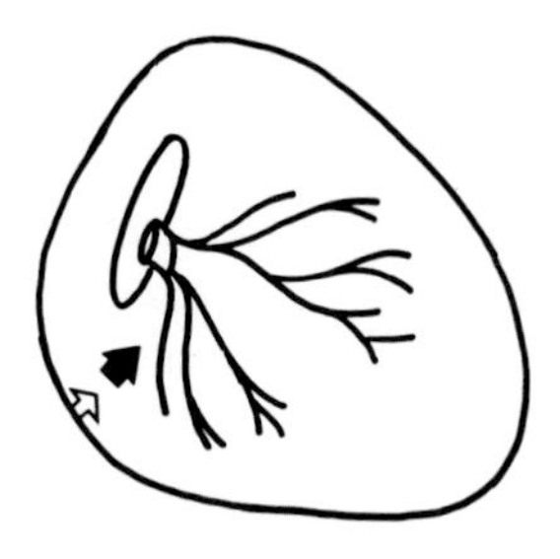

Diagram XII. *Explanation of the marked correlation between the direction of the delta wave and the rest of the QRS complex in cases with pre-excitation (see Chapter II). (Modified from Grant et al.[8])*

livered to the ventricular conduction system, the ventricles will be depolarized from below upwards, producing left axis deviation of the QRS vectors, without QRS prolongation (Diagram XII). On the other hand, if the pre-excitation starts at the upper margin of the septum, its direction of spread will be inferior, producing an inferiorly directed delta vector. It will spread towards the territory of the superior division of the left bundle and, if it reaches these fibers, ventricular depolarization will be from above downward within the ventricular conduction system, producing right axis deviation of the QRS vectors, without QRS prolongation (Diagram XII). Should pre-excitation start at a point in the septum between these two extremes, it will spread leftward and inferiorly, producing a leftward and inferior vector. With this direction it will enter both divisions of the left bundle almost simultaneously with no change in QRS vectors. Grant et al[8] wrote: "This hypothesis explains why certain delta vector directions are more frequently associated with QRS vector changes than others, and why the direction the QRS vectors take are specific for those delta vector directions; it also explains why, with a relatively small part of the ventricles involved in pre-excitation, QRS vector changes are so common; and it explains why there may be no prolongation of the QRS interval with the QRS vector changes."

Grant et al[8] mainly considered the changes observed in the frontal plane; but, with a bit of imagination we can envision similar development occurring also in the horizontal plane, producing an anterior (type A) or left posterior (type B) pre-excitation pattern. This is especially true if we assume that the premature ventricular depolarization vector first reaches the area of the branching part of the His bundle, which is an important crossroad region, at the beginning of the ventricular conduction system. An early advance of such a depolarization front into the septal divisions of the left bundle branch or even the right bundle branch, not opposed by vectors in opposite directions, could easily produce the anteriorly directed vectors, producing type A depolarization. This explanation is, in our opinion, preferable to another which draws straight lines from the assumed location of the pre-excitation to the place of breakthrough of the first electrical stimulus to the surface of the heart, and which does not take into consideration at all the existence of a ventricular conduction system.[14]

The hypothesis of Grant et al[8] can easily explain most of the events of pre-excitation which occur in the vicinity of the septal area. It is more difficult, however, to explain the electrical events in the cases where the anomalous structural AV connection is located between the free walls of the right or left atria and ventricles. It is our belief that Grant et al's principle can be applied here as well, but more experimental studies are required, such as electrical pacing of known specific and strategic points along the ventricular conduction system, to illuminate the directions of myocardial vectors during pre-excitation.

REFERENCES

1. **Bishop LF Jr:** Bundle branch block with short P-R interval in individuals without organic heart disease. *Am J Med Sci* 194:794, 1937.
2. **Roberts GH, Abramson DI:** Ventricular complexes of the bundle branch block type associated with short P-R intervals. *Ann Intern Med* 9:983, 1936.
3. **Wolff L:** Wolff-Parkinson-White syndrome: historical and clinical features. *Prog Cardiovasc Dis* 2:677, 1960.
4. **Holzmann M, Scherf D:** Über Elektrokardiogramme mit verkürtzter Vorhof-Kammer-Distanz und positiven P-Zacken. *Z Klin Med* 121:404, 1932.
5. **Hunter A, Papp C, Parkinson J:** The syndrome of short P-R interval, apparent bundle branch block and associated paroxysmal tachycardia. *Br Heart J* 2:107, 1940.
6. **Sodi-Pallares D:** Fusion beats and anomalous atrioventricular excitation in anomalous atrioventricular excitation: panel discussion. *Ann NY Acad Sci* 65:845, 1957.
7. **Sodi-Pallares D, Galder RM:** *New Bases of Electrocardiography,* St. Louis: Mosby, 1956.
8. **Grant RP, Tomlinson FB, Van Buren JK:** Ventricular activation in pre-excitation syndrome (Wolff-Parkinson-White). *Circulation* 18:355, 1958.
9. **Wolff L, Parkinson J, White PD:** Bundle branch block with short P-R interval in healthy young people prone to paroxysmal tachycardia. *Am Heart J* 5:685, 1930.
10. **Wolferth CC, Wood FC:** Further observations on the mechanism of production of a short P-R interval in association with prolongation of the QRS complex. *Am Heart J* 22:450, 1941.
11. **Wolff L, White PD:** Syndrome of short P-R interval with abnormal QRS complexes and paroxysmal tachycardia. *Arch Intern Med* 82:446, 1948.
12. **Averill KH, Fosmoe RJ, Lamb LE:** Electrocardiographic findings in 67,375 asymptomatic subjects. IV. Wolff-Parkinson-White syndrome. *Am J Cardiol* 6:108, 1960.
13. **Schiebler GL, Adams P Jr, Anderson RC:** The Wolff-Parkinson-White syndrome in infants and children. A review and a report of 28 cases. *Pediatrics* 24:585, 1959.
14. **Boineau JP, Moore EN:** Evidence for propagation of activation across an accessory atrioventricular connection in types A and B pre-excitation. *Circulation* 41:375, 1970.

INDEX

Please bear in mind that this index is based on major subject headings and does not include scattered comments. Remember also that the main topic for all listings is pre-excitation.

Aberration, QRS complex, 42, 56, 76, 217
Accelerated conduction, theory of, 2, 233
Accessory AV connection, 89-113, 230
  location by intracardiac recordings, 149, 150
  multiple, 148, 149
  terminology of, 230
Acetylcholine, 239
Acquired forms of pre-excitation, 244
Age, 8, 9
  and P-R interval, 38
A-H interval, 144-147
  in short P-R–normal QRS syndrome, 146
Airplane pilots, 163, 164
Ajmaline, 153, 180, 182, 183
  and intermittency, 151
  in normalization of pre-excitation conduction, 210
Allergic diseases, 23
Ambulatory monitoring (Holter), 62, 68, 74, 127-130
Amiodarone, 180
Amyl nitrite, in normalization of pre-excitation conduction, 210
Anatomical changes, postnatal, 231
Anesthesia, 164
Angina pectoris, 16
Animals, 111, 112
Anxiety, 163, 171
Aortic insufficiency, 20
Apexcardiography, 130
Arrhythmias, 61-81
  in infants (see infants, 166, 182)
  sinus, 75
Arteriosclerotic heart diseases (see atherosclerotic heart diseases, 16)
Artificial pacemaker, 185
Associated heart diseases, 163
Atherosclerotic heart diseases (ASHD), 16
Atresia, tricuspid, 12
Atrial fibrillation, 65, 180, 181
Atrial flutter, 68
Atrial pacing (see intracardiac recordings, 143-154)
Atrial parasystole, 61
Atrial premature beats, 61, 62
Atrial septal defect (ASD), 13
Atrial tachycardia, paroxysmal (PAT), 65, 179
Atropine, in normalization of pre-excitation conduction, 210

Atypical cases, 211, 212
Autonomic nervous system, 239
AV block
  complete, 80, 252
  first degree, 78, 252
  second degree, 252
  Wenckebach phenomenon in 80, 252
AV conduction anomalies, 78
AV connection, accessory, 89-113, 230
  location by intracardiac recordings, 149, 150
  multiple, 148, 149
  terminology of, 230
AV dissociation, 75
AV node, 246
  reciprocating tachycardia, 151

Bigeminy, 21, 61
Body surface isopotential mapping, 138
Bradycardia, sinus, 72-75, 253
Bradycardia-tachycardia syndrome, 74, 253
Bundle branch block (BBB)
  left (LBBB), 205
  right (RBBB), 204, 205
  and left axis deviation, 205, 206
  and vectorcardiography, 124
Bundle of His, electrogram (see intracardiac recordings, 143-154)
Bundle of Kent, theory of, 1, 230
"Burned out" cases, 172
Bypass fibers of James, 231, 233

Cardiac arrest (see sudden death, 174, 175)
Cardiac surgery (see surgery, 185-191)
Cardiomyopathy, familial, 9, 13
Cardioversion, electrical, 184, 185
Carotide pulse tracing, 131
Catecholamine, 239, 245
Cells
  P, 248
  Purkinje, 248
  transitional, 248
Central nervous system, disturbances, 23, 246
Chagas' disease, 89-113, 245
Cinematography, high speed, 132
Classification
  of pathological findings, 30
  of pre-excitation, 4
  of QRS patterns, 48-55
  use of VCG in, 124
Complete AV block, 80, 252
Complete transposition of great vessels, 12
Concertina effect, 55
Concomitant P and QRS changes, 247-249
Conduction, theory of
  accelerated, 2, 233
  synchronized sino-ventricular, 233
Congenital heart disease, 12-15
Coronary artery diseases (see atherosclerotic heart diseases, 16)
Coronary insufficiency, 16, 134
Countershock, electrical, 184, 185
Cryosurgery, 191

Death (see mortality, 172-175)
  sudden, 174, 175
Definition of pre-excitation, 3-5
Delta wave, 45
Diagnosis, 5, 198-212
  errors in, 198
Digitalis, 153, 181
  and sudden death, 174, 175
Digitoxin, in infants, 182
Digoxin, 153, 183
  in infants, 182

Distribution by sex, 7, 8
Disturbances
  in central nervous system, 23, 246
  electrolyte, 239
Dizziness, 171
Drugs, in treatment of tachycardias, 179-184
Dyspnea, 171

Ebstein's anomaly, 12
  P-R interval in, 78
Echocardiography, 130, 133
Ectopic beats (see premature beats, 61-63, 246)
Ectopic rhythms (see junctional escape beats, 75)
Electrical cardioversion, 184, 185
Electrical countershock, 184, 185
Electrogram, of His bundle (see intracardiac recordings, 143-154)
Electrokymography, 130, 132
Electrolyte disturbances, 239
Electrophysiological studies (see intracardiac recordings, 143-154)
Endocardial fibroelastosis, 15
Epicardial mapping, 154-156
  and surgery, 155
Ergometry, 134, 135
Escape beats (see junctional escape beats, 75)
Esophageal leads, 62
"Excitable center," theory of, 1, 232
Exercise tests, 133-138
  and normalization of pre-excitation conduction, 137
Extrasystoles, 61-63, 246
Eyeball pressure, to stop PAT, 66

Familial cardiomyopathy, 9, 13
Familial occurrence, 9
Fibrillation
  atrial, 65, 180, 181
  ventricular (see sudden death, 174, 175, 181)
Fibers
  James' bypass 231, 233
  Mahaim, 146, 147, 231
Fibroelastosis, endocardial, 15
First degree AV block, 78, 252
First heart sound (see phonocardiography, 130, 131)
Flutter, atrial, 68
"Fusion beats," theory of, 255

General anesthesia (see anesthesia, 164)
Genetic factors, 9
Great vessels, complete transposition of, 12

Heart block, 78, 80, 252
Heart diseases, associated, 16, 163
  arteriosclerotic, 16
  atherosclerotic, 16, 20, 164
  congenital, 12-15
  hypertensive, 20, 164
  rheumatic, 16, 164
Heart sounds (see phonocardiography, 130, 131)
Heredity (see genetic factors, 9)
High speed cinematography, 132
His bundle, electrogram, 143-154
History of pre-excitation, 1-3
Holter ambulatory monitoring, 62, 68, 74, 127-130
H-V interval, 144-147
  normal values, 144
  in short P-R–normal QRS syndrome, 146
Hypertensive cardiovascular disease, 20
Hyperthyroidism, 23, 246
Hypertrophic subaortic stenosis, 13

Hypertrophy
left ventricular, 20, 204
misdiagnosis of, 203
right ventricular, 14
misdiagnosis of, 202, 203

Idioventricular rhythm, 252
Incidence of pre-excitation (see prevalence, 5, 6)
Infants, 166, 182
and treatment of tachycardias, 182
Infarction, myocardial, 16, 198-200, 208-211
Inflammatory processes, 245
Insurance (see life insurance, 162)
Intermittency, of pre-excitation conduction, 55, 56
Internodal pathways, 233
Intracardiac recordings, 143-154
and determination of effective refractory period (ERP), 152, 153
and electrical pacing, 143-154
and location of accessory AV connection, 149, 150
and postoperative evaluation, 154
and tachycardias, 151, 152
Intracavitary leads, 130
Invasive methods of investigation, 143-156
Ischemia, myocardial, 16
Isolated pre-excitation, 163
Isopotential mapping, body surface, 138

James' bypass fibers, 231, 233
Jugular pulse recordings, 131
Junctional escape beats, 75
Junctional pacemaker, 75
Junctional premature beats (see supraventricular premature beats, 61-63)
Junctional rhythm, 42
Junctional tachycardia, 68

Kent bundle, theory of, 230

Lactic acid, 239
Left axis deviation and RBBB, 205, 206
Left bundle branch block (LBBB), 205
Left ventricular hypertrophy, 20, 204
misdiagnosis of, 203
Lidocaine, in treatment of tachycardias, 153, 181
Life insurance, 162
and mortality in pre-excitation, 173
Lown-Ganong-Levine syndrome (LGL), 217

Mahaim fibers, 146, 147, 231
Mapping
body surface isopotential, 138
epicardial, 154-156
and surgery, 155
Master 2-step test (see exercise tests, 133-138)
Mechanical precontraction, 130
Mental disorders (see central nervous system, disturbances, 23, 246)
Mitral stenosis, 18, 19
Monitoring, ambulatory (Holter), 62, 68, 74, 127-130
Morbidity, 163-172
Mortality, 172-175
and life insurance, 173
associated with tachycardia, 173
Multiple accessory AV connections, 148, 149

Myocardial infarction, 16, 198-200, 208-211
Myocardial ischemia, 16
Myocarditis, 22, 23, 245

Newborns (see infants, 166, 182)
Nodal premature beats (see supraventricular premature beats, 61-63)
Noninvasive methods of investigation, 117-144

Oculocardiac reflex (see eyeball pressure, 66)

Pacemaker
 artificial, 185
 junctional, 75
 wandering, 75, 77
P-A interval, 144, 147, 148
 normal values, 144
 in short P-R–normal QRS syndrome, 146
Palpitations of the heart, 171
Parasystoles (see atrial parasystole, 61)
Paroxysmal atrial fibrillation, 65, 180, 181
Paroxysmal atrial tachycardia (PAT), 65, 179
Paroxysmal sinus tachycardia, 68
Patent ductus arteriosus (PDA), 13
Pathological findings, 89-113
P cells, 248
Phonocardiography, 130, 131
Physical fitness, 134
Postnatal anatomical changes, 231
Potassium (see electrolyte disturbances, 239)
Precontraction, mechanical, 130
P-R interval, 36, 41
 and age, 38
 in Ebstein's anomaly, 78
Premature beats, 61-63, 246
 atrial, 61, 62
 supraventricular, 61-63
Prevalence of pre-excitation, 5, 6
Procainamide, in treatment of tachycardias, 153, 180
Prognosis of pre-excitation, 162-175
Propranolol, in treatment of tachycardias, 153, 181
Pseudoventricular tachycardia, 69
Pulse recordings, jugular, 131
Pulse tracing, carotide, 131
Purkinje cells, 248
P waves, changes in configuration, 41, 42

QRS complex, 42-56
 aberration, 42, 56, 76, 217
 characteristic of, 42-47
 classification of patterns, 48-55
 duration of, 47
 variation in form, 55, 56
Quinidine, in treatment of tachycardias, 153, 180, 184

Rheumatic fever, acute, 22, 245
Rheumatic heart diseases, 16, 164
Right bundle branch block (RBBB), 204, 205
 and left axis deviation, 205, 206
Right ventricular hypertrophy, 14
 misdiagnosis of, 202, 203
Roentgenkymography, 130, 132

Second degree AV block, 252
Second heart sound (see phonography, 130, 131)
Section, surgical
 of accessory bundle, 186, 187
 of His bundle, 187
Septal defect, ventricular (VSD)

(see congenital heart disease, 12-15)
Sex distribution, 7, 8
Short P-R–normal QRS syndrome, 212-219
in A-H, H-V, and P-A intervals, 146
and ventricular aberration, 217
Sick sinus syndrome (SSS), 74, 253
Sino-ventricular conduction (see synchronized sino-ventricular conduction, 233)
Sinus arrhythmias, 75
Sinus bradycardia, 72-75
Sinus node, arrest, 75
Sinus tachycardia (see paroxysmal sinus tachycardia, 68)
Sodium (see electrolyte disturbances, 239)
ST-T segment
depression (see exercise tests, 133-138)
elevation, 209
Sudden death, 174, 175
and digitalis, 174, 175
Supraventricular premature beats, 61-63
Supraventricular tachycardias, 65-69
Surgery, 185-191
indications for, 190, 191
results of, 188-190
Surgical section
of accessory bundle, 186, 187
of His bundle, 187
Surgical techniques, 185, 186
Synchronized sino-ventricular conduction, theory of, 233
Syncopal attacks, 171
Systemic diseases, 246

Tachycardias, 63-72, 164-172
age at onset, 165-166
atrial, paroxysmal (PAT), 65, 179
duration of suffering, 167
frequency and duration of paroxysms, 168, 169
and intracardiac recordings, 151, 152
junctional, 68
and mortality, 173
pseudoventricular, 69
reciprocating (AV node), 151
sinus, paroxysmal, 68
supraventricular, 65-69
treatment, 179-191
use of drugs in, 153, 179-184
in infants, 182
preventive, 182, 183
Valsalva's maneuvers in, 182, 183
symptoms and signs of, 170
types of, 170
ventricular, 72
Terminology, 3-5
of accessory AV connection, 230
Tetralogy of Fallot (see congenital heart disease, 12-15)
Theories of pre-excitation, 227-261
accelerated conduction, 2, 233
"excitable center," 1, 232
"fusion beats," 255
Kent bundle, 1, 230
synchronized sino-ventricular conduction, 233
Transitional cells, 248
Transposition of great vessels, 12
Treatment of tachycardias, 179-191
use of drugs in, 153, 179-184
in infants, 182
preventive, 182, 183
Valsalva's maneuvers in, 182, 183

Tricuspid atresia, 12
T wave changes, 208

Valsalva's maneuvers, in treatment of tachycardias, 182, 183
Vectorcardiography, 117-126
  and bundle branch block, 124
  classification of cases, 124
Ventricular fibrillation (see sudden death, 174, 175, 181)
Ventricular premature beats (see premature beats, 61-63, 246)
Ventricular septal defect (VSD), (see congenital heart disease, 12-15)
Ventricular tachycardia, 72
Verapamil, in treatment of tachycardias, 153, 181

Wandering pacemaker, 55, 77
Wenckebach phenomenon, in AV block, 80, 252
Work capacity, 134

TCH CARDIOLOGY